NUCLEAR REACTIONS

Daphne F. Jackson

READER IN NUCLEAR PHYSICS,
UNIVERSITY OF SURREY

METHUEN & CO LTD
11 NEW FETTER LANE, LONDON EC4

First published 1970

Printed in The Netherlands by
Nederlandse Boekdruk Industrie N.V.

SBN: 416 11780 5

Distributed in the U.S.A. by
Barnes & Noble, Inc.

Preface

This book has developed from a series of lectures on collision theory and reactions given as part of the postgraduate lecture course at the University of Surrey, and its aim is to provide an introduction to the study of nuclear reactions at a level suitable for first year research students. At the same time it is hoped that selected sections of the book could serve as suitable background reading for an advanced undergraduate course in nuclear physics. In either case, it is assumed that a student working through this book would also be studying, or have studied, some advanced quantum mechanics and nuclear physics.

The main emphasis in the book is on the standard models, theories and approximations which are used in the study of direct nuclear reactions in the medium and high energy region, their validity, relation to each other, and use in the interpretation of the data. No attempt is made to review the literature or to evaluate recent data, although many current problems are mentioned and important recent papers and review articles are listed in the bibliography. It is my hope that students who work through this book will emerge with the impression that the subject of nuclear reactions is a remarkably extensive one which nevertheless possesses an underlying unity, that it is closely connected with other branches of physics, and that it is very much alive.

I am indebted to C. J. Batty, E. J. Burge, and L. R. B. Elton, who read and criticized the whole of the first draft of the manuscript, to R. C. Barrett and R. C. Johnson who commented on Chapters 4 and 3 respectively, and to a number of Surrey research students who struggled with early versions as lecture notes. I am particularly grateful to J. S. Blair, both for his stimulating criticism of the manuscript and also for the many illuminating discussions on nuclear reactions we have had in Seattle and Guildford.

I would also like to thank Miss M. I. Nicholson for her rapid and accurate typing of the manuscript, C. E. Dear for the preparation of the figures, and D. C. Martin, R. W. Oakley and V. H. Rajaratnam for their help with the proofs.

Daphne F. Jackson
June 1969

Acknowledgements

I am indebted to the following authors and publishers for permission to reproduce or redraw figures: the editor of The Physical Review and The Physical Review Letters, Professor W. C. Barber, Professor L. C. Biedenharn, Dr. B. Buck, Professor H. Feshbach, Professor S. Ghoshal, Dr. N. K. Glendenning, Dr. T. A. Griffy, Dr. R. M. Haybron, Dr. D. L. Hendrie, Dr. D. S. Onley, Dr. H. G. Pugh, Dr. J. T. Reynolds, Dr. E. Rost, Dr. G. R. Satchler and Professor V. F. Weisskopf, for figures in *Physical Review* and *Physical Review Letters*; the North-Holland Publishing Company, Professor G. Backenstoss, Dr. C. J. Batty, Professor L. R. B. Elton, Dr. B. K. Jain, Professor I. E. McCarthy, Dr. G. Matthiae, Dr. D. S. Onley, Dr. J. T. Reynolds, Dr. G. R. Satchler, Dr. J. C. Sens and Dr. I. S. Towner for figures in *Nuclear Physics* and *Physics Letters*; the Physical Society of Japan and Professor J. P. Schiffer for a figure from the *Proceedings of the International Conference on Nuclear Structure* (Tokyo 1967); Longmans, Green & Co. Ltd. and Professor W. E. Burcham for figures from *W. E. Burcham, Nuclear Physics* (Longmans 1963); the Institute of Physics and The Physical Society for a figure from *Physics Education*; Taylor and Francis, Ltd. and Dr. I. S. Towner for figures from *Advances in Physics*; the editors and Dr. R. C. Barrett for a figure from the *Proceedings of the International Conference on Electromagnetic Sizes of Nuclei* (Carleton University 1968); Dr. F. E. Throw, the Argonne National Laboratory, Dr. G. R. Satchler, Dr. R. R. Spencer and Dr. K. T. Faler for figures from Argonne National Laboratory Reports; the Goldsmiths Librarian of the University of London and Dr. I. S. Towner for a figure from a Ph.D. Thesis.

Contents

1 Introduction

1.1 Definitions

The investigation of the scattering of low energy α-particles by the nuclei in a thin gold foil, and the interpretation of these results by Rutherford, brought about the beginning of nuclear physics. Subsequently, many of the basic properties of nuclei were established as a result of experiments on the interaction of suitable projectiles with nuclei. Important examples of this early work are the discovery of the neutron, and the experiments by Cockcroft and Walton which demonstrated the feasibility of nuclear reactions and artificial transmutation of the elements. The study of nuclear scattering and reactions now provides much of the available information about nuclear properties and nuclear structure [1,2,3].

When a projectile undergoes elastic scattering, the internal energy of the target is unchanged but the projectile is scattered out of the incident beam in a manner which depends on the interaction between the projectile and the target. Since the nucleus is a quantized system it may exist in one of a discrete spectrum of excited states, each one characterized by a set of quantum numbers, and the higher states may be excited in the process of inelastic scattering in which the projectile transfers a definite amount of energy to the nucleus. Alternatively, the projectile may be captured and a different particle may be emitted, or the original particle may reappear accompanied by other particles. In each of these processes the residual nucleus is left in a well-defined state. Thus the study of elastic scattering provides information about the nucleus in its normal or ground state, while the study of inelastic scattering and reactions provides information on the existence, location and properties of its excited states. The strength with which these states are excited in different reactions indicates the presence of selection rules and may also be related to the mode of excitation.

Each of the possible scattering or reaction processes is usually represented by the notation $A(a, b)B$ where a is the projectile, A is the target nucleus, and b and B are respectively the lightest and heaviest of the reaction products [4,5,6]. In many reactions there are more than two reaction products and in these cases the symbol b is replaced by two or more symbols. Each process is characterized by a *cross-section* σ which may be defined as the probability that the process will occur if the incident beam carries one particle per second and the target contains one nucleus per unit area. (The use of the term cross-section is carried over from classical billiard-ball scattering in which this probability clearly depends on the cross-sectional areas of the projectile and target [4]). The cross-section $d\sigma$ measures the probability that the reaction product b will appear within a solid angle $d\Omega$ about a given direction. The variation of this cross-section as a function of the energy of the particles b for a fixed energy of the particles a or equivalently, as a function of the energy transferred to the nucleus, is called the *energy spectrum* for the process and is denoted by the symbol $d^2\sigma/d\Omega\, dE$. (If two light products were emitted the symbol would be $d^4\sigma/d\Omega_1\, d\Omega_2\, dE_1\, dE_2$ and so on.) If this quantity is summed over all solid angles it yields the *excitation function* $d\sigma/dE$. The presence of peaks in the energy spectrum indicates the existence of discrete states in the residual nucleus and gives a direct measurement of their energies. Examples of energy spectra are given in figures 7.5 and 9.1, and these show that the amount of information obtainable depends on the extent to which the individual peaks in the spectrum are resolved. This is analogous to the resolving power of an optical instrument. The *angular distribution* of the reaction product b as a function of the scattering angle, i.e. the angle between the final direction of motion of the particle and the direction of the incident beam (see figure 1.1(a)), is normally measured at a fixed energy corresponding to one of the peaks in the energy spectrum. This measurement gives the *differential cross-section* $d\sigma/d\Omega$ for excitation of a particular final state. Examples of differential cross-sections are shown in many figures in Chapters 4, 6, 7, 8, 9, and 10. Some information can be deduced directly from the appearance of the various cross-sections but more detailed interpretation requires a comparison with theoretical predictions for the cross-sections. The formulation of these theoretical predictions and the deductions which can be made by comparison with the experimental data are the subject matter of this book.

In order to establish the basic formulae for the description of nuclear

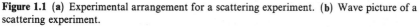

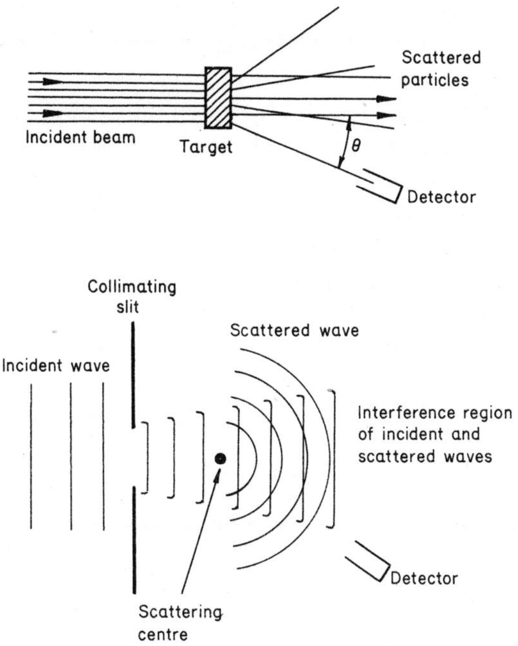

Figure 1.1 (**a**) Experimental arrangement for a scattering experiment. (**b**) Wave picture of a scattering experiment.

scattering we first take note of some standard results from non-relativistic classical mechanics pertaining to elastic scattering. In a billiard-ball collision the target, if not infinitely heavy, will acquire kinetic energy and momentum and will recoil in a direction determined by the conservation laws. The interaction $V(\boldsymbol{r})$ between the projectile and the target is a function of their relative positions, i.e. $\boldsymbol{r}=\boldsymbol{r}_1-\boldsymbol{r}_2$, and it is therefore convenient to transform from the laboratory frame of reference in which the target is at rest before collision to the centre-of-mass frame in which the centre-of-mass of the whole system is permanently at rest, since this yields an equation of motion for a single particle with co-ordinate $\boldsymbol{r}$ moving in the potential $V(\boldsymbol{r})$. This particle now has a *reduced mass* μ given by

$$\mu = \frac{m_1 m_2}{m_1+m_2} \tag{1.1}$$

and a reduced energy

$$E = \tfrac{1}{2}\mu v^2 = \frac{m_2}{m_1+m_2} E_0 \tag{1.2}$$

where E_0, v are the initial energy and velocity of the projectile of mass m_1, and m_2 is the mass of the target. The same general result holds in quantum mechanics and we have therefore to solve the Schrödinger equation for the particle of reduced mass for a suitably chosen interaction potential and to obtain particular solutions by applying the appropriate boundary conditions [4,5]. We shall, in fact, seek stationary-state solutions of the time-independent Schrödinger equation, and the argument for doing this is as follows. In the laboratory frame the incident beam passes through a collimating slit before reaching the target and so forms a wave packet but the size of the wave packet is so large compared with the size of region in which the scattering potential is effective that it is a good approximation to replace the wave packet by a plane wave of infinite extent. Thus, returning to the centre-of-mass frame, the complete time-dependent wavefunction is given by

$$\Psi = \psi e^{-iEt/\hbar} \tag{1.3}$$

and substituting this into the time-dependent equation

$$\left[-\frac{\hbar^2}{2\mu}\nabla^2 + V\right]\Psi = i\hbar\frac{\partial\Psi}{\partial t}$$

yields the usual time-dependent equation

$$\nabla^2\psi + \frac{2\mu}{\hbar^2}[E - V]\psi = 0\,. \tag{1.4}$$

The probabilities and cross-sections are now independent of time, and if $V \to 0$ as $r \to \infty$ the energy E is just the kinetic energy of the particle of reduced mass μ as given by equation (1.2).

Before the scattering occurs we have a collimated beam of particles moving along the z-direction with momentum $\hbar k = \mu v$. According to the argument given above, this beam is represented by the plane wavefunction $\psi = e^{ikz}$ which is the solution of equation (1.4) with $V = 0$. For this incident beam the probability density is

$$P_{\text{in}} = \psi^*\psi = 1$$

and the probability current density is

$$j_{\text{in}} = \frac{\hbar}{2\mu i}(\psi^*\nabla\psi - \psi\nabla\psi^*) = \frac{\hbar k}{\mu}$$

$$= v \text{ particles crossing unit area per sec.}$$

If the beam is scattered by a spherically symmetric potential $V(r)$ the scattered wave must have the form at large distances from the potential of an outgoing spherical wave with axial symmetry (see figure 1.1(b)). The wave-function of such a scattered wave is $(e^{ik'r}/r)f(\theta)$ where $f(\theta)$ is some function of the scattering angle θ which depends on the exact form of the potential. Hence for the scattered beam the probability density is

$$P_{\text{out}} = \frac{|f(\theta)|^2}{r^2}$$

and the probability current density is

$$j_{\text{out}} = \frac{\hbar k'}{\mu r^2}|f(\theta)|^2 .$$

On the surface of a sphere the element of area is $dS = r^2 d\Omega$, so that the current passing through an element of solid angle $d\Omega$ is $(\hbar k'/\mu)|f(\theta)|^2 d\Omega$, but the cross-section $d\sigma$ is just the ratio of the flux through the solid angle $d\Omega$ to the incident flux. Hence

$$d\sigma = \frac{k'}{k}|f(\theta)|^2 d\Omega$$

and for elastic scattering $k' = k$ so that

$$\frac{d\sigma}{d\Omega} = |f(\theta)|^2 .$$

Thus in order to calculate the cross-section we have to solve equation (1.4) with a given potential and find a solution ψ which has the asymptotic behaviour of an outgoing scattered wave and an incoming plane wave, i.e.

$$\psi \underset{r\to\infty}{\longrightarrow} e^{ikz} + f(\theta)\frac{e^{ik'r}}{r} .$$

The methods used to obtain these solutions are described in detail in Chapter 3.

The differential cross-section in the laboratory system is related to the differential cross-section in the centre-of-mass system by the expression [5]

$$\frac{d\sigma}{d\Omega}(\theta_{\text{lab}}) = \frac{(1+\gamma^2+2\gamma\cos\theta)^{\frac{3}{2}}}{|1+\gamma\cos\theta|}\frac{d\sigma}{d\Omega}(\theta)$$

where γ is the ratio of the speed of the centre-of-mass in the laboratory system to the speed of the scattered particle in the centre-of-mass system, and

$$\tan\theta_{\text{lab}} = \sin\theta/(\gamma+\cos\theta)\,.$$

The total cross-section, which is given by

$$\sigma = 2\pi\int_0^\pi \frac{d\sigma}{d\Omega}\sin\theta\,d\theta\,,$$

is the same in both systems since the probability per second for an incident particle to make a collision must be independent of the description of the scattering process.

1.2 Survey of nuclear reactions

The interactions between particles may be listed in order of decreasing strength as follows: strong or nuclear interaction, electromagnetic interaction, weak interaction, gravitational interaction [1,3,6]. The gravitational interaction is completely negligible in the context of nuclear reactions. The weak interaction plays an important role in nuclear physics since it is the cause of the slow decays such as the β-decay of the neutron but does not play a significant role in the processes which are generally classified under the heading of nuclear reactions. We are here concerned with scattering and reactions initiated by projectiles which interact with the target nuclei through the strong and/or electromagnetic interactions.

The electromagnetic interaction is in principle completely known, so that those projectiles which interact with the nucleus solely through the electromagnetic interaction should give information about nuclear structure which is fairly easy to interpret. The most important aspect of the electromagnetic interaction between charged particles is usually the Coulomb interaction due to the charges. This means that electron scattering from nuclei, for example, is sensitive primarily to the distribution of charge in the nucleus, but if the experimental conditions are chosen so that the Coulomb scattering is small, electron scattering can also be used to investigate the distribution of magnetic moment in the nucleus. One of the difficulties arising in the study of inelastic scattering and reactions due to electrons is that the energy spectra for these processes are superimposed on a bremsstrahlung or radiation tail due to the emission of radiation, hence loss of energy, by the electron during the scattering process or in the electromagnetic field of a different nucleus before and after scattering. An energy spectrum of electrons scattered from ^{12}C, showing the effect of the radiation tail, is given in figure 1.2.

Nuclear reactions and scattering have so far been considered in terms of the interaction of a beam of projectiles with a target, and it has been an implicit assumption that the energy of the projectiles is sufficiently high so that the extra-nuclear structure of the target may be ignored, i.e. the atomic structure of the target does not affect the nuclear reactions. The orbital electrons do, however, interact with the nuclear magnetic and quadrupole moments giving rise to the hyperfine structure seen in atomic spectra, so that information about the nucleus can be obtained from atomic proper-

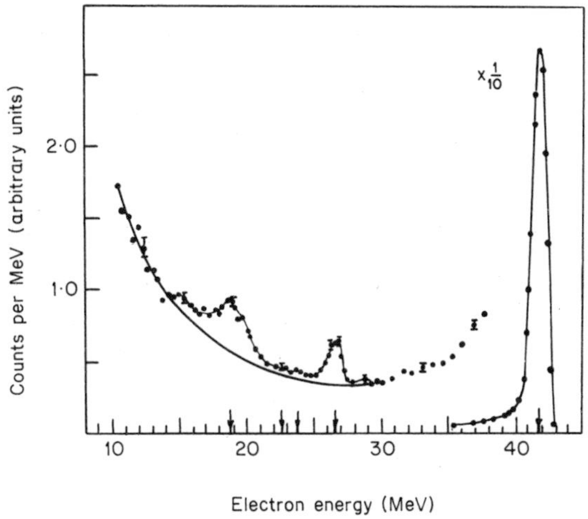

Figure 1.2 Energy spectrum of electrons of initial energy 42·5 MeV after 160° scattering from a carbon target. [From W. C. Barber, F. Berthold, G. Fricke and F. E. Gudden, *Phys. Rev.* **120,** 2081 (1960).]

ties [3]. Similarly, information can be obtained from the properties of muonic or μ-mesic atoms which are formed when a muon, which is essentially a heavy electron, is captured by an atom. The muon is usually captured into a state with high quantum number and cascades down to a lower, more tightly-bound, state with the emission of x-rays. Since, in the language of old quantum theory, the Bohr radius of an orbit is inversely proportional to the mass of the particle, the orbit of the heavier muon passes much closer to the nucleus than that of an electron with the same quantum numbers and the properties of the muonic atoms are much more sensitive to the details of the nuclear charge distribution.

Mesic atoms can also be formed by negatively charged pions and kaons [3]. In this case the meson also interacts with the nucleus through the strong interaction so that the meson can be captured from its extra-nuclear orbit and undergo a nuclear reaction. In these capture processes, the nuclear transition in question can often be studied under very different energy and momentum conditions from those arising in a reaction due to a projectile with considerable kinetic energy. More importantly, these capture processes involve both protons and neutrons in the nucleus and hence yield information on the nuclear matter distribution which is complementary to the information on the nuclear charge distribution obtained from electromagnetic interactions.

Nuclear scattering and reactions initiated by nucleons and other nuclear projectiles, such as deuterons and α-particles, still provide the major source of information on nuclear interactions and nuclear reaction mechanisms [6,7]. Our present understanding of these processes is illustrated in figure 6.3. The interaction between the nucleon and the nucleus can be represented by a potential which varies smoothly with mass number A and projectile energy E, and a residual interaction. Scattering due to the potential is called shape elastic or potential scattering. When the incident nucleon interacts with the nucleons in the nucleus through the residual interaction the target nucleus becomes excited. If the nucleon has moderately high energy, the major part of this energy is quickly carried away by an emitted particle, and a direct reaction or direct inelastic scattering occurs. If, however, the incident nucleon makes repeated interactions inside the nucleus, the structure of the $A+1$ system becomes very complicated and the energy of the incident nucleon becomes distributed among many particles. At this point it is said that a compound nucleus has been formed, and examination of the cross-section as a function of the energy of the incident nucleon shows sharp peaks or resonances which correspond to the formation of a quantized state of the $A+1$ system. Eventually the compound nucleus must decay and the incident nucleon may reappear to give compound elastic scattering or, more usually, a compound nucleus reaction occurs. The occurence of the various non-elastic processes means that nucleons are lost from the incident beam and hence the number of nucleons in the elastically scattered beam is reduced. This effect is taken into account in calculations by including in the potential a part which causes a reduction in the intensity of the scattered beam, and this total potential which describes the elastic scattering is called the optical potential by analogy with the refraction and absorption of light. (At an early stage in its development this approach to scattering was called the cloudy-crystal-ball model.) The same view may be taken of the scattering of composite particles.

Many experiments on direct inelastic scattering and reactions have been carried out, particularly with proton, deuteron, and α-particle beams. The deuteron breaks up very readily into its constituent proton and neutron or is stripped of its neutron by the target nucleus; because of this, deuteron stripping reactions were the first direct reactions to be studied in detail and the analysis of the data by Butler and others revealed the value of direct reactions in the study of the properties of low-lying states of nuclei. The analysis of these and similar reactions by more modern methods has strengthened this conclusion.

1.3 Guide to the organisation of the book

An indication of the function of the various chapters in this book is given in Table 1.1. Chapters 1–3 can be regarded as the introductory part of the book,

Table 1.1 Function of chapters.

Chapter	Purpose
1, 2	Introduction of definitions, formulae and vocabulary
3	Derivation and explanation of basic formalism of scattering theory
4	Study of electromagnetic processes from point of view of reaction theory
5, 6, 7, 8, 9	Study of standard theory of nuclear reactions and interpretation of data in terms of nuclear structure
10	Reviews of new types of reactions and theories so far applied to them

while Chapters 4–10 constitute the main text. Thus the main body of the work is concerned with nuclear reaction theory and its interpretation; the emphasis is on direct processes at ‘medium’ energies, which are defined here to be roughly between 25 MeV and 400 MeV. Resonance processes are included to give the necessary theoretical and experimental background but are not treated in the same detail as direct processes. Similarly, electromagnetic processes are dealt with in a limited way, from the point of view of their relation to nuclear reactions. For each process, the aim is to give an outline of the standard theoretical approach followed by a discussion of the way this is applied to obtain information about nuclear structure. A critical review of existing information and data is in general not included. However, Chapter 10 is much more in the form of a review of the current situation, as this chapter deals with new and rapidly developing topics.

The main part of the introduction is contained in Chapter 3, and is designed to give an outline of basic scattering formalism; this should be self-contained for any reader who has taken a preliminary course in quantum mechanics. Chapter 2 is not self-contained in the sense that it does not pretend to give an introduction to the theory of nuclear models; it should serve to explain and define the formulae and concepts required later in the book, but full understanding will require additional reading or a course in nuclear models or nuclear structure. Suitable texts are recommended in the bibliography.

The bibliography is given at the end of the book. The references are by chapters but with some cross-referencing. As far as possible, references are given to standard works and to review articles which contain lists of the important original papers. Footnotes to the main text are used to draw attention to alternative notation or definitions.

2 Nuclear models

2.1 The role of models in nuclear physics

The main objective of nuclear theory is the eventual description of all nuclear properties in terms of the interactions of the constituent nucleons. Unfortunately, the nucleus consists of a relatively small number (<250) of strongly interacting particles, and unlike the atom it is not obvious that there is any dominant central interaction such that the residual interaction between the particles can be treated as a small correction to the independent motion of the particles in the central field. Nor are there sufficient particles to make plausible the use of statistical methods to describe the bulk properties of the system. Furthermore, the nucleon–nucleon interaction, which is attractive within a separation distance of approximately 2 fm but becomes strongly repulsive within a separation distance of about 0.4 fm, and has velocity-dependent, spin-dependent, and non-central terms [8], is of a sufficiently complicated form that the theoretical techniques necessary for the description of such a system and the capacity and speed of computing systems have only recently advanced to the point where realistic calculations are feasible. It is therefore not surprising that much of the interpretation of nuclear structure during the past thirty years has proceeded through the study of nuclear models.

When we make a model of the nucleus, or of any other physical system, we deliberately simplify our study of the system and concentrate on particular features by constructing the model in such a way that it closely represents only these chosen features of the system. If the model is to be useful it must lead to a simplification of the mathematics of the problem but its primary purpose is to provide insight into the system and to provide a framework for the analysis and correlation of the data. We do not expect any given model to be completely successful or even uniformly successful but instead we study situations in which the model succeeds or fails in order to assess its relevance and validity. By isolating particular features of the system in this way, it is possible to study the relation between these features and the experimental data. It is essential that the set of assumptions on which the model is based should be expressible in mathematical terms and should lead to precise predictions for observable quantities. It is not essential that these assumptions should lead to a precise visual picture of the system although if such a visual representation is possible it is often very helpful.

There is no reason why several models developed from different assumptions should not be used to describe the same system and, in nuclear physics, both collective models and independent-particle models have been used. The term *collective* implies that a large group of nucleons in the nucleus are participating in some coherent motion. Thus the collective models emphasize the features of the system related to this coherent or co-operative motion, and they give a description of the macroscopic or bulk properties of the nucleus, such as shape, size and collective modes of excitation. The independent-particle models emphasize the opposite view of the nucleus as a system of individual nucleons each moving essentially independently of the others. Such models describe the *microscopic* or individual behaviour of the nucleons which underlies the behaviour of the group. In this chapter, the basic features of these models and some essential formulae are introduced in preparation for their use in the study of nuclear reactions. For a detailed discussion the reader must refer to texts on nuclear models and review articles [9–16].

2.2 Independent-particle models

The observation that nuclei possess exceptional stability when the proton number Z or the neutron number N has the value 2, 8, 20, 28, 50, 82, or 126, led to the suggestion that nuclei have a shell structure similar to the electron shells in atoms [16]. This implies that the nucleons in the nucleus move in some central spherically-symmetric potential and that they do so essentially independently of each other. The states in this shell model potential are called single-particle states and can be characterized by a set of quantum numbers. They can be filled in order of increasing energy according to the Pauli exclusion principle, exactly as in the case of electron bound states in atoms. In the case of atoms the Coulomb attraction of the nucleus provides an evident and usually dominant central interaction with reference to which the electron–electron repulsions may be treated as perturbations. In contrast, it is far from obvious that the concept of independent-particle motion in the nucleus is a useful one, even as a lowest order approximation, in view of the strong short-range nature of the nucleon–nucleon interaction, but consideration of the effect of the exclusion principle shows that such independent motion is indeed possible. When two nucleons collide inside the nucleus their energies and momenta are changed by the collision so that after the collision each nucleon has a different set of quantum numbers from that possessed before collision; however, all the states with energies near to those of the colliding nucleons are already occupied and it follows from the exclusion principle that these states are not available to the colliding nucleons. Hence collisions are largely prevented and nucleons can maintain their independent motions with long mean free paths. The same argument applies when an incident nucleon is scattered from a nucleus and leads to the optical model of nuclear scattering (see section 3.6 and Chapter 6) but in this case the exclusion principle is less effective because the incident nucleon has extra energy and can excite the target nucleons into empty states above the highest filled state.

The shell model potential in which a nucleon moves is derived from the interactions of the nucleon with its nearest neighbours. Inside the nucleus, the potential will therefore be essentially constant, but it must diminish in magnitude in the surface where some of the nearest neighbours are missing, and somewhere outside the nucleus the potential must go to zero. The precise choice of the radial shape of the potential is rarely of crucial importance, but a popular choice is the Saxon-Woods form shown in figure 2.1 which has a radial shape similar to that of the nuclear density distribution. To this nuclear potential $V(r)$ must be added the Coulomb potential $V_c(r)$ arising from the electrostatic repulsion between the protons. This potential does not, however, lead to shell closure at the correct number of nucleons unless an additional term, the spin-orbit potential $V_{so}(r)\boldsymbol{l}.\boldsymbol{\sigma}$, is added to the other terms. The effect of this term is to cause a splitting between the energies of single particle states with total angular momentum $j=l+\frac{1}{2}$ and $j=l-\frac{1}{2}$, (where $\boldsymbol{j}=\boldsymbol{l}+\boldsymbol{s}$ and $\boldsymbol{s}=\frac{1}{2}\boldsymbol{\sigma}$), and with a suitable strength for the spin-orbit term the shell closure occurs at the required number of nucleons, as shown in figure 2.2. The position of the single particle levels and the corresponding wavefunctions $\phi_{nlj}(\boldsymbol{r})$ are determined by solving the Schrödinger equation for a given set of potential parameters. Alternatively, if the single-particle energies are known (see section 4.2), the potential parameters may be adjusted to give the required energies and the corresponding wavefunctions. In either case,

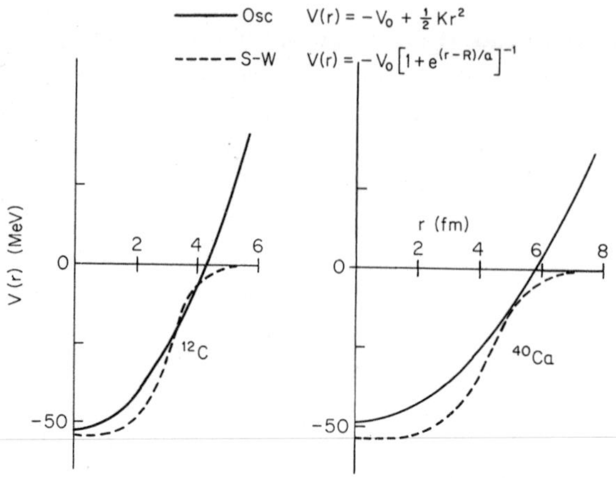

Figure 2.1 The Saxon-Woods and oscillator forms of the shell model potential. In each case the parameters are chosen to give the correct binding energy for the least bound nucleon.

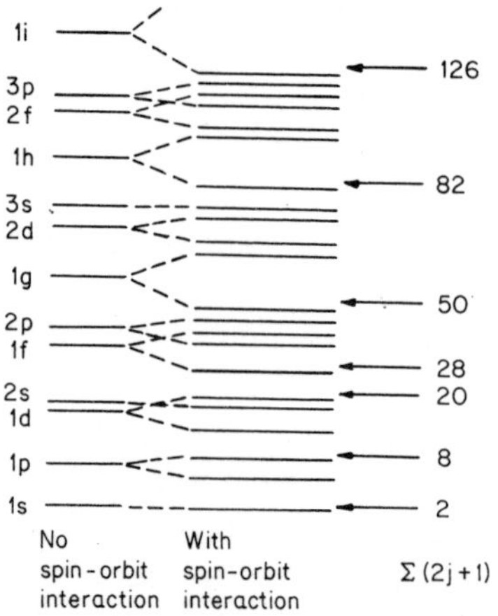

Figure 2.2 Single-particle levels in a realistic shell model potential.

the Schrödinger equation must be solved numerically if the Saxon-Woods or similar radial shapes are used. It is sometimes permissible to neglect the spin-orbit potential and to use nuclear potentials which yield analytic forms for the wavefunctions. For this purpose, the square well potential is sometimes used despite its unrealistic sharp boundary, but the interior and exterior solutions must be matched at the boundary and this removes much of the simplicity. More useful analytic functions are obtained from the harmonic oscillator potential, also shown in figure 2.1, the radial solutions of which can be written in the form

$$R_{nl}(r) = A_{nl}(r/a)^l \, {}_1F_1[-(n-1); l+\tfrac{3}{2}; (r/a)^2] e^{-\frac{1}{2}(r/a)^2} \tag{2.1}$$

where A_{nl} is the normalization constant, and ${}_1F_1(\alpha; \beta; x)$ is the confluent hypergeometric function [17]. The energy levels are given by

$$E_{nl} = [2(n-1)+l+\tfrac{3}{2}]\hbar\omega \tag{2.2}$$

$$\hbar\omega = \hbar^2/ma^2 = \hbar(K/m)^{\frac{1}{2}}. \tag{2.3}$$

Thus the oscillator potential gives rise to analytic functions of $(r/a)^2$ which are particularly convient for algebraic manipulation (see sections 2.5, 4.2, 9.4). Because these functions fall as $e^{-\frac{1}{2}(r/a)^2}$ with large r instead of $e^{-\alpha r}$ where α is determined by the single-particle energy, they give unreliable results for any matrix elements which are sensitive to the detailed behaviour of the wavefunction. Many standard computer codes now exist for the solution of the Schrödinger equation with realistic shell model potentials and it is always possible to expand the corresponding wave functions in terms of a complete set of oscillator functions.

If we consider the sum of the two-body interactions V_{ij} which can take place between the nucleons in a nucleus, the total potential can be written in the form

$$\begin{aligned} V &= \tfrac{1}{2} \sum_{i \neq j} V_{ij} \\ &= \sum_i V_i + \tfrac{1}{2} \sum_{i \neq j} W_{ij} \end{aligned} \tag{2.4}$$

where V_i is the shell model potential for the ith nucleon and ΣW_{ij} represents all the residual interactions which cannot be incorporated into the shell model potential. The complete neglect of these residual interactions leads to the *extreme independent-particle model* which is unlikely to have very wide validity. The term *independent-particle model* is usually used to describe the situation in which the residual interactions between the nucleons in the last unfilled shell are taken into account while the remaining nucleons are considered to form an inert core [16,18]. The residual interactions in the unfilled shell are likely to be the most important since the exclusion principle is least effective here in preventing collisions, and the effect of including them is to remove some of the degeneracy of the shell model states. For example, the states formed by nucleons within a given shell model configuration rearranging themselves to different total angular momenta J are degenerate in the extreme model but are split by the residual interactions to give a ground state and several low-lying excited states. Further, the residual interactions will lead to the admixture into each of these states of components arising from more complicated shell model configurations in which some of the nucleons are promoted to higher unfilled single-particle levels. This situation is known as *configuration mixing*.

The total angular momentum of the group of nucleons in the unfilled shell is obtained by coupling together the angular momenta of the individual nucleons [16, 18]. If there is no spin-orbit potential and the two-body interactions are central the total orbital angular momentum L is a good quantum number and the L–S coupling sheme is appropriate, but in the presence of a pure spin-orbit potential the total angular momentum is the only good quantum number so that the j–j coupling scheme must be used. In practice, as noted earlier, the shell model potential must contain both central and spin-orbit terms so that the true coupling situation is one of *intermediate coupling*, but the true nuclear wavefunctions may be expanded in terms of basis states in either coupling scheme

In addition to the single-particle quantum numbers which characterize the spin and spatial state of the nucleon, it is convenient to introduce the quantum numbers of I-spin (isotopic, or more accurately *isobaric spin*) to characterize the proton and neutron as different charge states of the nucleon [7, 16, 18]. This is done through the I-spin operator $\boldsymbol{t}$ with eigenvalues $t_z = \pm\frac{1}{2}$. The total I-spin of a nucleus is then given by $\boldsymbol{T} = \Sigma \boldsymbol{t}_i$ where the sum runs over all the nucleons in the nucleus. There are, unfortunately, several different conventions for the I-spin. In high energy physics the convention is to allocate quantum numbers (t, t_z) as follows,

$$\text{proton } (\tfrac{1}{2}, +\tfrac{1}{2}), \qquad \text{neutron } (\tfrac{1}{2}, -\tfrac{1}{2}). \tag{2.5}$$

This will be referred to as *convention A* and used in Chapter 10 in connection with pion-nucleon interactions. In nuclear physics the more common usage is

$$\text{proton } (\tfrac{1}{2}, -\tfrac{1}{2}), \qquad \text{neutron } (\tfrac{1}{2}, +\tfrac{1}{2}) \tag{2.6}$$

which will be referred to as *convention B* and used in Chapters 6, 7, and 9*.

* The use of two conventions in a short book is unfortunate. However, it is hoped that the explicit discussion of these conventions may help to remove confusion when students progress to original papers.

In the latter case we have

$$T_z = \sum t_z = \tfrac{1}{2}(N-Z) \tag{2.7}$$

so that $2T_z$ is equal to the neutron excess. Thus if T is a good quantum number we would expect a nucleus with given T, T_z to be one of $(2T+1)$ states of an I-spin multiplet. In practice, even assuming charge independence of nuclear forces, T is in principle not a good quantum number because of the Coulomb interaction of the protons, but nevertheless it is found to be a very useful quantum number for light nuclei for which the neutron excess is small, and even for heavy nuclei is a useful concept in particular circumstances, such as the interpretation of isobaric analogue states (see section 7.5).

2.3 Collective models

In table 2.1 a comparison is made of some basic properties of nuclei, atoms, and liquids. It appears that in the nucleus these properties are essentially independent of the number of nucleons since the addition of more nucleons does not change the density or the binding energy per nucleon. Thus the nucleus, like the liquid drop, shows this phenomenon of *saturation*, and in the nucleus this is due to the combined effect of the short range repulsive part of the nucleon–nucleon interaction and of the exclusion principle which limits the number of nucleon pairs which can interact strongly. The static *liquid drop model* of the nucleus, based on this similarity, was the first successful nuclear model and was used to describe the bulk properties of nuclei such as nuclear masses and binding energies and to give a qualitative explanation of the process of nuclear fission.

The dynamical theory of the collective motion of a liquid drop is developed in terms of the collective oscillations of the surface. The behaviour of the surface may be described by the expression [10, 12]

$$R(\theta, \phi) = R_0\left[1 + \sum_{\lambda\mu} \alpha_{\lambda\mu} Y_\lambda^\mu(\theta, \phi)\right] \tag{2.8}$$

where θ, ϕ are the polar angles with respect to an arbitrary space-fixed axis, and Y_λ^μ is a spherical harmonic of order λ and projection μ. If the drop is incompressible R_0 is the radius of the undisturbed spherical surface. Collective motion is described by allowing the coefficients $\alpha_{\lambda\mu}$ to vary with time so that they become generalized collective co-ordinates, and for small oscillations and irrotational flow the kinetic energy of the collective motion is given by [10, 12]

Table 2.1

	Atoms	Liquids	Nuclei
Density	The density of electrons in atoms is not uniform. The density is much higher near the nucleus than in the outer regions.	The density has a unique value independent of the number of molecules in the drop, provided this is sufficiently large.	The density of nucleons in nuclei is approximately constant from the centre out to a thin surface region where it falls off rapidly.
Volume	The volume occupied by the electrons does not increase systematically with Z, but tends to be related to the position in the Periodic Table.	The volume increases with the number of molecules.	To a good approximation a size parameter for nuclei can be defined as $R = r_0 A^{\ddagger}$ where r_0 is a constant. This gives the volume as $(4\pi/3^-)r_0^3 A$.
Total binding energy	The total binding energy is determined by the inner electrons which experience almost the full Coulomb attraction of the nucleus, and increases roughly as $Z^{7/3}$.	The total binding energy increases with the number of molecules.	To a good approximation the total binding energy is proportional to A.

$$T = \tfrac{1}{2}\sum_{\lambda\mu} B_\lambda |\dot{\alpha}_{\lambda\mu}|^2 = \tfrac{1}{2}\sum_{\lambda\mu} \frac{1}{B_\lambda} |\pi_{\lambda\mu}|^2 , \tag{2.9}$$

where
$$B_\lambda = \rho R_0^5/\lambda \tag{2.10}$$

is the mass parameter for constant density ρ, and

$$\pi_{\lambda\mu} = B_{\lambda\mu}\dot{\alpha}_{\lambda\mu} \tag{2.11}$$

is the generalized momentum conjugate to the co-ordinate $\alpha_{\lambda\mu}$. The potential energy is

$$V = \tfrac{1}{2}\sum_{\lambda\mu} C_\lambda |\alpha_{\lambda\mu}|^2 \tag{2.12}$$

where C_λ is the restoring force parameter. For a classical liquid, C arises only from the surface tension but for a charged system there is an additional Coulomb term, so that

$$C_\lambda = SR_0^2(\lambda-1)(\lambda+2) - \frac{3Z^2e^2}{2\pi R_0}\frac{\lambda-1}{2\lambda+1} \tag{2.13}$$

where S is the coefficient of surface tension. For oscillations about a spherical shape C_λ, B_λ are independent of μ. The Hamiltonian for the system is thus

$$H = T+V = \tfrac{1}{2}\sum_{\lambda\mu}\left[\frac{1}{B_\lambda}|\pi_{\lambda\mu}|^2 + C_\lambda|\alpha_{\lambda\mu}|^2\right] \tag{2.14}$$

which is just the Hamiltonian for a set of uncoupled harmonic oscillators with frequencies

$$\omega_\lambda = [C_\lambda/B_\lambda]^{\frac{1}{2}} . \tag{2.15}$$

From equations (2.10) and (2.13) it follows that $\omega_\lambda = 0$ for $\lambda = 0, 1$. The term $\lambda = 0$ would describe density oscillations of a spherical nucleus, which could occur but at much higher excitation energies than incompressible vibrations. The terms with $\lambda = 1$ describe motion of the centre of mass of the system.

The Hamiltonian (2.14) is quantized through the commutator requirement that

$$[\pi_{\lambda\mu}, \alpha_{\lambda'\mu'}] = -i\hbar\delta_{\lambda\lambda'}\delta_{\mu\mu'}$$

and it is convenient to introduce creation and destruction operators $b^\dagger_{\lambda\mu}$, $b_{\lambda\mu}$ which create or destroy a *phonon** in the state $\lambda\mu$. In terms of these operators the generalized co-ordinates are

* (Since the phonon is the quantum of the surface oscillations it is sometimes called a *surfon*.)

$$\alpha_{\lambda\mu} = (\hbar\omega_\lambda/2C_\lambda)^{\frac{1}{2}}[b_{\lambda\mu} + (-1)^\mu b^\dagger_{\lambda-\mu}] \tag{2.16}$$

and the excitation energies of the system are $\Sigma_\lambda n_\lambda \hbar\omega_\lambda$, where n_λ is the number of phonons of order λ in the excited state. The parity of the phonon is given by $(-1)^\lambda$. This means that the lowest one-phonon states will be a 2^+ state ($\lambda = 2$) and a 3^- state ($\lambda = 3$) and above these will lie the two-phonon states 0^+, 2^+, 4^+ formed by coupling two $\lambda = 2$ phonons and the states $0^+, 2^+, 4^+, 6^+$ formed by coupling two $\lambda = 3$ phonons. The degeneracy of the two-phonon states should be removed by residual interactions so that these states would appear as multiplets.

If the energy of deformation is much larger than the zero-point vibrational energies the nucleus can take up a permanent non-spherical shape, and can in the extreme case be treated as a non-spherical rigid rotator [10, 12]. It is convenient in this case to define the surface in the body-fixed system as

$$R(\theta', \phi') = R_0[1 + \sum_{\lambda q} \alpha'_{\lambda q} Y^q_\lambda(\theta', \phi')] \tag{2.17}$$

where θ', ϕ' are the polar angles in the body-fixed system, and by comparison with equation (2.8) we have

$$\sum_{\lambda\mu} \alpha_{\lambda\mu} Y^\mu_\lambda(\theta, \phi) = \sum_{\lambda q} \alpha'_{\lambda q} Y^q_\lambda(\theta', \phi') .$$

The spherical harmonics transform as

$$Y^\mu_\lambda(\theta, \phi) = \sum_q Y^q_\lambda(\theta', \phi') \mathscr{D}^\lambda_{\mu q}(\hat{s})$$

where $\hat{s}$ denotes the three Euler angles of the body-fixed system with respect to the space-fixed system and $\mathscr{D}$ is a rotation matrix* so that

$$\alpha_{\lambda\mu} = \sum_q \alpha'_{\lambda q} \mathscr{D}^{\lambda *}_{\mu q}(\hat{s}) , \quad \alpha'_{\lambda q} = \sum_\mu \alpha_{\lambda\mu} \mathscr{D}^\lambda_{\mu q}(\hat{s}) . \tag{2.18}$$

In the rotational model the state vectors are denoted by $|JMK\rangle$ where M and K are the projections of the total angular momentum J on the z-axis in the space-fixed and body-fixed systems respectively (see figure 2.3).

$$J^2|JMK\rangle = J(J+1)|JMK\rangle \tag{2.19}$$

$$J_z|JMK\rangle = M|JMK\rangle , \quad J_3|JMK\rangle = K|JMK\rangle . \tag{2.20}$$

* For the equation of the surface and the definition of the rotation matrix we use the notation of Preston [12]. The relation between these rotation matrices and those of other authors is given by Preston in an appendix.

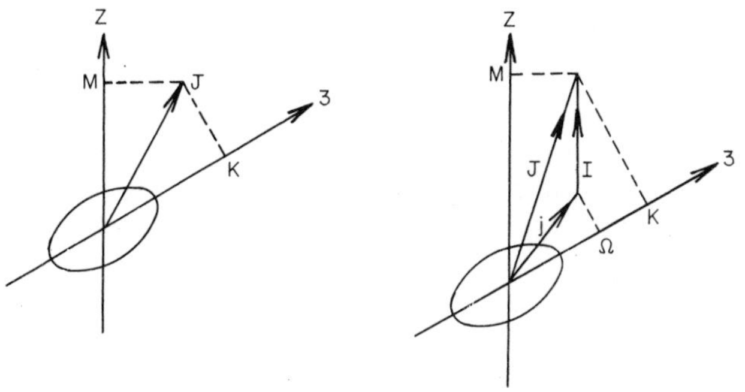

Figure 2.3 (**a**) Definition of quantum numbers for a defined even-even nucleus. (**b**) Definition of quantum numbers for a nucleus with one nucleon outside a deformed core.

The Hamiltonian for a deformed rigid rotator is

$$H = \frac{\hbar^2}{2} \sum_{k=1}^{3} \frac{J_k^2}{\mathscr{I}_k}$$

where the $\mathscr{I}_k$ are the moments of inertia, and for an axially symmetric rotator with the 3-axis as the symmetry axis we have

$$\mathscr{I}_1 = \mathscr{I}_2 = \mathscr{I}$$

$$H = \frac{\hbar^2}{2}\left[\frac{\boldsymbol{J}^2}{\mathscr{I}} + \left(\frac{1}{\mathscr{I}_3} - \frac{1}{\mathscr{I}}\right) J_3^2\right]$$

$$E = \frac{\hbar^2}{2}\left[\frac{J(J+1)}{\mathscr{I}} + \left(\frac{1}{\mathscr{I}_3} - \frac{1}{\mathscr{I}}\right) K^2\right],$$

and the wavefunction are just the rotation matrices $\mathscr{D}^J_{MK}$. The experimental evidence indicates the absence of rotational states in spherical nuclei at least for low excitations, i.e. there is no rigid rotation about a symmetry axis, and from this it follows that the projection of J along the 3-axis is zero. Thus the simplest low-lying states have $K=0$ and energies given by

$$E_J = \frac{\hbar^2}{2\mathscr{I}} J(J+1), \quad J = 0, 2, 4, \dots . \tag{2.21}$$

The absence of states with $K=0$ and $J=1, 3, 5, \dots$ arises from the symmetry with respect to a plane perpendicular to the symmetry axis. When the conditions of complete rigidity are relaxed, corrections to the energy expression (2.21) arise.

The simplest model for deformed even-even nuclei is the *collective model* due to Bohr and Mottelson [14] which restricts the description to quadrupole deformation. It is assumed that the nucleus can rotate and vibrate but that the vibrations are small so that the rotating nucleus has a reasonably permanent shape. The system thus has five degrees of freedom which are represented by the $\alpha_{2\mu}$ in the space-fixed system, or by the three Euler angles and α'_{20}, α'_{22}, in the body-fixed system. (The axes in the body-fixed system can be chosen so that $\alpha'_{21}=\alpha'_{2-1}=0$ and $\alpha'_{22}=\alpha'_{2-2}$.) It is usual to replace the $\alpha'_{2\mu}$ co-ordinates through the relations

$$\alpha'_{20}=\beta_2 \cos\gamma\,, \quad \alpha'_{22}=\alpha'_{2-2}=\beta_2 \frac{1}{\sqrt{2}} \sin\gamma\,.$$

Since

$$\sum_{\mu} |\alpha'_{2\mu}|^2 = \beta_2^2\,,$$

and from equation (2.8) $\beta_2 \simeq \delta R/R$, the quantity β_2 is a measure of the mean deformation and is usually referred to as the *deformation parameter*. The states of this system can be grouped into a number of *rotational bands* the lowest of which has $K=0$ and energies given by equation (2.21) with the value of $\mathscr{I}$ determined by the values of β_2 and γ. There can be additional bands with $K=0$ which correspond to larger values of β_2. These bands are associated with vibrations in the β degree of freedom and are referred to as *β-vibrations.* Vibrations in the γ degree of freedom gives rise to bands which are referred to as *γ-vibrations*, and the change in γ leads to loss of axial symmetry in the excited states of these bands which have $K \geqslant 2$. For the case of axial symmetry, the wavefunctions for these states can be written in the product form

$$|JMK; xy\rangle = f_{JKxy}(\beta) g_{JKx}(\gamma) |JMK\rangle \tag{2.22}$$

$$|JMK\rangle = \left[\frac{2J+1}{16\pi^2(1+\delta_{Ko})}\right]^{\frac{1}{2}} \{\mathscr{D}^J_{MK} + (-1)^{J+K} \mathscr{D}^J_{M-K}\} \tag{2.23}$$

where x, y are the vibration quantum numbers.

In the extension of this model to odd-even and odd-odd nuclei it is assumed that the odd protons and/or neutrons are coupled to a deformed even-even core so that the extra-core nucleons move in the potential due to the core. Thus if there is a single nucleon outside the core, the total angular momentum $\boldsymbol{J}$ of the nucleus is obtained by coupling the total angular momentum $\boldsymbol{j}$ of the nucleon to the total angular momentum of the core $\boldsymbol{I}$ (see figure 2.3). Since the core is deformed, $\boldsymbol{j}^2$ is not a constant of the motion although the projection Ω of j on the 3-axis will be a good quantum number if there is axial symmetry.

2.4 Microscopic models of collective effects

The collective models described in section 2.3 give a description of collective effects but do not provide any information about the behaviour of the individual nucleons. Since the collective effects must be generated by the microscopic motions of the nucleons, it should be possible to describe them by microscopic models, although such models will inevitably be more complicated in structure than the macroscopic models. The basis of these microscopic models lies in an understanding of the way that nucleons influence the motion of other nucleons [13, 14, 19].

If one nucleon is in a single-particle state outside a closed shell core its density distribution is non-spherical and so it yields a non-spherical contribution to the average potential. A second nucleon added to the unfilled shell will tend to correlate its motion with the average potential due to the first, and so on as more nucleons are added. This situation is described by a wavefunction constructed in the *aligned coupling* scheme from single-particle states derived from the average and non-spherical potential which they themselves generate. Such a scheme takes into account the long range part of the nucleon–nucleon interactions which produces the average potential and allows a microscopic description of an intrinsically non-spherical shape. The low-lying excited states of the system are just the rotational states described in section 2.3, and the nuclear wavefunctions can be expressed in a form similar to that given by equations (2.22) and (2.23), but with the inclusion of intrinsic wavefunctions constructed from the single-particle wavefunctions generated in the deformed potential. The calculation of these single-particle states was first described by Nilsson [20] using a potential of oscillator type, and has since been extended to more realistic potentials.

It may be said that the aligned coupling scheme describes the long-range correlations between nucleons in the nucleus arising from the correlation of the individual nucleon motions with the average potential. A different type of correlation is associated with the short-range part of the nucleon–nucleon interaction which gives rise to *pairing effects* [13, 21]. Because the Pauli principle precludes the possibility of more than two protons or two neutrons coming very close together, the optimum wavefunction in the presence of short-range forces consists of a pair-wise correlation of the nucleons of the same type. These nucleon pairs have total angular momentum zero and a spherically symmetric density distribution, and therefore give rise to a spherically symmetric average potential. Thus there is a competition between the pairing effects and the aligned coupling effects. It is found that pairing is dominant when the number of nucleons in the unfilled shell is small and the energy spectra show the large gap between the lowest $J=0$ state and the higher states of the j^2 configuration which is characteristic of a short range force. As the number of nucleons increases the average potential influences the time-variation or fluctuation of the pairs giving rise to collective vibrational states. Finally, when the number of nucleons is sufficiently large the deformed potential becomes stable and the rotational behaviour is apparent.

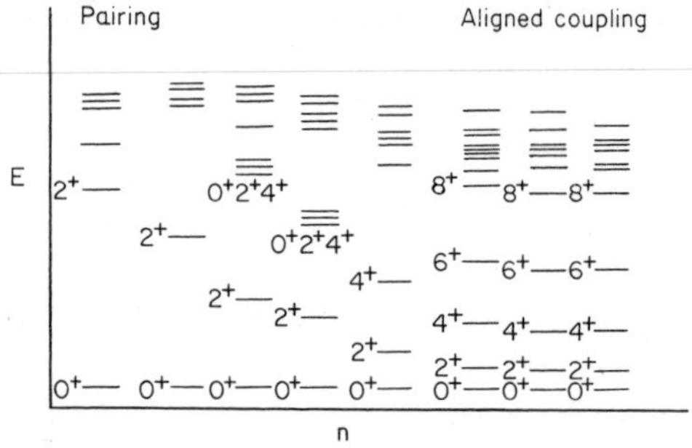

Figure 2.4 The behaviour of the spectra of even-even nuclei as a function of the number of nucleons n in the unfilled shell.

The trend of the spectra through the shell is shown schematically in figure 2.4. One important feature of pairing is due to the fact that the correlated pair of nucleons exist in a mixture of single particle states, so that there is an effect similar to configuration mixing and particles from the highest filled levels are promoted to adjacent unfilled levels.

The existence of pairing in nuclei was assumed from the early days of the shell model, when Meyer and Jensen [16] proposed that the nucleons coupled in pairs to zero spin so that the total spin of an odd nucleus was given by the last unpaired nucleon, and this idea was formalized into a set of rules for determining the spins of nuclei. Also, Racah developed the formal theory of nucleons interacting through a δ-function force, called the *seniority scheme* [18]. But the full importance of the pairing effect in nuclei was not realized until the theory of superconductivity was developed by Bardeen, Cooper and Schrieffer.

It is clear from the preceding discussion that the relation between the single-particle states of individual nucleons and the average potential which these nucleons generate is of crucial importance, and the quantitative development of the microscopic approach requires that we should be able to generate the average potential and the single-particle states in a self-consistent way. This can be done through the *Hartree-Fock approximation.* In this method the many-body wavefunction is written as an antisymmetrized product, or determinant, of single-particle wavefunctions. These wavefunctions are then determined through a variational procedure by minimizing the energy of the ground state E_0. This means that the wavefunction of the Hartree-Fock ground state is that single-particle product wavefunction which gives the best (minimum) value of the total energy. The variational procedure leads to a set of single-particle equations of the form [13, 22]

$$T\phi_i(\mathbf{r})+V(\mathbf{r})\,\phi_i(\mathbf{r})-\int U(\mathbf{r},\mathbf{r}')\,\phi_i(\mathbf{r}')d\mathbf{r}'=\varepsilon_i\phi_i(\mathbf{r}) \tag{2.24}$$

where the potentials are given in terms of the two-body interaction $V(\mathbf{r}-\mathbf{r}')$ as

$$\left.\begin{aligned} V(\mathbf{r}) &= \sum_{j=1}^{A}\int \phi_j^*(\mathbf{r}')\,V(\mathbf{r}-\mathbf{r}')\,\phi_j(\mathbf{r}')d\mathbf{r}' \\ U(\mathbf{r},\mathbf{r}') &= \sum_{j=1}^{A}\phi_j^*(\mathbf{r}')\,V(\mathbf{r}-\mathbf{r}')\,\phi_j(\mathbf{r})\,. \end{aligned}\right\} \tag{2.25}$$

These equations can be solved to find the ϕ_i and ε_i, and from these the total energy

$$\begin{aligned} E_0 &= \sum_i \langle T\rangle_i + \tfrac{1}{2}\sum_{i\neq j}\langle V\rangle_{ij} \\ &= \sum_i \varepsilon_i - \tfrac{1}{2}\sum_{i\neq j}\langle V\rangle_{ij} \end{aligned}$$

can be calculated. Thus the ith nucleon moves in an average field generated by the other nucleons which is non-local and state-dependent. The Hartree-Fock method has been used to calculate properties of nuclear ground states, including the equilibrium shape. The repulsive core in the nucleon–nucleon interaction can be incorporated in Hartree-Fock type calculations using a modified approach [13] developed from the work of Brueckner.

2.5 Centre of mass motion in the shell model

The law of conservation of linear momentum imposes the condition that the Hamiltonian for a system of nucleons is translation invariant, i.e. it must depend only on relative co-ordinates. In the shell model, however, the use of a single-particle potential introduces an arbitrary origin of co-ordinates, namely the origin of the potential, and the corresponding wavefunctions describe an additional, spurious, collective motion in which the centre of mass of the nucleus oscillates about the origin of the potential [22–25]. As shell model wavefunctions are used very frequently in later chapters, we include here a discussion of the effects of centre of mass motion.

In order to see how this centre of mass motion comes about we consider first the Hamiltonian

$$H = \sum_i \frac{\hbar^2 p_i^2}{2m} + \sum_{i>j} V_{ij}(\boldsymbol{r}_i - \boldsymbol{r}_j) \tag{2.26}$$

where $\boldsymbol{r}_i$ represents the co-ordinates of the ith nucleon relative to some arbitrary origin, p_i is the conjugate momentum and the potential V_{ij} is a function of the relative co-ordinate $\boldsymbol{r}_i - \boldsymbol{r}_j$. This Hamiltonian can be rewritten using the *Gartenhaus and Schwartz transformation*

$$\boldsymbol{\rho}_i = \boldsymbol{r}_i - \boldsymbol{R}\,, \quad \boldsymbol{q}_i = \boldsymbol{p}_i - \frac{1}{A}\boldsymbol{P} \tag{2.27}$$

where $\boldsymbol{R}$ and $\boldsymbol{P}$ are the position vector and momentum of the centre of mass, so that

$$\boldsymbol{R} = \frac{1}{A}\sum_i \boldsymbol{r}_i\,, \quad \boldsymbol{P} = \sum_i \boldsymbol{p}_i \tag{2.28}$$

and

$$\sum r_i^2 = \sum \rho_i^2 + AR^2\,, \quad \sum p_i = \sum q_i^2 + \frac{1}{A}P^2\,. \tag{2.29}$$

(It may be noted that the co-ordinates $\boldsymbol{\rho}_i$ are not independent, since $\sum_{i=1}^{A} \boldsymbol{\rho}_i = 0$, so that the number of independent relative co-ordinates is less than the number of nucleons.) The Hamiltonian now becomes

$$H = \frac{\hbar^2 P^2}{2Am} + H_{\text{int}}(\boldsymbol{q}_1 \ldots \boldsymbol{q}_{A-1}, \boldsymbol{\rho}_1 \ldots \boldsymbol{\rho}_{A-1})$$

and the wavefunction of the system is

$$\Psi_{Pn}(\boldsymbol{R}, \boldsymbol{\rho}_1 \ldots \boldsymbol{\rho}_{A-1}) = A^{-\frac{1}{2}}(2\pi)^{-\frac{3}{2}} e^{i\boldsymbol{P}\cdot\boldsymbol{R}} \Phi_n(\boldsymbol{\rho}_1 \ldots \boldsymbol{\rho}_{A-1})\,. \tag{2.30}$$

We may now choose a frame of reference in which the centre of mass of the system is at rest, i.e. the c.m. system, but this is not necessary. In an arbitrary frame of reference the matrix element of an operator, such as $e^{i\boldsymbol{q}\cdot\boldsymbol{r}_i}$, is given by

$$X_{mn}=A^{-1}(2\pi)^{-3}\int e^{-i\boldsymbol{P}'.\boldsymbol{R}}\Phi_m^*(\rho_1\,\ldots\,\rho_{A-1})e^{i\boldsymbol{q}.(\rho_1+\boldsymbol{R})}$$

$$\times\Phi_n(\rho_1\,\ldots\,\rho_{A-1})e^{i\boldsymbol{P}.\boldsymbol{R}}A\,d\boldsymbol{R}\,d\rho\,\ldots\,d\rho_{A-1}$$

$$=\delta(\boldsymbol{P}-\boldsymbol{P}'+\boldsymbol{q})\int\Phi_m^*(\rho_1\,\ldots\,\rho_{A-1})e^{i\boldsymbol{q}.\rho_1}\Phi_n(\rho_1\,\ldots\,\rho_{A-1})$$

$$\times\,d\rho_1\,\ldots\,d\rho_{A-1} \tag{2.31}$$

where the δ-function simply expresses momentum conservation, so that the matrix element involves only the internal wavefunctions.

If we now assume that the interaction ΣV_{ij} can be replaced in part by a shell model potential of oscillator form, the Hamiltonian (2.26) becomes

$$H=\sum_i\frac{\hbar^2p_i^2}{2m}+\sum_i\tfrac{1}{2}Kr_i^2+\sum_{i>j}W_{ij}=H^{\text{sm}}+\sum_{i>j}W_{ij},$$

where ΣW_{ij} represents the sum of the residual interactions and H^{sm} is the oscillator form for the shell model Hamiltonian. Using the Gartenhaus and Schwartz transformation H^{sm} becomes

$$H^{\text{sm}}=\frac{\hbar^2P^2}{2Am}+\tfrac{1}{2}AKR^2+H_{\text{int}}^{\text{sm}}(q_1,\,\ldots\,q_{A-1},\rho_1\,\ldots\,\rho_{A-1}) \tag{2.32}$$

and the wavefunction of the system is

$$\Psi_{\alpha n}^{\text{sm}}(\boldsymbol{R},\rho_1,\,\ldots\,\rho_{A-1})=\phi_\alpha(\boldsymbol{R})\Phi_n(\rho_1,\,\ldots\,\rho_{A-1}) \tag{2.33}$$

where ϕ_α is now not a plane wave but an oscillator function describing the motion of the centre-of-mass in an oscillator potential with energy spacing $\hbar(AK/Am)^{\frac{1}{2}}=\hbar\omega$, which is the same as the spacing of the single-particle states (see equations (2.2), and (2.3)). The energy of the system is

$$E_{\alpha n}^{\text{sm}}=(\alpha+\tfrac{3}{2})\hbar\omega+E_n^{\text{int}}\,. \tag{2.34}$$

The lowest allowed energy state of the system is that in which the centre-of-mass is in a $1s$ ($\alpha=0$) state with energy $\frac{3}{2}\hbar\omega$. Thus even when the centre-of-mass is in its lowest state there is some 'zero-point' energy due to its motion in the oscillator potential [25]. The internal, or true, energy of the system is

$$E_n^{\text{int}}=E_{0n}^{\text{sm}}-\tfrac{3}{2}\hbar\omega\,. \tag{2.35}$$

It can be seen from equations (2.33) and (2.34) that a wavefunction $\Psi_{\alpha n}^{\text{sm}}$ with $\alpha>0$ represents an excited state of the centre-of-mass motion, i.e. it describes the spurious oscillation of the nucleus as a whole about the origin of the potential. A spurious state can occur only among states which have at least 1 $\hbar\omega$ of excitation energy because this is the energy required to raise the centre-of-mass into its first excited state, and it follows that there can be no spurious states if the only unfilled shell in the shell model configuration is the lowest allowed by the exclusion principle. If this condition is not satisfied it is necessary to eliminate the spurious states by constructing a suitable linear combination of the shell model states [24, 26].

We choose the normalization

$$\int |\Psi_{\alpha n}^{\text{sm}}|^2 d\boldsymbol{r}_1 \ldots d\boldsymbol{r}_A = 1\,, \quad \int |\Phi|^2 d\boldsymbol{\rho}_1 \ldots d\boldsymbol{\rho}_{A-1} = 1\,,$$

$$\int |\phi_\alpha|^2 A\, d\boldsymbol{R} = 1 \tag{2.36}$$

so that Ψ_{0n}^{sm} becomes

$$\Psi_{0n}^{\text{sm}} = A^{+\frac{1}{4}} \pi^{-\frac{3}{4}} a^{-\frac{3}{2}} e^{-AR^2/2a^2} \Phi_n(\boldsymbol{\rho}_1 \ldots \boldsymbol{\rho}_{A-1})$$

where a is the oscillator length parameter defined in equation (2.3). The matrix element of $e^{i\boldsymbol{q}\cdot\boldsymbol{r}_i}$ is given by

$$\begin{aligned} X_{mn}^{\text{sm}} &= A^{\frac{1}{2}} \pi^{-\frac{3}{2}} a^{-3} \int e^{-AR^2/a^2} \Phi_m^*(\boldsymbol{\rho}_1 \ldots \boldsymbol{\rho}_{A-1}) e^{i\boldsymbol{q}\cdot(\boldsymbol{\rho}_1 + \boldsymbol{R})} \\ &\quad \times \Phi_n(\boldsymbol{\rho}_1 \ldots \boldsymbol{\rho}_{A-1}) A\, d\boldsymbol{R}\, d\boldsymbol{\rho}_1 \ldots d\boldsymbol{\rho}_{A-1} \\ &= e^{-q^2a^2/4A} \int \Phi_m^*(\boldsymbol{\rho}_1 \ldots \boldsymbol{\rho}_{A-1}) e^{i\boldsymbol{q}\cdot\boldsymbol{\rho}_i} \Phi_n(\boldsymbol{\rho}_1 \ldots \boldsymbol{\rho}_{A-1}) d\boldsymbol{\rho}_1 \ldots d\boldsymbol{\rho}_{A-1} \\ &= e^{-q^2a^2/4A} X_{mn}^{\text{int}}\,. \end{aligned} \tag{2.37}$$

In this case the true matrix element X^{int} can be obtained from the shell model matrix element X^{sm} by multiplying [27] by the correcting factor $e^{q^2a^2/4A}$ which may be important for light nuclei but tends to unity for large A.

In order to obtain a relationship of the form $X^{\text{sm}} = C(A) X^{\text{int}}$ it is necessary that the operator and the wavefunctions can be expressed as functions of the centre-of-mass co-ordinate and the relative coordinates. This occurs for a limited choice of operators and only for the oscillator wavefunctions. It is usual in most calculations of shell model wavefunctions to replace the nucleon mass by the appropriate reduced mass and then to ignore the difference between co-ordinates referred to the origin of the potential and those referred correctly to the centre-of mass.

2.6 Distribution and correlation functions

In order to test the various nuclear models and to determine the extent to which a given model describes the properties of particular nuclei or a particular scattering process, we have to make predictions of observable quantities by evaluating the matrix elements of the relevant operators taken between the model wavefunctions for the initial and final nuclear states. As an intermediate step in these calculations it is often convenient to use the nuclear wavefunctions to construct a variety of distribution functions.

We take $\Psi_n(\boldsymbol{r}_1, \boldsymbol{r}_2, \ldots \boldsymbol{r}_A)$ to be a normalized, anti-symmetrized wavefunction for a nucleus in state n. As discussed in section 2.5, translation invariance requires that the co-ordinates $\boldsymbol{r}_i$ are relative co-ordinates obeying the condition $A^{-1} \sum_{i=1}^{A} \boldsymbol{r}_i = 0$. This requirement will be ignored in this section and the co-ordinates $\boldsymbol{r}_i$ will be regarded as independent. The errors arising in the matrix elements are assumed to be of order $1/A$. The spatial distribution of protons in the ground state of the nucleus can then be defined as

$$\rho_{\text{p}}(\boldsymbol{r}) = \frac{1}{Z} \sum_{i=1}^{Z} \left[\int \Psi_0^*(\boldsymbol{r}_1, \boldsymbol{r}_2, \ldots \boldsymbol{r}_A)\, \Psi_0(\boldsymbol{r}_1, \ldots \boldsymbol{r}_A)\, d\boldsymbol{r}_1 \ldots d\boldsymbol{r}_{i-1}\, d\boldsymbol{r}_{i+1} \ldots d\boldsymbol{r}_A \right]_{\boldsymbol{r}=\boldsymbol{r}_i}$$

$$= \frac{1}{Z} \langle 0 \left| \sum_{i=1}^{Z} \delta(\boldsymbol{r} - \boldsymbol{r}_i) \right| 0 \rangle . \tag{2.38}$$

and the normalization of the wavefunction is usually chosen to be

$$\int |\Psi_0|^2\, d\boldsymbol{r}_1 \ldots d\boldsymbol{r}_n = 1$$

so that the normalization* of the proton distribution defined in equation (2.38) is

$$\int \rho_{\text{p}}(\boldsymbol{r})\, d\boldsymbol{r} = 1 . \tag{2.39}$$

The distribution ρ_{p} represents the distribution of point protons in the nucleus or, equivalently, the distribution of the proton centres-of-mass. If the

* An alternative normalization is obtained by omitting the factor $1/Z$ in equation (2.38) and similarly omitting the factor $1/A$ in equations (2.42) and (2.46). The normalization given in the text is most commonly used in the analysis of electron scattering while the alternative normalization is more frequently used in the analysis of nuclear reactions.

true charge distribution in the nucleus is required it is necessary to take account of the fact that there is a spatial distribution of the charge of each proton (i.e. the proton is not a point but has a finite size) by folding in the known distribution function for the proton charge*. This gives the *nuclear charge distribution*

$$\rho_{\mathrm{ch}}(\boldsymbol{r}) = \int \rho_{\mathrm{p}}(\boldsymbol{r}')\,\rho_{\mathrm{proton}}(|\boldsymbol{r}-\boldsymbol{r}'|)\,d\boldsymbol{r}' , \tag{2.40}$$

and the Fourier transform of this is the *charge form factor*

$$F_{\mathrm{ch}}(q) = \int \rho_{\mathrm{ch}}(\boldsymbol{r})\,e^{i\boldsymbol{q}\cdot\boldsymbol{r}}\,d\boldsymbol{r} . \tag{2.41}$$

By comparison with equation (2.38) it is possible to define a neutron distribution and a *nuclear matter distribution.* The latter is given by

$$\rho(\boldsymbol{r}) = \frac{1}{A}\langle 0 \left| \sum_{i=1}^{A} \delta(\boldsymbol{r}-\boldsymbol{r}_i) \right| 0\rangle \tag{2.42}$$

and may also be called the *one-particle density function.* The Fourier transform of this function is the *nuclear momentum distribution*

$$F(q) = \int \rho(\boldsymbol{r})\,e^{i\boldsymbol{q}\cdot\boldsymbol{r}}\,d\boldsymbol{r} . \tag{2.43}$$

If we require the matrix element of a single-particle or one-body operator or a sum of such operators, this can always be expressed in terms of the one-particle density function. Thus, for an operator $\Sigma V(\boldsymbol{r}_i)$ such that $V(\boldsymbol{r}_i)$ depends only on the co-ordinate $\boldsymbol{r}_i$, the matrix element is given by

$$X = \langle 0 \left| \sum_{i=1}^{A} V(\boldsymbol{r}_i) \right| 0\rangle = A \int \rho(\boldsymbol{r})\,V(\boldsymbol{r})\,d\boldsymbol{r} . \tag{2.44}$$

The same situation arises if the one-body operator causes excitation of the nucleus to a definite final state, in which case

$$X_{n0} = \langle n \left| \sum_{i=1}^{A} V(\boldsymbol{r}_i) \right| 0\rangle = A \int \rho_{n0}(\boldsymbol{r})\,V(\boldsymbol{r})\,d\boldsymbol{r} \tag{2.45}$$

where ρ_{n0} is the *transition density*

* This difference between the point proton distribution and the charge distribution is the cause of a surprising amount of error and confusion. It should be noted that the function which appears in formulae describing interactions with nuclear projectiles is always the point distribution, and not the charge distribution which is relevant to electromagnetic interactions.

$$\rho_{n0}(\boldsymbol{r}) = \frac{1}{A}\langle n \left| \sum_{i=1}^{A} \delta(\boldsymbol{r}-\boldsymbol{r}_i) \right| 0\rangle \tag{2.46}$$

$$\int \rho_{n0}(\boldsymbol{r})d\boldsymbol{r} = \delta_{n0} \,. \tag{2.47}$$

In contrast, the matrix elements of many-body operators involve the higher density functions. The two-particle density function is defined as [28]

$$C(\boldsymbol{r}, \boldsymbol{r}') = \frac{1}{A(A-1)}\langle 0 \left| \sum_{i\neq j} \delta(\boldsymbol{r}-\boldsymbol{r}_i)\,\delta(\boldsymbol{r}'-\boldsymbol{r}_j) \right| 0\rangle \tag{2.48}$$

and if $C(\boldsymbol{r}, \boldsymbol{r}')=\rho(\boldsymbol{r})\,\rho(\boldsymbol{r}')$ the system is said to be uncorrelated, but in general $C(\boldsymbol{r}, \boldsymbol{r}')$ can be written in the form

$$C(\boldsymbol{r}, \boldsymbol{r}') = D(\boldsymbol{r}, \boldsymbol{r}') + \rho(\boldsymbol{r})\rho(\boldsymbol{r}') \tag{2.49}$$

where $D(\boldsymbol{r}, \boldsymbol{r}')$ is the *pair correlation function.* A typical situation involving the two-particle density function arises if we consider the summed effect of transitions to all final states $n\neq 0$ due to the one-body operators $\Sigma V(\boldsymbol{r}_i)$. In this case we have to evaluate the expression

$$\sum_{n\neq 0} |X_{n0}|^2 = \sum_{n=0}^{\infty} \left|\langle n|\sum_i V(\boldsymbol{r}_i)|\,0\rangle\right|^2 - \left|\langle 0|\sum_i V(\boldsymbol{r}_i)|\,0\rangle\right|^2$$

and the form of the first term suggests that we use the closure relation [5]

$$\sum_n |n\rangle\langle n| = \delta(\boldsymbol{r}-\boldsymbol{r}')$$

so that

$$\begin{aligned}\sum_n |\langle n|\sum_i V(\boldsymbol{r}_i)|\,0\rangle|^2 &= \sum_n \langle 0|\sum_i V^*(\boldsymbol{r}_i)|n\rangle\langle n|\sum_j V(\boldsymbol{r}_j)|0\rangle \\ &= \langle 0|\sum_i V^*(\boldsymbol{r}_i)\,V(\boldsymbol{r}_i)|\,0\rangle \\ &\quad + \langle 0|\sum_{i\neq j} V^*(\boldsymbol{r}_i)\,V(\boldsymbol{r}_j)|\,0\rangle\,,\end{aligned}$$

and

$$\begin{aligned}\sum_{n\neq 0} |X_{n0}|^2 = A\int \rho(\boldsymbol{r})\,V^*(\boldsymbol{r})\,V(\boldsymbol{r})d\boldsymbol{r} + A(A-1)\iint C(\boldsymbol{r}, \boldsymbol{r}')\,V^*(\boldsymbol{r})\,V(\boldsymbol{r}')d\boldsymbol{r}\,d\boldsymbol{r}' \\ - |A\int \rho(\boldsymbol{r})\,V(\boldsymbol{r})d\boldsymbol{r}|^2\,.\end{aligned} \tag{2.50}$$

In the special case when $V(\boldsymbol{r}_i) = e^{i\boldsymbol{q}\cdot\boldsymbol{r}_i}$ we have

$$\sum_{n\neq 0} |X_{n0}|^2 = A + A(A-1)F(q)F(-q) + A(A-1)D(q) - A^2F^2(q)$$

$$= A[1 - F^2(q) + (A-1)D(q)] \tag{2.51}$$

where $F(q)$ is given by equation (2.43), $F(q) = F(-q)$ and $D(q)$ is the Fourier transform of the pair correlation function*

$$D(q) = \iint D(\boldsymbol{r}, \boldsymbol{r}')e^{i\boldsymbol{q}\cdot(\boldsymbol{r}-\boldsymbol{r}')}d\boldsymbol{r}\,d\boldsymbol{r}' . \tag{2.52}$$

In the single-particle model the one- and two-particle density functions for closed-shell nuclei have very simple forms, since in this case the pair correlation function arises solely from the exclusion principle [29]. The additional correlations not explicitly included in a single-particle model are the short-range correlations arising from the repulsive behaviour of the nucleon–nucleon force at very short distances, and the long-range correlations associated with collective behaviour. This point is discussed further in section 4.2.

* A wide variety of notations have been used for the correlation function and its transform. Some authors use $C(\boldsymbol{r}, \boldsymbol{r}')$ to denote the function we have called $D(\boldsymbol{r}, \boldsymbol{r}')$, some use $C(q)$ to denote the transform of $C(\boldsymbol{r}, \boldsymbol{r}')$ and others use $C(q, -q)$ to denote the transform of $D(\boldsymbol{r}, \boldsymbol{r}')$. See, for example, references [28–30].

3 Basic scattering theory

3.1 Introduction

In this chapter an outline is given of the formalism used in the analysis and interpretation of nuclear scattering and reactions. This outline is severely restricted to those parts of scattering theory which are most frequently used for the description of nuclear reactions, and for a more complete study the reader is referred to the works on quantum mechanics and scattering theory listed in the bibliography. In this chapter and in most other chapters we are concerned with non-relativistic processes, but some processes which require a relativistic theory are described in Chapters 4 and 10.

We begin with a relatively old-fashioned and elementary approach to the phase shift analysis of elastic potential scattering of spinless projectiles. This approach is also used to describe the scattering of spin $\frac{1}{2}$ projectiles, and the results are then expressed in the much more general formalism of the scattering matrix which does not assume the existence of a potential. The description of non-elastic processes is studied first through potential scattering from a complex potential and then through the S-matrix formalism which is a generalization including elastic, inelastic and rearrangement collisions. In the final sections, the formal theory of elastic potential scattering is developed and Green's function techniques are used to extend this formalism for application to inelastic scattering and reactions.

3.2 Scattering of spin zero projectiles from a spin zero target

As explained in Chapter 1, in the theory of non-relativistic potential scattering we use the centre-of-mass system and seek a steady-state solution of the Schrödinger equation describing the scattering of a projectile of reduced mass μ. The interaction between the projectile and the target is represented by a potential which depends on the relative co-ordinate $\boldsymbol{r}$ which relates the position of the centre of mass of the projectile to the centre-of-mass of the target. The potential may also depend on other co-ordinates of the system, such as the spin co-ordinates. The solution of this Schrödinger equation must satisfy the boundary conditions that it is finite at the origin and that it has the asymptotic behaviour of an incoming plane wave and an outgoing spherical wave. In this section we use this approach [4, 5] to obtain formulae for the scattering amplitude for elastic scattering of spin zero projectiles.

We first assume that the potential describing the interaction between the target and projectile is spherically symmetric, so that the Schrödinger equation is

$$\nabla^2 \psi + [k^2 - U(r)]\psi = 0 \tag{3.1}$$

where

$$k^2 = \frac{2\mu}{\hbar^2} E, \quad U(r) = \frac{2\mu}{\hbar^2} V(r), \tag{3.2}$$

and E is the energy in the centre of mass system. We take the direction of the incident beam to be the z-axis, so that the most general solution of this equation with axial symmetry is

$$\psi = \sum_{l=0}^{\infty} B_l f_l(kr) P_l(\cos\theta) \tag{3.3}$$

where $P_l(\cos\theta)$ is a Legendre polynomial and f_l is the solution of the radial equation

$$\frac{d^2 f_l}{dr^2} + \frac{2}{r}\frac{df_l}{dr} + \left[k^2 - U(r) - \frac{l(l+1)}{r^2}\right] f_l = 0 \tag{3.4}$$

which is bounded at the origin, and the constants B_l are to be determined by comparison with the asymptotic form

$$\psi \xrightarrow[r\to\infty]{} e^{ikz} + f(\theta)\frac{e^{ikr}}{r}. \tag{3.5}$$

The plane wavefunction can be expressed in spherical co-ordinates as

$$e^{ikz} = \sum_l i^l (2l+1) j_l(kr) P_l(\cos\theta) \tag{3.6}$$

where j_l is the spherical Bessel function which is regular at the origin and satisfies the radial equation

$$\frac{d^2 j_l}{dr^2} + \frac{2}{r}\frac{dj_l}{dr} + \left[k^2 - \frac{l(l+1)}{r^2}\right] j_l = 0 \,. \tag{3.7}$$

Equations (3.4) and (3.7) differ only through the presence of the potential $U(r)$ and if this term falls to zero sufficiently rapidly at large r, the asymptotic forms of f_k and j_l can differ only by a constant phase factor. (The restriction on the behaviour of the potential at large distances will be specified more precisely at the end of this section.) The asymptotic behaviour of the spherical Bessel function is known to be [5, 17]

$$j_l(kr) \to \frac{1}{kr} \sin(kr - \tfrac{1}{2} l\pi) \tag{3.8}$$

so that the asymptotic behaviour of f_l must be

$$f_l(kr) \to \frac{1}{kr} \sin(kr - \tfrac{1}{2} l\pi + \delta_l) \tag{3.9}$$

where δ_l is called the *phase shift* due to the potential. There may of course be a very substantial difference in the forms of j_l and f_l in the interior region where the behaviour of f_l depends on the nature of the potential. The coefficients B_l and the scattering amplitude $f(\theta)$ can now be obtained from a comparison of the asymptotic behaviour of the scattering wavefunction ψ defined by equation (3.3) with the required behaviour given by equation (3.5). Thus

$$\psi \to \sum_l P_l(\cos\theta)\, B_l \frac{1}{kr} \sin(kr - \tfrac{1}{2} l\pi + \delta_l)$$

$$\to (2ikr)^{-1} \sum_l P_l(\cos\theta)\, B_l \{e^{i(kr - \frac{1}{2}l\pi)} e^{i\delta_l} - e^{-i(kr - \frac{1}{2}l\pi)} e^{-i\delta_l}\} \tag{3.10}$$

and

$$e^{ikz} \to \sum_l P_l(\cos\theta)\, i^l (2l+1) \frac{1}{kr} \sin(kr - \tfrac{1}{2} l\pi)$$

$$\to (2ikr)^{-1} \sum_l P_l(\cos\theta)\, i^l (2l+1) \{e^{i(kr - \frac{1}{2}l\pi)} - e^{-i(kr - \frac{1}{2}l\pi)}\} \,.$$

Now by comparison with equation (3.5) it can be seen that the asymptotic form of $\psi - e^{ikz}$ must be such that the coefficient of e^{-ikr} is zero and the coefficient of e^{ikr}/r is by definition the scattering amplitude $f(\theta)$. Hence

$$B_l = i^l (2^l + 1) e^{i\delta_l} \tag{3.11}$$

$$f(\theta) = \frac{1}{2ik}\sum_i (2l+1)(e^{2i\delta_l}-1)\, P_l(\cos\theta) \tag{3.12}$$

and

$$\psi = \sum_l i^l (2l+1)\, e^{i\delta_l} f_l(kr)\, P_l(\cos\theta)\,. \tag{3.13}$$

The differential cross-section for elastic scattering from the potential $V(r)$ can now be expressed directly in terms of the phase shifts δ_l and is given by

$$\frac{d\sigma}{d\Omega} = \frac{1}{4k^2}\left|\sum_l (2l+1)(e^{2i\delta_l}-1)\, P_l(\cos\theta)\right|^2. \tag{3.14}$$

The total cross-section is given by

$$\sigma = 2\pi\int_0^\pi |f(\theta)|^2 \sin\theta\, d\theta$$

and in order to evaluate this it is convenient to write the scattering amplitude in the form

$$\begin{aligned} f(\theta) = {} & \frac{1}{2k}\sum_l (2l+1)\sin 2\delta_l P_l(\cos\theta) \\ & - \frac{i}{2k}\sum_l (2l+1)(\cos 2\delta_l - 1) P_l(\cos\theta)\,. \end{aligned} \tag{3.15}$$

The square modulus of this function will contain terms of the form P_l^2 and terms of the form $P_m P_n$, but the latter will give a zero contribution to the total cross-section because of the orthogonality condition [5, 17]

$$\int_{-1}^{+1} P_n P_m d(\cos\theta) = 0 \qquad m \neq n\,,$$

and the total cross-section reduces to

$$\begin{aligned} \sigma &= \frac{\pi}{2k^2}\sum_l \int_0^\pi (2l+1)^2 \{\sin^2 2\delta_l + (\cos 2\delta_l - 1)^2\} P_l^2(\cos\theta) \sin\theta\, d\theta \\ &= \frac{4\pi}{k^2}\sum_l (2l+1)\sin^2\delta_l\,. \end{aligned} \tag{3.16}$$

By comparison of equations (3.15) and (3.16) we can deduce the relation

$$\operatorname{Im} f(0) = \frac{k\sigma}{4\pi} \tag{3.17}$$

which is known as the *optical theorem.*

Equations (3.13) and (3.6) give the *partial wave expansions* for the scattering wavefunction and the plane wavefunction, and the description of elastic scattering given above is referred to as a *partial wave analysis* of the scattering. It can be seen from equation (3.14) that the contribution from the lowest partial wave with $l=0$, i.e. from *s*-wave scattering, is isotropic but the contributions from the higher partial waves give the differential cross-section a distinctive angular variation. Further, there is interference between the contributions of different partial waves to the differential cross-section but not to the total cross-section, and this is to be expected as the total cross-section is simply a measure of the total number of particles scattered per unit incident flux. However, the fact that these scattered particles appear at angles $\theta > 0$ implies that there is a depletion of the beam in the forward direction, i.e. at $\theta \approx 0$, and this redistribution is produced by destructive interference between the incident plane wave and the scattered wave in the forward direction. There is, therefore, a close connection between the scattering amplitude

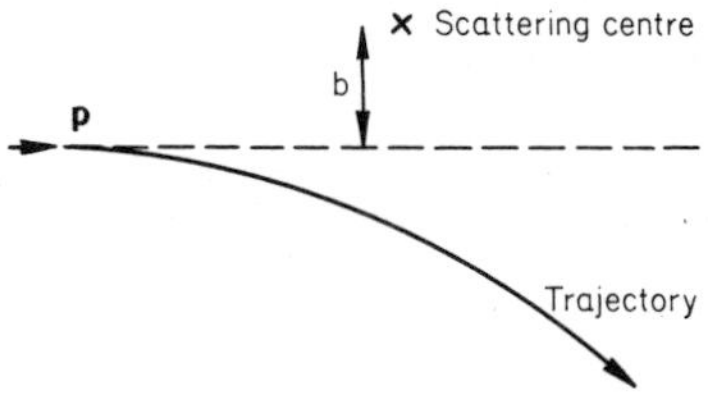

Figure 3.1 Definition of the impact parameter b for scattering of a classical particle.

in the forward direction and the total cross-section which is expressed by the optical theorem. In fact, the optical theorem can also be derived from the continuity equation which expresses the conservation of particles. It is a matter of some importance to know whether the summations indicated in equations (3.12)–(3.16) can be restricted to a finite number of terms, and a semi-classical argument can conveniently be used to show that this is so [31]. If the scattering of a particle is described classically, the trajectory may be characterized by the impact parameter b as shown in figure 3.1 and the angular momentum of the particle with this trajectory is pb, where p is the magnitude of the momentum. If the potential has a finite range r_0, scattering will occur only if $b < r_0$. The connection with quantum mechanical scattering is made through the substitutions

$$pb \rightarrow \hbar[l(l+1)]^{\frac{1}{2}}, \qquad p \rightarrow \hbar k,$$

so that a contribution may be expected from those partial waves for which $[l(l+1)]^{\frac{1}{2}} < kr_0$.

The arguments leading to equation (3.5) and the derivation of equation (3.12) are valid provided the potential goes to zero more rapidly than $1/r$ at large r, and are therefore not valid in the case when the potential includes a Coulomb term. When the Coulomb potential is included, it is still possible to make a partial wave expansion of the scattering wavefunction but the asymptotic behaviour is now [32]

$$\psi \to \psi_i + f_c(\theta)\psi_s \tag{3.18}$$

where
$$\psi_i = [1-\gamma^2/ik(r-z)]\exp[ikz+i\gamma \log k(r-z)]$$
$$\psi_s = r^{-1}\exp[ikr-i\gamma \log 2kr]$$

and γ is the Coulomb parameter

$$\gamma = Z_1Z_2e^2/\hbar v = \mu Z_1Z_2e^2/\hbar^2 k\,. \tag{3.19}$$

For scattering by a pure Coulomb potential the scattering amplitude and scattering wavefunction are given by

$$f_c(\theta) = -\frac{\gamma}{2k\sin^2\frac{1}{2}\theta}\exp[-i\gamma\log(\sin^2\tfrac{1}{2}\theta)+2i\sigma_0] \tag{3.20}$$

$$\psi = \sum_l i^l(2l+1)e^{i\sigma_l}L_l(kr)P_l(\cos\theta) \tag{3.21}$$

where the asymptotic behaviour of L_l is given by

$$L_l(kr) \to \frac{1}{kr}\sin(kr-\tfrac{1}{2}l\pi+\sigma_l-\gamma\log 2kr) \tag{3.22}$$

where σ_l is the Coulomb phase shift defined by

$$e^{2i\sigma_l} = \Gamma(1+l+i\gamma)/\Gamma(1+l-i\gamma)\,. \tag{3.23}$$

Because the Coulomb potential has such a long range, it affects the wavefunction even in the asymptotic region and f_l contains a logarithmic contribution to the phase which can not be neglected. However, the current density $\boldsymbol{j}_{\text{in}}$ calculated from ψ_i is directed along the z-axis and has magnitude v at large negative values of z, and the current density calculated from ψ_s is directed radially outwards from the scattering centre and has magnitude $v|f_c(\theta)|^2/r^2$ at large values of r. This means that ψ_i and ψ_s can be interpreted as incident and scattered waves and the differential cross-section is given by $|f_c(\theta)|^2$.

When the potential $V(r)$ contains both nuclear (short-range) and Coulomb (long-range) terms the asymptotic behaviour of the radial part of the scattering wavefunctions must be compared, not with the spherical Bessel function j_l, but with the Coulomb function L_l. This comparison yields the phase shift δ_l which is now the additional phase shift due to the nuclear potential and the difference between the Coulomb potential due to a nucleus of finite size and that due to a point charge (see below). Using the result

$$e^{2i(\sigma_l+\delta_l)}-1=(e^{2i\sigma_l}-1)+e^{2i\sigma_l}(e^{2i\delta_l}-1)$$

the scattering amplitude can be written in the form

$$f(\theta)=f_c(\theta)+f_N(\theta) \tag{3.24}$$

where $f_c(\theta)$ is given by equation (3.20) and

$$f_N(\theta)=\frac{1}{2ik}\sum_l(2l+1)e^{2i\sigma_l}(e^{2i\delta_l}-1)P_l(\cos\theta)\,. \tag{3.25}$$

It may be helpful to examine exactly what the previous paragraph implies concerning the additivity of phase shifts. To do this we consider the scattering from two short-range potentials V and U. The asymptotic behaviour of the wavefunctions for the case when these two potentials act together or separately are given by

$$F_l^U(kr)\to\frac{1}{kr}\sin\left[kr-\tfrac{1}{2}l\pi+\delta_l(U)\right]$$

$$F_l^V(kr)\to\frac{1}{kr}\sin\left[kr-\tfrac{1}{2}l\pi+\delta_l(V)\right]$$

$$F_l^{V+U}(kr)\to\frac{1}{kr}\sin\left[kr-\tfrac{1}{2}l\pi+\delta_l(V+U)\right]$$

$$\to\frac{1}{kr}\sin\left[kr-\tfrac{1}{2}l\pi+\delta_l^U(V)+\delta_l(U)\right]$$

If we compare the solution F_l^{U+V} with the plane wave solution (3.8) we obtain the total phase shift $\delta_l(U+V)$ due to the potential $U+V$, but if we compare the solution F_l^{U+V} with the solution F_l^U we obtain the additional phase shift $\delta_l^U(V)$ due to potential V acting in the presence of potential U, and this is clearly not the same as the phase shift $\delta_l(V)$ due to potential V acting alone. Thus

$$\delta_l(V+U) = \delta_l^U(V) + \delta_l(U) \neq \delta_l(V) + \delta_l(U) \tag{3.26}$$

The same conclusion applies if U is a long-range Coulomb potential so that the phase shift δ_l used in the previous paragraph is the additional phase shift due to scattering from a short-range nuclear potential and a modified Coulomb potential in the presence of a pure (point) Coulomb potential.

3.3 Scattering of spin $\frac{1}{2}$ projectiles from a spin zero target

If the projectile has non-zero spin the interaction between the projectile and the target may be spin dependent. It is, therefore, necessary to modify the formalism to include explicitly the spin wavefunctions of the projectile and to allow for the possibility that the projectile changes its spin state as a result of the interaction. We denote the spin wavefunctions for a projectile of spin s as χ_s^μ, so that the asymptotic form of the scattering wavefunction must now be written as

$$\psi_{s,\mu} \to \chi_s^\mu e^{ikz} + \sum_\nu \chi_s^\nu f_{\nu\mu}(\theta, \phi) \frac{e^{ikr}}{r}, \tag{3.27}$$

where $f_{\nu\mu}$ is the scattering amplitude for scattering from spin state (s, μ) to state (s, ν). For simplicity we have taken the projectile to be uncharged, but the modification of the formalism for a charged particle proceeds in a completely similar manner to that outlined here. The partial cross-section for scattering from one spin state to another is therefore given by

$$\frac{d\sigma}{d\Omega}\,\nu\mu = |f_{\nu\mu}|^2 . \tag{3.28}$$

If the final spin state is not determined and the incident beam has no preferred spin direction, the cross-section must be obtained by taking the sum over the final spin states and the average over initial states. This gives

$$\frac{d\sigma}{d\Omega} = \frac{1}{g}\sum_{\nu\mu}\frac{d\sigma}{d\Omega}\,\nu\mu = \frac{1}{g}\sum_{\nu\mu}|f_{\nu\mu}|^2 \tag{3.29}$$

where g is the total number of spin states. In general, if the projectile has spin s and the target has spin S there are $(2s+1)(2S+1)$ spin states. In the present case we have $s=\frac{1}{2}$, $S=0$, and $g=2$.

We assume that the interaction conserves the total angular momentum $\boldsymbol{j}=\boldsymbol{l}+\boldsymbol{s}$ and also conserves parity. For the case of $s=\frac{1}{2}$, these assumptions imply conservation of orbital angular momentum since states with $l=j+\frac{1}{2}$ and $l=j-\frac{1}{2}$ have opposite parity. The scattering wavefunction can then be

written as
$$\psi_{s,\mu}=\sum_{jl} i^l[4\pi(2l+1)]^{\frac{1}{2}} f_{l_j}(kr)(l\;0\;s\;\mu|jm)\psi_j^m \tag{3.30}$$

where $(l\;0\;s\;\mu|jm)$ is the Clebsch-Gordan coefficient [33, 34] which couples the angular momenta l and s to a definite j with projection m. The eigenfunction ψ_j^m is constructed from a spherical harmonic and a spin-function [33], i.e.

$$\psi_j^m=\sum_{\lambda}(l\;\lambda\;s\;\nu|jm)\,Y_l^{\lambda}(\theta,\phi)\chi_s^{\nu}\,, \tag{3.31}$$

and the spherical harmonics are related to the Legendre polynomials and Associated Legendre functions by the relations [5]

$$Y_l^0(\theta,\phi)=\left[\frac{2l+1}{4\pi}\right]^{\frac{1}{2}}P_l(\cos\theta)\,, \tag{3.32}$$

$$Y_l^m(\theta,\phi)=(-1)^m\left[\frac{2l+1}{4\pi}\,\frac{(l-m)!}{(l+m)!}\right]^{\frac{1}{2}}P_l^m(\cos\theta)e^{im\phi}\quad(m\geqslant 0) \tag{3.33}$$

$$Y_l^{-m}(\theta,\phi)=(-1)^m\,Y_l^m(\theta,\phi)^*\,. \tag{3.34}$$

If we now put $s=\frac{1}{2}$, $\mu=\pm\frac{1}{2}$ and substitute the expansion (3.31) into equation (3.30) we have

$$\begin{aligned}\psi_{\frac{1}{2},\pm\frac{1}{2}}=&\sum_{jl} i^l[4\pi(2l+1)]^{\frac{1}{2}}f_{lj}(kr)\,(l\;0\;\tfrac{1}{2}\pm\tfrac{1}{2}|j\pm\tfrac{1}{2})^2\,Y_l^0\chi_{\frac{1}{2}}^{\pm\frac{1}{2}}\\&+\sum_{jl} i^l[4\pi(2l+1]^{\frac{1}{2}}f_{lj}(kr)(l\;0\;\tfrac{1}{2}\pm\tfrac{1}{2}|j\pm\tfrac{1}{2})\\&\qquad\times(l\pm1\;\tfrac{1}{2}\mp\tfrac{1}{2}|j\pm\tfrac{1}{2})\,Y_l^{\pm1}\chi_{\frac{1}{2}}^{\mp\frac{1}{2}}\end{aligned} \tag{3.35}$$

where the sum over j reduces to two terms with $j=l+\frac{1}{2}$ and $j=l-\frac{1}{2}$. By comparison with equations (3.9) and (3.13) it can be seen that the factor $e^{i\delta_{lj}}$ has been absorbed into the f_{lj} whose asymptotic form must therefore be

$$\begin{aligned}f_{lj}(kr)&\to e^{i\delta_{lj}}\frac{1}{kr}\sin(kr-\tfrac{1}{2}l\pi+\delta_{lj})\\&\to(2ikr)^{-1}\left[e^{2i\delta_{lj}}e^{i(kr-\frac{1}{2}l\pi)}-e^{-i(kr-\frac{1}{2}l\pi)}\right],\end{aligned} \tag{3.36}$$

and the relevant Clebsch-Gordan coefficients are

$$\begin{aligned}(l\;0\;\tfrac{1}{2}\pm\tfrac{1}{2}|j\pm\tfrac{1}{2})&=[(l+1)/(2l+1)]^{\frac{1}{2}}\quad&&\text{for } j=l+\tfrac{1}{2}\\&=\mp[l/(2l+1)]^{\frac{1}{2}}&&\text{for } j=l-\tfrac{1}{2}\end{aligned}$$

$$(l \pm 1\ \tfrac{1}{2} \mp \tfrac{1}{2} | j \pm \tfrac{1}{2}) = [l/(2l+1)]^{\frac{1}{2}} \qquad \text{for } j = +\tfrac{1}{2}$$
$$= \pm [(l+1)/(2l+1)]^{\frac{1}{2}} \text{ for } j = l - \tfrac{1}{2}\,.$$

Finally, by substituting these coefficients and the expansion (3.36) into equation (3.35) and subtracting the plane wave expansion we obtain the asymptotic behaviour of the scattering wavefunction

$$(2ikr)(\psi_{\frac{1}{2},\pm\frac{1}{2}} - \chi_{\frac{1}{2}}^{\pm\frac{1}{2}} e^{ikz}) \rightarrow$$
$$\chi_{\frac{1}{2}}^{\pm\frac{1}{2}} \sum_l [(l+1)\exp(2i\delta_{lj}^+) + l\exp(2i\delta_{lj}^-) - (2l+1)]\, e^{ikr} P_l(\cos\theta)$$
$$\mp \chi_{\frac{1}{2}}^{\pm\frac{1}{2}} \sum_l [\exp(2i\delta_{lj}^+) - \exp(2i\delta_{lj}^-)]\, e^{ikr} P_l^1(\cos\theta) e^{\pm i\phi}\,, \tag{3.37}$$

where δ_{lj}^+, δ_{lj}^- are the phase shifts for $j = l + \frac{1}{2}$ and $j = l - \frac{1}{2}$, respectively. From equation (3.37) we can pick out the scattering amplitudes

$$f_{\pm\frac{1}{2}\pm\frac{1}{2}}(\theta, \phi) = \frac{1}{2ik} \sum_l [(l+1)\ \exp(2i\delta_{lj}^+) + l\exp(2i\delta_{lj}^-) - (2l+1)]\, P_l(\cos\theta) \tag{3.38}$$

and

$$f_{\mp\frac{1}{2}\pm\frac{1}{2}}(\theta, \phi) = \pm \frac{i}{2k} \sum_l [\exp(2i\delta_{lj}^+) - \exp(2i\delta_{lj}^-)]\, P_l^1(\cos\theta) e^{\pm i\phi}\,. \tag{3.39}$$

Comparison of equations (3.38) and (3.39) with equation (3.12) shows immediately that if there is no spin-dependent term in the potential, so that $\delta_{lj}^+ \equiv \delta_{lj}^-$, the formalism for spin $\frac{1}{2}$ projectiles gives exactly the same results as the formalism for spin zero projectiles. However, if there is a spin-dependent term in the potential, such as a spin-orbit potential, then 'spin-flip' transitions from one spin state to another occur with scattering amplitudes given by equation (3.39).

Much of the complexity of these formulae may be removed by defining an operator M such that

$$M\chi_s^\mu = f_{\nu\mu}\chi_s^\nu\,,$$

and in terms of this operator equation (3.27) becomes

$$\psi_{s,\mu} \rightarrow \left\{ e^{ikz} + \frac{e^{ikr}}{r} M \right\} \chi_s^\mu\,. \tag{3.40}$$

Thus M is a matrix of order $g \times g$ whose elements are the $f_{\nu\mu}$. The partial cross-section for scattering from one spin state to another is given by

$$\frac{d\sigma}{d\Omega}\,\nu\mu = |\langle\chi_s^\nu|M|\chi_s^\mu\rangle|^2 \tag{3.41}$$

and the averaged cross-section is given by

$$\frac{d\sigma}{d\Omega} = \frac{1}{g}\sum_{\nu\mu}\frac{d\sigma}{d\Omega}\,\nu\mu = \frac{1}{g}\sum_{\mu}(M^\dagger M)_{\mu\mu} = \frac{1}{g}\,\text{Trace}(M^\dagger M)\,. \tag{3.42}$$

Returning to the case of spin $\frac{1}{2}$ projectiles we define the functions $g(\theta)$ and $h(\theta)$ through the relations

$$f_{\frac{1}{2}\frac{1}{2}} = f_{-\frac{1}{2}-\frac{1}{2}} = g(\theta) \tag{3.43}$$

$$f_{\frac{1}{2}-\frac{1}{2}} = ih(\theta)e^{i\phi}\,,\quad f_{-\frac{1}{2}\frac{1}{2}} = -ih(\theta)e^{-i\phi}\,, \tag{3.44}$$

so that the scattering matrix has the form

$$M = \begin{pmatrix} g(\theta) & -ih(\theta)e^{-i\phi} \\ ih(\theta)e^{i\phi} & g(\theta) \end{pmatrix} = g(\theta)\mathbf{1} + h(\theta)\boldsymbol{A} \tag{3.45}$$

with

$$A = \begin{pmatrix} 0 & -ie^{-i\phi} \\ ie^{i\phi} & 0 \end{pmatrix} = -\sin\phi\begin{pmatrix} 0 & 1 \\ 1 & 0 \end{pmatrix} + \cos\phi\begin{pmatrix} 0 & -i \\ i & 0 \end{pmatrix}$$

$$= -\sin\phi\,\sigma_x + \cos\phi\,\sigma_y\,, \tag{3.46}$$

where σ_x and σ_y are two of the three Pauli matrices.

The incident beam, represented by the plane wavefunction e^{ikz}, is moving along the z-axis towards the origin, and may therefore be indicated by a wave vector $\boldsymbol{k}_i$ along the z-axis with magnitude $k_i = \mu v/\hbar$. A wave vector $\boldsymbol{k}_f$ for the outgoing beam has polar co-ordinates (θ, ϕ) where θ is the angle of scattering and, for elastic scattering, the magnitude $k_f = k_i$. From these vectors we can construct the unit vector $\hat{\boldsymbol{n}}$ to be normal to the plane containing $\boldsymbol{k}_i$ and $\boldsymbol{k}_f$ through the relation

$$\hat{\boldsymbol{n}} = \boldsymbol{k}_i \wedge \boldsymbol{k}_f / |\boldsymbol{k}_i \wedge \boldsymbol{k}_f| \tag{3.47}$$

so that $\hat{\boldsymbol{n}}$ has components $(-\sin\phi, \cos\phi, 0)$. Since the vectors $\boldsymbol{k}_i$ and $\boldsymbol{k}_f$ change sign under reflection of co-ordinates, the vector $\hat{\boldsymbol{n}}$ is unchanged and is therefore an axial or pseudo vector. In this property $\hat{\boldsymbol{n}}$ resembles the spin operator $\boldsymbol{\sigma}$ and other angular momentum vectors, since a vector of the form $\boldsymbol{r} \wedge \boldsymbol{p}$ transforms to $(-\boldsymbol{r}) \wedge (-\boldsymbol{p})$. From these properties of the vector $\hat{\boldsymbol{n}}$ and from the nature of the matrix $\boldsymbol{A}$ defined by equation (3.46) it appears that the scattering matrix M can be expressed in a more general form as

$$M = g(\theta)\mathit{1} + h(\theta)\boldsymbol{\sigma}.\hat{\boldsymbol{n}}\,. \tag{3.48}$$

This form for M can be justified by the following general arguments. Since M is a 2×2 matrix it can always be expressed in terms of the unit matrix and the three Pauli matrices σ_x, σ_y, σ_z. It must be independent of any external direction and dependent only on the geometry of the collision, so that the coefficients g and h must be functions of the scalar $\boldsymbol{k}_i.\boldsymbol{k}_f$, i.e. of scattering angle. Since it is assumed that parity is conserved in the interaction, M must be invariant under reflections, and this is satisfied by constructing the second term from the scalar product of two pseudo vectors. Thus equation (3.45) gives a particular form of the scattering matrix defined by equation (3.48) which arises from our choice for the axis of quantization.

By using equation (3.48) or equation (3.45) for the scattering matrix M we can calculate the differential cross-section from equation (3.42), and this gives

$$\frac{d\sigma}{d\Omega} = |g|^2 + |h|^2\,. \tag{3.49}$$

We can also calculate the expectation value $\langle\boldsymbol{\sigma}\rangle$ of the spin of a particle scattered in the direction $\boldsymbol{k}_f$, which is given by

$$\langle\boldsymbol{\sigma}\rangle = \mathrm{Trace}(M^\dagger\boldsymbol{\sigma}M)/\mathrm{Trace}(M^\dagger M)\,. \tag{3.50}$$

To do this we take $\phi=0$ so that $\boldsymbol{k}_f$ is in the xz-plane and $\hat{\boldsymbol{n}}$ is along the y-axis, and find

$$\langle\sigma_x\rangle = \langle\sigma_z\rangle = 0$$

$$\langle\sigma_y\rangle = \frac{2\,\mathrm{Re}(g^*h)}{|g|^2+|h|^2}\,.$$

Thus as a result of the scattering there is a net alignment of spins in the y-direction, and the unpolarized incident beam which had no preferred spin direction has been converted into a polarized scattered beam. Since $\langle\sigma_y\rangle=0$ if $h=0$, this polarization has arisen from the spin-flip transitions. For scattering with $\phi=180°$, the vector $\hat{\boldsymbol{n}}$ and the direction of polarization are reversed, but the cross-section is unchanged since this is independent of ϕ. The polarization can, however, be detected and measured by performing a second scattering. For the arrangement shown in figure 3.2 we have $\phi_1=0$, $\phi_2=0$ or $180°$ so that the complete scattering matrix is

$$M = M_2 M_1$$

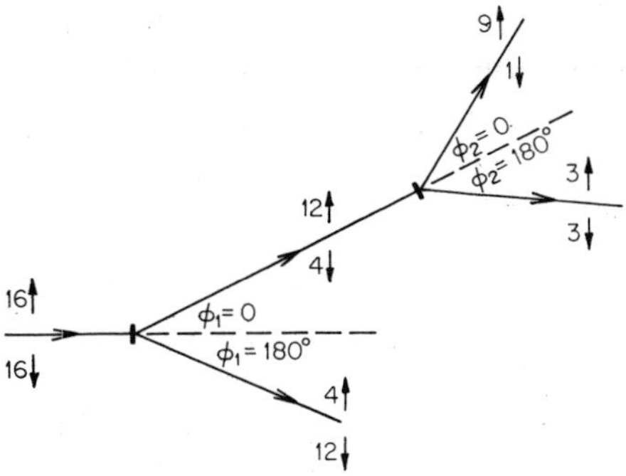

Figure 3.2 Double scattering of an initially unpolarized beam of spin $\frac{1}{2}$ particles. The scattering angles θ on either side of the incident direction are the same for each scattering. For this example it is assumed that $P_1 = P_2 = \frac{1}{2}$.

where $$M_1 = \begin{pmatrix} g_1 & -ih_1 \\ ih_1 & g_1 \end{pmatrix}, \qquad M_2 = \begin{pmatrix} g_2 & -ih_2 e^{-i\phi_2} \\ ih_2 e^{i\phi_2} & g_2 \end{pmatrix}$$

and the differential cross-section measured after the second scattering is given by

$$\frac{d\sigma}{d\Omega}2 = \text{Trace}(M^+ M)/\text{Trace}(M_1^+ M_1)$$

$$= \{|g_2|^2 + |h_2|^2\}[1 + P_1(\theta_1)P_2(\theta_2)\cos\phi_2] \tag{3.51}$$

where

$$P_1(\theta) = \frac{2\,\text{Re}(g_1^* h_1)}{|g_1|^2 + |h_1|^2}, \quad P_2(\theta) = \frac{2\,\text{Re}(g_2^* h_2)}{|g_2|^2 + |h_2|^2}. \tag{3.52}$$

Equation (3.51) shows clearly that the polarization induced by the first scattering can be detected through the ϕ-dependence of the cross-section for the second scattering. The simplest case arises for $P_1 = P_2$ when measurements of the cross-section for scattering to the left and to the right of the incident beam yield a measure of the polarization from the relation

$$P_1^2 = \left(\frac{d\sigma^L}{d\Omega}2 - \frac{d\sigma^R}{d\Omega}2\right) \bigg/ \left(\frac{d\sigma^L}{d\Omega}2 + \frac{d\sigma^R}{d\Omega}2\right) \tag{3.53}$$

Results identical to equations (3.51) and (3.53) would be obtained for the single scattering of a beam initially polarized by an amount P_1.

Measurements of the differential cross-section and polarization determine the quantities $(|g|^2+|h|^2)$ and $\mathrm{Re}\,(g^*h)$, and a triple-scattering experiment is necessary to completely determine $|g|$, $|h|$, and their relative phase. Since [17]

$$P_l^1(\cos\theta) = \sin\theta\,\frac{dP_l}{d(\cos\theta)}$$

it follows from equation (3.39) that $h(\theta)=P(\theta)=0$ at $\theta=0°$ and $\theta=180°$.

3.4 Scattering of spin $\frac{1}{2}$ projectiles from a spin $\frac{1}{2}$ target

In this case the total number of spin states is 4 so that the scattering matrix M is a 4×4 matrix. The matrix M must be a scalar constructed from the spin operators $\boldsymbol{\sigma}_1$ and $\boldsymbol{\sigma}_2$, the pseudo vector $\hat{\boldsymbol{n}}$ and the unit vectors $\hat{\boldsymbol{P}}$ and $\hat{\boldsymbol{q}}$ defined by the relations [35, 36]

$$\hat{\boldsymbol{q}} = (\boldsymbol{k}_i-\boldsymbol{k}_f)/|\boldsymbol{k}_i-\boldsymbol{k}_f| \tag{3.54}$$

$$\hat{\boldsymbol{P}} = (\boldsymbol{k}_i+\boldsymbol{k}_f)/|\boldsymbol{k}_i+\boldsymbol{k}_f| \tag{3.55}$$

$$\hat{\boldsymbol{n}} = \hat{\boldsymbol{q}} \wedge \hat{\boldsymbol{P}}\,.$$

Under time reversal, $\boldsymbol{k}_i\to-\boldsymbol{k}_f$, $\boldsymbol{k}_f\to-\boldsymbol{k}_i$, so that $\hat{\boldsymbol{n}}$ and $\hat{\boldsymbol{P}}$ change sign but $\hat{\boldsymbol{q}}$ does not, and $\boldsymbol{\sigma}\to-\boldsymbol{\sigma}$. The simplest form for M which can be constructed from these components which is invariant under reflections, rotations and time reversal is given by

$$\begin{aligned} M = {} & A+B(\boldsymbol{\sigma}_1\,.\,\hat{\boldsymbol{n}})(\boldsymbol{\sigma}_2\,.\,\hat{\boldsymbol{n}})+C(\boldsymbol{\sigma}_1+\boldsymbol{\sigma}_2)\,.\,\hat{\boldsymbol{n}}+D(\boldsymbol{\sigma}_1-\boldsymbol{\sigma}_2)\,.\,\hat{\boldsymbol{n}} \\ & +E(\boldsymbol{\sigma}_1\,.\,\hat{\boldsymbol{q}})(\boldsymbol{\sigma}_2\,.\,\hat{\boldsymbol{q}})+F(\boldsymbol{\sigma}_1\,.\,\hat{\boldsymbol{P}})(\boldsymbol{\sigma}_2\,.\,\hat{\boldsymbol{P}})\,. \end{aligned} \tag{3.56}$$

The coefficients are functions of energy and scattering angle. The invariance requirements imposed on M imply that the interaction giving rise to the scattering is such that total angular momentum and parity are conserved.

In equation (3.56) the term containing $\boldsymbol{\sigma}_1-\boldsymbol{\sigma}_2$ is the only one which changes sign under exchange of particles 1 and 2. Now if 1 and 2 are identical particles, for example in the case of proton–proton scattering, the matrix M must be symmetric under exchange and the coefficient D must therefore be zero. If the charge independence of nuclear forces is assumed, the same must be true for neutron–proton scattering. The determination of these parameters is more complicated than in the previous case when one of the particles had spin zero, and requires a series of single, double and triple scattering experiments [3, 35].

3.5 Scattering from a non-symmetric potential

As an example of a non-symmetric potential we consider a potential which has axial but not spherical symmetry. Such a potential can be expanded in terms of Legendre polynomials to give

$$\frac{2\mu}{\hbar^2} V(r) = U(r) = \sum_{\lambda=0}^{\infty} U_\lambda(r) P_\lambda(\cos\theta) . \tag{3.57}$$

The scattering wavefunction can still be expanded in the form

$$\psi = \sum_l i^l (2l+1) f_l(kr) P_l(\cos\theta) ,$$

if the direction of the incident beam is along the symmetry axis and we take this to be the z-axis. In order to obtain the radial equation for the f_l, we multiply the Schrödinger equation

$$(\nabla^2 + k^2)\psi = U(r)\psi$$

on the left by $P_n(\cos\theta)$ and integrate over the angular co-ordinates. This gives

$$\left[\frac{d^2}{dr^2} + \frac{2}{r}\frac{d}{dr} - \frac{n(n+1)}{r^2} + k^2\right] f_n(kr) = \sum_{l\lambda} U_{ln}^\lambda(r) f_l(kr) \tag{3.58}$$

where

$$U_{ln}^\lambda(r) = \tfrac{1}{2} i^{l-n}(2l+1)\, U_\lambda(r) \int_{-1}^{+1} P_n(\cos\theta)\, P_\lambda(\cos\theta)\, P_l(\cos\theta)\, d(\cos\theta) .$$

For a spherically symmetric case we have $\lambda = 0$ only, and

$$U_{ln}^0(r) = U(r)\delta_{l,n}$$

so that equation (3.58) reduces to equation (3.4), but in general the radial functions obey the set of coupled differential equations (3.58).

The example described above is a rather artificial one. However, if the potential does not possess axial symmetry or if the symmetry axis and the incident beam direction do not coincide, more general expansions in terms of spherical harmonics and a more general set of coupled equations can be derived by the same procedure. A realistic problem which gives rise to such equations is described in section 7.4.

3.6 Scattering from a complex potential

We consider a complex potential of the form $V(r) + iW(r)$. From the Schrödinger equation

$$(\nabla^2 + k^2)\psi = \frac{2\mu}{\hbar^2}(V + iW)\psi \tag{3.59}$$

and its complex conjugate we can construct the divergence of the probability current density,

$$\begin{aligned} \operatorname{div} \boldsymbol{j} &= \frac{\hbar}{2i\mu} \nabla . (\psi^* \nabla\psi - \psi\nabla\psi^*) \\ &= \frac{2}{\hbar} W\psi^*\psi . \end{aligned} \tag{3.60}$$

Thus, in the presence of a complex potential, the current is no longer divergenceless. The imaginary part of the potential acts as a source or a sink of particles depending on the sign of W and if W is negative, particles are absorbed or removed from the incident beam at a rate proportional to the local probability density. It is illuminating to evaluate div $\boldsymbol{j}$ for a beam of particles moving along the z-direction in a region of uniform potential $-(V_0 + iW_0)$. The wave number in this region is given by

$$k_1^2 = \frac{2\mu}{\hbar^2}[E + V_0 + iW_0] = \frac{2\mu}{\hbar^2}(E + V_0)\left[1 + \frac{iW_0}{E + V_0}\right]$$

so that for $W_0 \ll E + V_0$ and $K^2 = 2\mu(E + V_0)/\hbar^2$,

$$k_1 \approx K\left[1 + \frac{iW_0}{2(E + V_0)}\right],$$

and the wavefunction is

$$\psi \approx e^{ik_1 z} \approx e^{ikz} e^{-z/2\lambda},$$

where
$$\lambda = (E + V_0)/KW_0 . \tag{3.61}$$

Hence
$$\psi^*\psi = e^{-z/\lambda} \tag{3.62}$$

$$\operatorname{div} \boldsymbol{j} = -\frac{\hbar K}{\lambda\mu}\psi^*\psi \tag{3.62}$$

and comparing equation (3.62) with equation (3.60), we find

$$W = -\tfrac{1}{2}\frac{\hbar v}{\lambda} \tag{3.63}$$

where $v = \hbar K/\mu$ is the velocity in the medium. Thus we see that in a region described by a negative complex potential the beam is attenuated with an attenuation coefficient $1/\lambda$, or equivalently, the particles have a mean free path λ, and equation (3.60) is related to the classical continuity equation for the steady state

$$\operatorname{div} \boldsymbol{j} = -v\rho/\lambda\,.$$

In a real scattering process removal of particles from the beam occurs when other processes, such as inelastic scattering or rearrangement collisions, compete with elastic scattering. In order to see what effect this has on the cross-section and other quantities we use the asymptotic expansion (3.10) for the scattering wavefunction which, using equation (3.11), can be rewritten in the form

$$\psi \to \frac{-i}{2kr}\sum_l i^l(2l+1)\,P_l(\cos\theta)\left[e^{2i\delta_l}e^{i(kr-\frac{1}{2}l\pi)}-e^{-i(kr-\frac{1}{2}l\pi)}\right]$$

$$\to \tfrac{1}{2}\sum_l i^{l+1}(2l+1)\,P_l(\cos\theta)\left[\mathscr{I}_l(kr)-\eta_l\mathscr{O}_l(kr)\right] \tag{3.64}$$

where $\mathscr{I}_l$ and $\mathscr{O}_l$ represent incoming and outgoing spherical waves respectively and the *reflection coefficient* η_l is defined as

$$\eta_l = e^{2i\delta_l}\,. \tag{3.65}$$

Now if $|\eta_l| = 1$ the amplitude of the outgoing wave is equal to the incoming wave and the reflection is complete. There may be redistribution of intensity as in optical or electron diffraction but there is no overall loss of intensity. However, if absorption occurs the intensity of the outgoing elastically scattered wave must be less than that of the incident beam, so that $|\eta_l| < 1$. In this case the phase shift δ_l must be complex with Im $\delta_l > 0$. If $|\eta_l| = 0$ there is complete absorption and no outgoing wave emerges for this partial wave. The total cross-section for elastic scattering is still given by equation (3.16) which can be expressed in terms of η_l as

$$\sigma_{\text{sc}} = \frac{\pi}{k^2}\sum_l (2l+1)\;|1-\eta_l|^2\,. \tag{3.66}$$

The total cross-section for absorption is defined as the number of particles removed from the beam per second per unit incident flux. Since the difference between the intensities of the lth incoming and outgoing waves is $1-|\eta_l|^2$ the absorption cross-section is given by*

* In the literature, the total cross-sections for elastic scattering and absorption are often denoted by σ_{el} and σ_{re}, respectively.

$$\sigma_{\text{abs}} = \frac{\pi}{k^2} \sum_l (2l+1)(1-|\eta_l|^2) \tag{3.67}$$

and the total cross-section for all processes is given by

$$\sigma_{\text{tot}} = \sigma_{\text{sc}} + \sigma_{\text{abs}} = \frac{2\pi}{k^2} \sum_l (2l+1)(1-\text{Re}\,\eta_l)\,. \tag{3.68}$$

From equations (3.12), (3.65) and (3.68), it follows that the optical theorem now becomes

$$\text{Im}\, f(0) = \frac{k\sigma_{\text{tot}}}{4\pi}\,. \tag{3.69}$$

The differential cross-section for elastic scattering is still given by equation (3.14).

3.7 Reaction channels and the *S*-matrix

In the formalism described in the previous section all processes which cause removal of particles from the elastically scattered beam are treated together as a single absorption effect. It would not make sense, therefore, to define a differential cross-section for absorption. It is of considerable interest, however, to study the differential cross-section for a particular non-elastic process and to do this it is helpful to use the concept of *reaction channels* [6, 7]. We consider the possible processes which can follow from the interaction of two particles

$a+A \rightarrow$	$a+A$	elastic scattering
	$a+A^*$	inelastic scattering
	$b+B$	rearrangement collision
	etc.	

The system $a+A$ is referred to as the *entrance channel* and *the systems* $a+A^*$, $b+B$, etc. are referred to as *exit channels.* For simplicity, we assume that only two-particle exit channels are important. The wavefunction in a given channel must be constructed by coupling the internal states of the two particles to the state of their relative motion, and this can be done using several different coupling schemes [33]. One way is to couple the total angular momenta of the two particles, excluding the angular momentum due to their relative motion, to give the *channel spin* $\boldsymbol{Z}$, and then to add $\boldsymbol{Z}$ to the angular momentum $\boldsymbol{l}$ of the relative motion to give the total angular momentum j of the channel. The channel wavefunction which is an eigenfunction of j with projection m is then given by

$$\psi_j^m = \sum_{\nu} (l\nu Z\mu | jm)\, Y_l^{\nu}(\theta_x, \phi_x)\, U_l(k_x r_x)\, \Phi_Z^{\mu} \tag{3.70}$$

where Φ is a function of the internal co-ordinates of both particles including the spin co-ordinates and x labels the channel co-ordinates. In order to simplify the notation we write equation (3.70) as

$$U_\gamma = U_\gamma(r_x)\phi_\gamma$$

where γ represents the set of quantum numbers x, j, m, l, and Z. (In this representation two channels may have the same x, j, m and Z, and differ only in the relative angular momentum l.) The radial function U_γ is normalized to unit incident flux so that its asymptotic behaviour is

$$U_\gamma \to A_\gamma \mathscr{I}_\gamma - B_\gamma \mathscr{O}_\gamma$$

where $\mathscr{I}$ and $\mathscr{O}$ are the incoming and outgoing spherical waves

$$\mathscr{I}_\gamma = v_x^{-\frac{1}{2}} r_x^{-1} \exp[-i(k_x r_x - \tfrac{1}{2}l\pi)] \tag{3.71}$$

$$\mathscr{O}_\gamma = v_x^{-\frac{1}{2}} r_x^{-1} \exp[i(k_x r_x - \tfrac{1}{2}l\pi)]\,. \tag{3.72}$$

The amplitude A_γ of the incoming wave is given by

$$A_\gamma = i^{l+1}[\pi(2l+1)/k_x^2]^{\frac{1}{2}} \tag{3.73}$$

and the amplitude B_γ is to be determined.

The scattering initiated from a particular channel $\gamma = \alpha$ will be described by the wavefunction

$$\psi_\alpha = \sum_\beta U_{\alpha\beta}(r_\beta)\phi_\beta \tag{3.74}$$

where the sum over β runs over all reaction channels which can be reached from channel α, i.e. over all 'open' channels. In the channel $\beta = \alpha$ there is an incoming wave and an outgoing elastically scattered wave, but because of the depletion of the intensity in the elastic channel due to inelastic scattering and reactions the amplitude B_α is not equal to A_α. We can write

$$B_\alpha = S_{\alpha\alpha} A_\alpha\,,$$

where $S_{\alpha\alpha}$ is a constant, so that

$$U_{\alpha\alpha} \to A_\alpha[\mathscr{I}_\alpha - S_{\alpha\alpha}\mathscr{O}_\alpha]\,.$$

In the channels $\beta \neq \alpha$ there is no incoming wave but only an outgoing wave

$$U_{\alpha\beta} \to -A_\alpha S_{\alpha\beta} \tag{3.75}$$

so that the asymptotic behaviour of ψ_α is given by

$$\psi_\alpha \to A_\alpha [\mathscr{I}_\alpha \phi_\alpha - \sum_\beta S_{\alpha\beta} \mathscr{O}_\beta \phi_\beta] \, .$$

The complete scattering wavefunction can be written as a sum over the ψ_α. Since α specifies the incident particles and these are assumed to be in their ground states in the entrance channel, the additional summation required to give the complete wavefunction is a sum over l and the complete wavefunction is given by

$$\psi \to (\pi/k_x^2)^{\frac{1}{2}} \sum_{l\nu} i^{l+1} (2l+1)^{\frac{1}{2}} [\mathscr{I}_\alpha \phi_\alpha - \sum_\beta S_{\alpha\beta} \mathscr{O}_\beta \phi_\beta] \, . \tag{3.76}$$

Thus the asymptotic behaviour of the wavefunction is given by the $S_{\alpha\beta}$ which form the elements of the S-matrix*, and the diagonal elements are just the reflection coefficients defined in equation (3.65).

The cross-section for the process $\alpha \to \beta$ ($\beta \neq \alpha$) can now be determined by examining the asymptotic behaviour of the wavefunction ψ in the channel β. We take the axis of quantization along the direction of the incident beam so that $\nu = 0$, and associate the quantum numbers l', ν' and the channel co-ordinates y with the channel β. Thus, using equations (3.72) and (3.76) and integrating over the angular co-ordinates we obtain the total cross-section

$$\sigma_{\alpha\beta} = \frac{\pi(2l+1)}{k_x^2} |S_{\alpha\beta}|^2 , \quad \beta \neq \alpha \tag{3.77}$$

$$= \frac{\pi(2l+1)}{k_x^2} |S_{\alpha\beta} - \delta_{\alpha\beta}|^2 , \quad \text{all } \beta \, . \tag{3.78}$$

In practice we are usually more interested in the cross-sections for transitions between particular internal states of the colliding particles summed over all relative angular momentum states. We denote this process by $\bar{\alpha} \to \bar{\beta}$. The cross-section for this process can be obtained from equation (3.76) by retaining the summations over l and β, and using equation (3.72) this gives the differential cross-section

* This is also known as the U-matrix or collision matrix.

$$\frac{d\sigma}{d\Omega}(\bar{\alpha}\rightarrow\bar{\beta}) = \frac{\pi}{k_x^2}\left|\sum_{ll'v'} i^{l-l'}(2l+1)^{\frac{1}{2}}\, Y_{l'}^{v'}(\theta_y, \phi_y)(S_{\alpha\beta}-\delta_{\alpha\beta})\right|^2 \tag{3.79}$$

and the total cross-section

$$\sigma(\bar{\alpha}\rightarrow\bar{\beta}) = \frac{\pi}{k_x^2}\sum_{l'v'}\left|\sum_{l} i^{l}(2l+1)^{\frac{1}{2}}(S_{\alpha\beta}-\delta_{\alpha\beta})\right|^2 . \tag{3.80}$$

All information about the scattering processes is contained in the S-matrix since all measurable quantities can be determined from the elements $S_{\alpha\beta}$, and the $S_{\alpha\beta}$ can in principle be calculated from the Schrödinger equation for the system. In practice, the Hamiltonian for the whole system is a complicated many-body operator and exact solution of the Schrödinger equation is feasible only in certain special cases. In other cases the many-body interaction is replaced by a model two body interaction such that the Schrödinger equation can be solved and the S-matrix elements determined from the asymptotic behaviour of the radial solution as described in section 3.2. In the optical model (see Chapter 6), the effect of the reaction channels is described by means of a complex potential which takes account of the absorption from the elastic channels as discussed in section 3.6. In the strong coupling model (see section 7.4) a limited number of reaction channels are included explicitly and this leads to coupled radial equations. An alternative approach [37] is the parametrization of the S-matrix in terms of simple functions of energy and angular momentum, thus avoiding the need to solve the Schrödinger equation. This method works well for those processes in which the absorption effects are strong and is described in connection with strong absorption in Chapter 8. Yet another approach which is useful when the number of partial waves involved is small is the direct analysis of the data in terms of the phase shifts. This method is particularly important in the study of nucleon–nucleon scattering where the variation of particular phase shifts with energy reveals important features of the nucleon–nucleon interaction, and is also useful for the study of very low energy nucleon–nucleus scattering. For all these methods it is useful to have some knowledge of the basic characteristics of the S-matrix which are imposed, not by the particular problem under consideration, but by the basic principles of quantum mechanics and the conservation laws.

In the notation of this section the amplitude of the incoming wave in channel α is A_α. From equation (3.75) it can be seen that the amplitude of the outgoing wave in channel β due to the incoming wave in channel α is $-S_{\alpha\beta}A_\alpha$ so that the total outgoing amplitude in channel β is

$$B_\beta = \sum_\alpha S_{\alpha\beta} A_\alpha \,. \tag{3.81}$$

Conservation of flux requires that the outgoing intensity summed over all channels is equal to the incoming intensity summed over all channels. This gives

$$\sum_\beta |B_\beta|^2 = \sum_\alpha |A_\alpha|^2$$

$$\sum_{\alpha\beta\gamma} S^*_{\alpha\beta} A^*_\alpha S_{\gamma\beta} A_\gamma = \sum_\alpha A^*_\alpha A_\alpha$$

and, if this is to hold for arbitrary A_α,

$$\sum_\beta S^*_{\alpha\beta} S_{\gamma\beta} = \sum_\beta S_{\gamma\beta} S^+_{\beta\alpha} = \delta_{\alpha\gamma} \tag{3.82}$$

or

$$SS^\dagger = 1 \,.$$

Thus the requirement of conservation of flux or probability imposes the condition that the full S-matrix is unitary. Conservation of total angular momentum (rotational invariance) requires that j is preserved during the collision, i.e. Z' and l' combine to form the same j, which implies that S is diagonal in j. Since different values of m can be obtained by rotation of the coordinate system, rotational invariance also implies that S is independent of m.

The requirement of time reversal invariance also imposes a condition on the S-matrix. In order to investigate this we use the result that a time-reversed wavefunction is obtained by taking the complex conjugate of the wavefunction and reversing the spins [31]. From equations (3.71) and (3.72) this means that outgoing waves become incoming waves, and vice versa. Thus the time-reversed form of

$$U_\gamma \rightarrow A_\gamma \mathscr{I}_\gamma - B_\gamma \mathscr{O}_\gamma$$

is

$$U^T_\gamma \rightarrow A^*_{\gamma'} \mathscr{O}_\gamma - B^*_{\gamma'} \mathscr{I}_\gamma$$

where $A_{\gamma'}$, $B_{\gamma'}$ differ from A_γ, B_γ in that the signs of all magnetic quantum numbers are reversed, but by definition the time-reversed wavefunction obeys the same Schrödinger equation so that we may equate the coefficients of U_γ and U^T_γ. This gives

$$-B_\gamma = A^*_{\gamma'} \,, \quad -A_\gamma = B^*_{\gamma'}$$

and, from equation (3.81) we have

$$A^*_{\gamma'} = \sum_{\alpha} S_{\alpha\gamma} B^*_{\alpha'} = \sum_{\alpha\beta'} S_{\alpha\gamma} S^*_{\beta'\alpha'} A^*_{\beta'}$$

$$\delta_{\beta'\gamma'} = \sum_{\alpha} S_{\beta'\alpha'} S^*_{\alpha\gamma} \,.$$

Combining this with the unitary condition (3.82) yields the *reciprocity relation*

$$S_{\beta'\alpha'} = S_{\alpha\beta} \tag{3.83}$$

which implies that the amplitude for the process $\alpha \to \beta$ is the same as the amplitude for the time-reversed process $\beta \to \alpha$. If spin zero particles are involved, equation (3.83) requires that the S-matrix is symmetric. It follows from equations (3.83), (3.79) and (3.29) that

$$g_\alpha k_\alpha^2 \frac{d\sigma}{d\Omega}(\bar{\alpha} \to \bar{\beta}) = g_\beta k_\beta^2 \frac{d\sigma}{d\Omega}(\bar{\beta}' \to \bar{\alpha}') \tag{3.84}$$

where g_α, g_β are the total numbers of spin states in channels α and β. This result is known as the *principle of detailed balance.*

3.8 The integral equation for potential scattering

In the preceding sections we have approached the task of solving the Schrödinger equation for a scattering process by writing down solutions and then imposing the required asymptotic behaviour. In the formal theory which we develop in this and the following sections, more elegant mathematical techniques are used to write down formal solutions which contain the required asymptotic behaviour. The first step in the development of this formal theory is the conversion of the Schrödinger equation into an integral equation which incorporates the asymptotic boundary condition.

We extend the notation of the preceding sections as follows. The solution of the Schrödinger equation for the potential $V(\boldsymbol{r})$ and energy $E = \hbar^2 k^2 / 2\mu$ is denoted by $\psi^+(\boldsymbol{k}, \boldsymbol{r})$ where the plus sign indicates that the asymptotic behaviour is that of an outgoing spherical wave, i.e. $\psi^+(k, r)$ obeys equation (3.5). The wavefunction $\psi^-(\boldsymbol{k}, \boldsymbol{r})$ is a solution of the same Schrödinger equation but has the asymptotic behaviour of an incoming spherical wave. The behaviour of $\psi^-(\boldsymbol{k}, \boldsymbol{r})$ can be determined from the condition for time-reversal invariance

$$\psi^{-*}(\boldsymbol{k}, \boldsymbol{r}) = \psi^+(-\boldsymbol{k}, \boldsymbol{r})\,, \tag{3.85}$$

which gives $$\psi^-(\mathbf{k},\mathbf{r}) \to e^{i\mathbf{k}.\mathbf{r}} + f^*(\pi-\theta,\pi+\phi)\frac{e^{-ikr}}{r}. \tag{3.86}$$

If the sign is not given, $\psi^+(\mathbf{k},\mathbf{r})$ is to be understood. The plane wavefunction $e^{i\mathbf{k}.\mathbf{r}}$ will be denoted by $\phi(\mathbf{k},\mathbf{r})$ or by ϕ_E. The time-independent Schrödinger equation for the problem of potential scattering can now be written as

$$(E-H_0)\psi(\mathbf{k},\mathbf{r}) = V(\mathbf{r})\psi(\mathbf{k},\mathbf{r}) \tag{3.87}$$

where H_0 is the kinetic energy operator. The plane wavefunction obeys the equation

$$(E-H_0)\phi(\mathbf{k},\mathbf{r}) = 0\,. \tag{3.88}$$

It also obeys the equation

$$(E-H_0)\phi(\mathbf{k}',\mathbf{r}) = (E-E')\phi(\mathbf{k}',\mathbf{r})\,, \tag{3.89}$$

the *orthogonality relation*

$$\int \phi^*(\mathbf{k}',\mathbf{r})\phi(\mathbf{k},\mathbf{r})d\mathbf{r} = (2\pi)^3\delta(\mathbf{k}-\mathbf{k}')\,, \tag{3.90}$$

and the *closure relation*

$$\int \phi^*(\mathbf{k},\mathbf{r}')\phi(\mathbf{k},\mathbf{r})d\mathbf{k} = (2\pi)^3\delta(\mathbf{r}-\mathbf{r}')\,. \tag{3.91}$$

We now expand the scattering solution in terms of plane wave states

$$\psi(\mathbf{k},\mathbf{r}') = \int a(k'')\phi(\mathbf{k}'',\mathbf{r}')d\mathbf{k}'' \tag{3.92}$$

so that, using equation (3.89), equation (3.87) becomes

$$\int a(k'')(E-E'')\phi(\mathbf{k}'',\mathbf{r}')d\mathbf{k}'' = V(\mathbf{r}')\psi(\mathbf{k},\mathbf{r}')\,.$$

We then multiply on the left by $\phi^*(\mathbf{k}',\mathbf{r}')$ and integrate over $\mathbf{r}'$ and use the orthogonality relation so that the coefficient $a(k')$ is determined as

$$(2\pi)^3(E-E')a(k') = \int \phi^*(\mathbf{k}',\mathbf{r}')V(\mathbf{r}')\psi(\mathbf{k},\mathbf{r}')d\mathbf{r}'\,, \tag{3.93}$$

and substituting this back into equation (3.92) we have

$$\psi(\boldsymbol{k}, \boldsymbol{r}) = \int a(k')\phi(\boldsymbol{k}', \boldsymbol{r})d\boldsymbol{k}'$$

$$= \int G_0(\boldsymbol{r}, \boldsymbol{r}')V(\boldsymbol{r}')\psi(\boldsymbol{k}, \boldsymbol{r}')d\boldsymbol{r}' \tag{3.94}$$

where $G_0(\boldsymbol{r}, \boldsymbol{r}')$ is given by

$$G_0(\boldsymbol{r}, \boldsymbol{r}') = (2\pi)^{-3} \int \frac{\phi(\boldsymbol{k}', \boldsymbol{r})\phi^*(\boldsymbol{k}', \boldsymbol{r}')}{E - E'} d\boldsymbol{k}' . \tag{3.95}$$

By adding to equation (3.94) any solution of equation (3.88) we obtain the general solution of equation (3.87)

$$\psi(\boldsymbol{k}, \boldsymbol{r}) = \phi(\boldsymbol{k}, \boldsymbol{r}) + \int G_0(\boldsymbol{r}, \boldsymbol{r}')V(\boldsymbol{r}')\psi(\boldsymbol{k}, \boldsymbol{r}')d\boldsymbol{r}' . \tag{3.96}$$

We must now choose the form of $G_0(\boldsymbol{r}, \boldsymbol{r}')$ so that the wavefunction $\psi(\boldsymbol{k}, \boldsymbol{r})$ satisfies the required boundary condition. Equation (3.95) is unsatisfactory as it stands because of the presence of poles at $k' = \pm k$, and to eliminate the divergence we redefine $G_0(\boldsymbol{r}, \boldsymbol{r}')$ as

$$G_0^{\pm}(\boldsymbol{r}, \boldsymbol{r}') = \lim_{\varepsilon \to +0} (2\pi)^{-3} \frac{2\mu}{\hbar^2} \int \frac{\phi(\boldsymbol{k}', \boldsymbol{r})\phi^*(\boldsymbol{k}', \boldsymbol{r}')}{k^2 - k'^2 \pm i\varepsilon} d\boldsymbol{k}' \tag{3.97}$$

where ε is a small positive quantity which is allowed to go to zero after the integration has been performed. We can now perform the integration over the angular co-ordinates of k', followed by the integration in the complex k'-plane taking the residue at the poles $k' = \pm\sqrt{(k^2 \pm i\varepsilon)}$, and finally take the limit as $\varepsilon \to 0$. This gives [5]

$$G_0^{\pm}(\boldsymbol{r}, \boldsymbol{r}') = \frac{-\mu}{2\pi\hbar^2} \frac{\exp[\pm ik|\boldsymbol{r} - \boldsymbol{r}'|]}{|\boldsymbol{r} - \boldsymbol{r}'|} \tag{3.98}$$

where the plus and minus signs in the exponential arise from the residues at the poles in the k'-plane for $+\varepsilon$ and $-\varepsilon$ respectively. In order to examine the asymptotic behaviour of $G_0(\boldsymbol{r}, \boldsymbol{r}')$ and $\psi(\boldsymbol{k}, \boldsymbol{r})$ we assume that $V(\boldsymbol{r}')$ falls off rapidly with r' so that in the asymptotic region r is very much greater than the values of r' which contribute to the integrand. We then have

$$|\boldsymbol{r} - \boldsymbol{r}'|^{-1} \to \frac{1}{r}$$

$$|\boldsymbol{r} - \boldsymbol{r}'| \to r - \hat{\boldsymbol{r}} . \boldsymbol{r}', \quad \hat{\boldsymbol{r}} = \boldsymbol{r}/r ,$$

$$\exp[\pm ik|\boldsymbol{r} - \boldsymbol{r}'|] \to \exp[\pm i(kr - \boldsymbol{k}' . \boldsymbol{r}')] , \quad \boldsymbol{k}' = k\hat{\boldsymbol{r}} ,$$

so that from equations (3.96) and (3.98) the asymptotic behaviour of the scattering wavefunction is given by

$$\psi^{\pm}(\boldsymbol{k}, \boldsymbol{r}) \underset{r\to\infty}{\longrightarrow} e^{i\boldsymbol{k}.\boldsymbol{r}} - \frac{\mu}{2\pi\hbar^2}\frac{e^{\pm ikr}}{r}\int e^{\mp i\boldsymbol{k}'.\boldsymbol{r}'}V(\boldsymbol{r}')\psi^{\pm}(\boldsymbol{k}, \boldsymbol{r}')d\boldsymbol{r}' . \tag{3.99}$$

Here $\boldsymbol{k}'$ is a vector of magnitude k which has the direction given by the polar angles of the point at which the scattering amplitude is measured. Thus the wavefunctions $\psi^{\pm}(\boldsymbol{k}, \boldsymbol{r})$ correspond to the same energy $\hbar^2k^2/2\mu$ as the plane wave $\phi(\boldsymbol{k}, \boldsymbol{r})$. Further, if we associate the plus and minus signs in the expression (3.98) for $G_0^{\pm}(\boldsymbol{r}, \boldsymbol{r}')$ with the outgoing and incoming boundary conditions for the wavefunctions $\psi^{\pm}(\boldsymbol{k}, \boldsymbol{r})$, then these wavefunctions do indeed have the correct asymptotic behaviour. Thus the boundary conditions are associated with the sign of ε in the expression (3.97) for G_0. Finally, by comparison with equation (3.5) we obtain the expression for the scattering amplitude

$$f(\theta, \phi) = \frac{-\mu}{2\pi\hbar^2}\int e^{-i\boldsymbol{k}'.\boldsymbol{r}'}V(\boldsymbol{r}')\psi^{+}(\boldsymbol{k}, \boldsymbol{r}')d\boldsymbol{r}' . \tag{3.100}$$

3.9 Green's function techniques

The next step in the development of the formal theory is to express equation (3.96) in terms of an operator formalism. We first multiply both sides of equation (3.89) by $(E-H_0)^{-1}$ which gives

$$(E-H_0)^{-1}(E-H_0)\phi(\boldsymbol{k}', \boldsymbol{r}) = (E-E')(E-H_0)^{-1}\phi(\boldsymbol{k}', \boldsymbol{r})$$

from which we obtain the equation

$$(E-H_0)^{-1}\phi(\boldsymbol{k}', \boldsymbol{r}) = \frac{1}{E-E'}\phi(\boldsymbol{k}', \boldsymbol{r}) . \tag{3.101}$$

Thus the $\phi(\boldsymbol{k}', \boldsymbol{r})$ are eigenfunctions of the operator $(E-H_0)^{-1}$ with eigenvalues $1/(E-E')$ where $E'>0$ and $E\neq E'$. For $E=E'$ the operator $(E-H_0)^{-1}$ is not defined. If we now operate on equation (3.87) in a similar way we obtain

$$\psi(\boldsymbol{k}, \boldsymbol{r}) = (E-H_0)^{-1}V(\boldsymbol{r})\psi(\boldsymbol{k}, \boldsymbol{r})$$

to which we must add any solution of equation (3.88) and remove the divergence by inserting the small quantity ε. Hence we obtain the formal solution

$$\psi^{\pm}(\boldsymbol{k}, \boldsymbol{r}) = \phi(\boldsymbol{k}, \boldsymbol{r}) + (E-H_0 \pm i\varepsilon)^{-1}V(\boldsymbol{r})\psi(\boldsymbol{k}, \boldsymbol{r}) . \tag{3.102}$$

(This is a solution of the Schrödinger equation only in the limit $\varepsilon \to 0$.) By comparison with equation (3.96) we may identify the function $G_0^{\pm}(\boldsymbol{r}, \boldsymbol{r}')$ defined in equation (3.97) as the *Green's function* for the operator $(E - H_0 \pm i\varepsilon)^{-1}$. This can be checked using equation (3.101) which can be written in the more general form

$$\frac{1}{E - H + i\varepsilon} \Phi_n(r) = \frac{1}{E - E_n + i\varepsilon} \Phi_n(r)$$

where H is any Hamiltonian and Φ_n are the corresponding eigenfunctions. Using the closure relation we have

$$\frac{1}{E - H + i\varepsilon} = \sum_n \frac{1}{E - E_n + i\varepsilon} \Phi_n(r) \Phi_n^*(r') \tag{3.103}$$

where $\sum_n$ implies summation over discrete eigenstates and integration over continuous eigenstates. If H is replaced by the kinetic energy operator H_0 and the Φ_n are replaced by the continuous eigenstates $(2\pi)^{-\frac{3}{2}} \phi(\boldsymbol{k}', \boldsymbol{r})$ equation (3.103) reduces to

$$\frac{1}{E - H_0 + i\varepsilon} = (2\pi)^{-3} \int \frac{\phi(\boldsymbol{k}', \boldsymbol{r}) \phi^*(\boldsymbol{k}', \boldsymbol{r}')}{E - E' + i\varepsilon} d\boldsymbol{k}'$$

which is identical with equation (3.97). Equation (3.103) is referred to as the *spectral representation of the Green's function.*

In equation (3.96) the Green's function appears as the kernel in an integral equation, but it can also be regarded as the matrix representative of an operator G_0 such that

$$G_0^+(\boldsymbol{r}, \boldsymbol{r}') = \langle \boldsymbol{r} | G_0^+ | \boldsymbol{r}' \rangle$$

and similarly

$$G_0^-(\boldsymbol{r}, \boldsymbol{r}') = \langle \boldsymbol{r} | G_0^- | \boldsymbol{r}' \rangle .$$

Since $G_0(\boldsymbol{r}, \boldsymbol{r}')$ is symmetric in $\boldsymbol{r}$ and $\boldsymbol{r}'$ (see equation (3.98)), it follows that

$$G_0^- = (G_0^+)^\dagger . \tag{3.104}$$

Thus if we take a scattering state $|\psi^{\pm}\rangle$ whose corresponding wavefunction is

$$\langle \boldsymbol{r} | \psi^{\pm} \rangle = \psi^{\pm}(\boldsymbol{k}, \boldsymbol{r})$$

the effect of operating with G_0 is given by

$$\langle \boldsymbol{r}|G_0|\psi\rangle = \int G_0(\boldsymbol{r}, \boldsymbol{r}')\psi(\boldsymbol{k}, \boldsymbol{r}')d\boldsymbol{r}', \tag{3.105}$$

and we can now rewrite equations (3.96) and (3.102) in the form

$$|\psi^{\pm}\rangle = |\phi\rangle + G_0^{\pm} V|\psi^{\pm}\rangle \tag{3.106}$$

which is known as the *Lipmann-Schwinger equation.* It is also convenient to define an operator corresponding to the total Hamiltonian $H = H_0 + V$ as

$$G^{\pm} = \frac{1}{E - H \pm i\varepsilon}. \tag{3.107}$$

The relation between G and G_0 can be determined from the relation

$$\frac{1}{A} - \frac{1}{B} = \frac{1}{B}(B - A)\frac{1}{A}$$

which is true for operators as well as for numbers. Hence if $A = E - H \pm i\varepsilon$ and $B = E - H_0 \pm i\varepsilon$ we find

$$G^{\pm} = G_0^{\pm} + G_0^{\pm} V G^{\pm} \tag{3.108}$$

and reversing the definitions of A and B we find

$$G_0^{\pm} = G^{\pm} - G^{\pm} V G_0^{\pm}. \tag{3.109}$$

Substituting equation (3.109) into equation (3.106) yields

$$\begin{aligned} \psi^{\pm} &= \phi + G^{\pm} V(\psi^{\pm} - G_0^{\pm} V\psi^{\pm}) \\ &= \phi + G^{\pm} V\phi \end{aligned} \tag{3.110}$$

where for simplicity of notation we have dropped the bra and ket symbols. Equation (3.110) is the formal solution of the Lippmann-Schwinger equation and hence of the scattering problem. Unfortunately, it is solely a formal solution because the explicit form of $G^{\pm}$ is given in terms of the eigenfunctions of the total Hamiltonian H (see equation (3.103)) and these are just the solutions we are trying to find. The formal theory is extremely useful, however, because it lends itself to manipulation and to a careful study of various approximations. For example, the iterative solution of equation (3.108) is

$$G^{\pm} = G_0^{\pm} + G_0^{\pm} V G_0^{\pm} + \ldots = \sum_{n=0}^{\infty} (G_0^{\pm} V)^n G_0^{\pm}, \tag{3.111}$$

and substituting in equation (3.110) and using equation (3.105) we obtain the Born expansion for the scattering wavefunction

$$\psi^{\pm}(\boldsymbol{k}, \boldsymbol{r}) = \phi(\boldsymbol{k}, \boldsymbol{r}) + \int G_0^{\pm}(\boldsymbol{r}, \boldsymbol{r}') V(\boldsymbol{r}') \phi(\boldsymbol{k}, \boldsymbol{r}') d\boldsymbol{r}'$$
$$+ \int\int G_0^{\pm}(\boldsymbol{r}, \boldsymbol{r}'') V(\boldsymbol{r}'') G_0^{\pm}(\boldsymbol{r}'', \boldsymbol{r}') V(\boldsymbol{r}') \phi(\boldsymbol{k}, \boldsymbol{r}') d\boldsymbol{r}' d\boldsymbol{r}'' + \ldots \tag{3.112}$$

Thus we may interpret the Green's function $G_0(\boldsymbol{r}'', \boldsymbol{r}')$ as a propagator which takes the wave from its scattering at the point $\boldsymbol{r}'$ to its next scattering at the point $\boldsymbol{r}''$, and this is represented diagrammatically in figure 3.3. Between scatterings the propagation is that of a free wave.

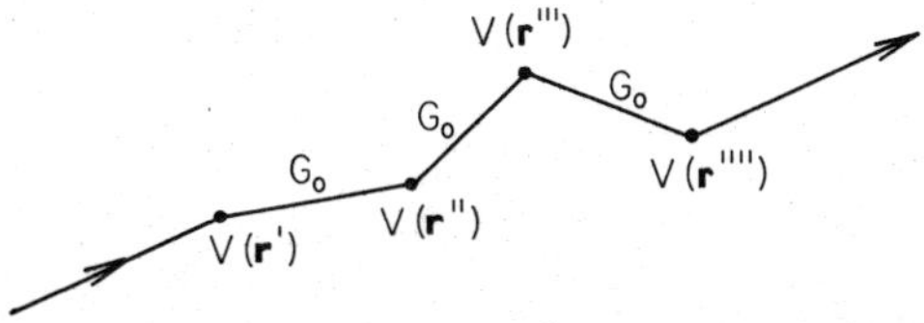

Figure 3.3 The propagation of a wave between successive scatterings.

The *Møller wave operators* $\Omega^{\pm}$ are defined by the equation

$$\psi^{\pm} = \Omega^{\pm}\phi = (1 + G^{\pm} V)\phi \tag{3.113}$$

and transform plane wave states into scattering states. The *transition operators* $T^{\pm}$ are defined through the relations

$$\begin{aligned} T^{\pm} &= V\Omega^{\pm} \\ &= V + VG^{\pm} V = V + VG_0^{\pm} T^{\pm}, \end{aligned} \tag{3.114}$$

and the transition matrix element is defined as

$$T_{fi} = \langle \phi | T^{+} | \phi \rangle = \langle \phi | V | \psi^{+} \rangle . \tag{3.115}$$

By comparison with equation (3.100) we have

$$f(\theta, \phi) = \frac{-\mu}{2\pi\hbar^2} T_{fi} . \tag{3.116}$$

These operators G, G_0, Ω, T are defined for a given incident energy E, and we say that E is the *energy shell* for these operators. A matrix element T_{fi} with $k'^2 = k^2 = 2\mu E/\hbar^2$ is said to be an 'on-shell' or 'on-energy-shell' matrix element, and conversely if $k'^2 \neq k^2$ the matrix element is 'off the energy-shell'.

In many problems of interest the potential can be separated into two terms,

$$V = U + W ,$$

such that the scattering problem with the potential U alone has a known solution χ. We use the notation

$$H_1 = H_0 + U\,, \qquad H = H_1 + W\,, \tag{3.117}$$

$$G_1^{\pm} = \frac{1}{E - H_1 \pm i\varepsilon}\,, \qquad G^{\pm} = \frac{1}{E - H \pm i\varepsilon}\,, \tag{3.118}$$

so that χ obeys the equation

$$(E - H_1)\chi^{\pm} = 0 \tag{3.119}$$

and can be expressed in terms of plane waves states using equations (3.106) and (3.110) to give

$$\chi^{\pm} = \phi + G_1^{\pm}\, U\phi \tag{3.120}$$

$$= \phi + G_0^{\pm}\, U\chi^{\pm}\,. \tag{3.121}$$

The solution to the complete problem obeys the equation

$$(E - H)\psi^{\pm} = 0 \tag{3.122}$$

and can be expressed in terms of $\chi^{\pm}$, again using equations (3.106) and (3.110),

$$\psi^{\pm} = \chi^{\pm} + G^{\pm}\, W\chi^{\pm} \tag{3.123}$$

$$= \chi^{\pm} + G_1^{\pm}\, W\psi^{\pm}\,. \tag{3.124}$$

The transition matrix element is given by

$$T_{fi} = \langle \phi | T^{+} | \phi \rangle = \langle \phi | U + W | \psi^{+} \rangle$$

$$= \langle \phi | U | \chi^{+} \rangle + \langle \phi | U G_1^{+}\, W | \psi^{+} \rangle + \langle \phi | W | \chi^{+} \rangle + \langle \phi | W G_1^{+}\, W | \psi^{+} \rangle$$

where we have used equation (3.124). Now using equations (3.104) and (3.120) we have

$$\langle \phi | U G_1^{+}\, W | \psi^{+} \rangle = \langle G_1^{-}\, U^{\dagger} \phi | W | \psi^{+} \rangle = \langle \chi^{-} | W | \psi^{+} \rangle - \langle \phi | W | \psi^{+} \rangle$$

and similarly, using equation (3.124),

$$\langle \phi | W G_1^{+}\, W | \psi^{+} \rangle = \langle \phi | W | \psi^{+} \rangle - \langle \phi | W | \chi^{+} \rangle\,.$$

Thus, collecting up the terms the transition matrix element is given by

$$T_{fi} = \langle \phi | U | \chi^{+} \rangle + \langle \chi^{-} | W | \psi^{+} \rangle \tag{3.125}$$

or, using the iterative solution of equation (3.124),

$$T_{fi} = \langle \phi | U | \chi^+ \rangle + \sum_{n=0}^{\infty} \langle \chi^- | W (G_1^+ W)^n | \chi^+ \rangle \,. \tag{3.126}$$

If the initial and final states of the target are not connected by the potential U the first term in equation (3.125) is zero, but the potential U still has an effect as it determines the function χ. This result, due to Gell-Mann and Goldberger, is the source of several approximate methods which are used in the description of inelastic scattering and reactions [38].

3.10 Approximate methods for elastic scattering

We have already obtained in equation (3.112) the Born series for the scattering wavefunction, and substituting this into equation (3.100) yields the Born series for the scattering amplitude

$$\begin{aligned} f(\theta, \phi) = \frac{-\mu}{2\pi\hbar^2} \Bigg[& \int e^{-i\boldsymbol{k}' \cdot \boldsymbol{r}} V(\boldsymbol{r}) e^{i\boldsymbol{k} \cdot \boldsymbol{r}} \, d\boldsymbol{r} \\ & + \int e^{-i\boldsymbol{k}' \cdot \boldsymbol{r}} V(\boldsymbol{r}) G_0^+ (\boldsymbol{r} . \boldsymbol{r}') V(\boldsymbol{r}') e^{i\boldsymbol{k} \cdot \boldsymbol{r}'} \, d\boldsymbol{r}' \, d\boldsymbol{r} \\ & + \int \ldots d\boldsymbol{r}'' \, d\boldsymbol{r}' \, d\boldsymbol{r} + \ldots \Bigg] . \end{aligned} \tag{3.127}$$

The first term in this series is the first *Born approximation* for the scattering amplitude and represents the simplest possible approximation obtained by terminating the series for the scattering wavefunction after the plane wave term. In terms of the momentum transfer $\boldsymbol{q} = \boldsymbol{k} - \boldsymbol{k}'$, the Born amplitude is

$$f_B(\theta, \phi) = \frac{-\mu}{2\pi\hbar^2} \int V(\boldsymbol{r}) e^{i\boldsymbol{q} \cdot \boldsymbol{r}} \, d\boldsymbol{r} \tag{3.128}$$

and is just proportional to the Fourier transform of the potential. This approximation usually leads to very simple calculations even for non-symmetric or spin-dependent potentials. Estimates of the validity of this approximation yield the condition [31, 32]

$$\frac{|V|}{E} \ll \frac{1}{kd}$$

where d is the range over which the potential changes significantly, and this is rarely satisfied for nuclear scattering processes.

A better approximation for the scattering wavefunction can be derived in the high energy limit

$$\frac{|V|}{E} \ll 1\,, \qquad kd \gg 1\,.$$

We follow the derivation of this *high energy* or *semi-classical approximation* given by Glauber [39] and start from the integral equation (see equations (3.96) and (3.98))

$$\psi^+(\boldsymbol{k}, \boldsymbol{r}) = e^{i\boldsymbol{k}\cdot\boldsymbol{r}} - \frac{\mu}{2\pi\hbar^2}\int \frac{e^{ik|\boldsymbol{r}-\boldsymbol{r}'|}}{|\boldsymbol{r}-\boldsymbol{r}'|} V(\boldsymbol{r}')\psi^+(\boldsymbol{k}, \boldsymbol{r}')d\boldsymbol{r}\,.$$

The wavefunction can also be written as the product of a plane wave and a modulation function

$$\psi^+(\boldsymbol{k}, \boldsymbol{r}) = e^{i\boldsymbol{k}\cdot\boldsymbol{r}}\rho(\boldsymbol{r}) \tag{3.129}$$

so that $\rho(\boldsymbol{r})$ is given by

$$\rho(\boldsymbol{r}) = 1 - \frac{\mu}{2\pi\hbar^2}\int \frac{e^{ik|\boldsymbol{r}-\boldsymbol{r}'|}}{|\boldsymbol{r}-\boldsymbol{r}'|} e^{-i\boldsymbol{k}\cdot(\boldsymbol{r}-\boldsymbol{r}')} V(\boldsymbol{r}')\rho(\boldsymbol{r}')d\boldsymbol{r}'$$

In the high-energy limit the product $V\rho$ should vary slowly within a distance of the order of the reduced wavelength $1/k$, so that contributions to the integral from regions in which the exponential varies rapidly should be small. The largest contribution to the integral will occur when $\boldsymbol{r}-\boldsymbol{r}'$ is roughly parallel to $\boldsymbol{k}$, since in this case $\exp[ik|\boldsymbol{r}-\boldsymbol{r}'| - i\boldsymbol{k}\cdot(\boldsymbol{r}-\boldsymbol{r}')]$ is almost stationary. We make the substitution

$$\boldsymbol{u} = \boldsymbol{r}-\boldsymbol{r}'\,, \quad d\boldsymbol{u} = u^2 du\, dw\, d\eta\,, \quad w = \boldsymbol{k}\cdot\boldsymbol{u}/ku\,,$$

and integrate by parts keeping only first-order terms. This gives

$$\rho(\boldsymbol{r}) = 1 + \frac{\mu}{2\pi\hbar^2}\int u^2 du\, d\eta \left[\frac{e^{iku(1-w)}}{iku^2} V(\boldsymbol{r}-\boldsymbol{u})\rho(\boldsymbol{r}-\boldsymbol{u})\right]_{-1}^{+1}.$$

Using the same argument as before, we take the contribution from the limit $w = -1$ to be negligible and evaluate the integrand at the limit $w = +1$ or $\boldsymbol{u}$ parallel to $\boldsymbol{k}$, which gives

$$\begin{aligned}\rho(\boldsymbol{r}) &= 1 - \frac{i}{\hbar v}\int_0^\infty V(\boldsymbol{r}-\boldsymbol{u})\rho(\boldsymbol{r}-\boldsymbol{u})du \qquad \boldsymbol{u}\,\|\,\boldsymbol{k} \\ &= 1 - \frac{i}{\hbar v}\int_0^\infty V(\boldsymbol{r}-\hat{\boldsymbol{k}}s)\rho(\boldsymbol{r}-\hat{\boldsymbol{k}}s)ds\,. \end{aligned} \tag{3.130}$$

The solution of this integral equation is

$$\rho(\boldsymbol{r}) = \exp\left[\frac{-i}{\hbar v}\int_0^\infty V(\boldsymbol{r}-\hat{\boldsymbol{k}}s)ds\right] \tag{3.131}$$

so that the approximation for the wavefunction (3.129) is

$$\psi^+(\boldsymbol{k}, \boldsymbol{r}) \simeq \exp\left[i\boldsymbol{k}.\boldsymbol{r} - \frac{i}{\hbar v}\int_0^\infty V(\boldsymbol{r}-\hat{\boldsymbol{k}}s)ds\right]. \tag{3.132}$$

If we choose the z-axis along the direction $\boldsymbol{k}$ of the incident beam, integrals along the path s can be written as

$$\int_0^\infty V(\boldsymbol{r}-\hat{\boldsymbol{k}}s)ds = \int_{-\infty}^{z} V(xyz')dz', \quad \int_0^\infty V(\boldsymbol{r}+\hat{\boldsymbol{k}}s)ds = \int_z^\infty V(xyz')dz', \tag{3.133}$$

or we can substitute $b^2 = x^2 + y^2$ so that equation (3.132) becomes

$$\psi^+(\boldsymbol{k}, \boldsymbol{r}) \simeq \exp\left[ikz - \frac{i}{\hbar v}\int_{-\infty}^{z} V(\boldsymbol{b}+\hat{\boldsymbol{k}}z')dz'\right]. \tag{3.134}$$

The scattering amplitude in this approximation is given by

$$f(\theta, \phi) = \frac{-\mu}{2\pi\hbar^2}\int e^{-i\boldsymbol{k}'.\boldsymbol{r}}V(\boldsymbol{r})\psi^+(\boldsymbol{k}, \boldsymbol{r})d\boldsymbol{r}$$

$$\simeq \frac{-\mu}{2\pi\hbar^2}\int e^{i(\boldsymbol{k}-\boldsymbol{k}').(\boldsymbol{b}+\hat{\boldsymbol{k}}z)}V(\boldsymbol{b}+\boldsymbol{k}z)\rho(\boldsymbol{b}+\hat{\boldsymbol{k}}z)b\,db\,dz\,d\eta\,.$$

Now for elastic scattering and small scattering angles

$$(\boldsymbol{k}-\boldsymbol{k}').\hat{\boldsymbol{k}} \approx 0\,, \quad (\boldsymbol{k}-\boldsymbol{k}').\boldsymbol{b} \approx 2kb\sin\tfrac{1}{2}\theta\cos\eta.$$

Also, since from equation (3.130),

$$\rho(\boldsymbol{b}+\boldsymbol{k}z) = 1 - \frac{i}{\hbar v}\int_{-\infty}^{z} V(\boldsymbol{b}+\hat{\boldsymbol{k}}z')\rho(\boldsymbol{b}+\hat{\boldsymbol{k}}z')dz'$$

$$\frac{d\rho}{dz} = \frac{-i}{\hbar v}V(\boldsymbol{b}+\hat{\boldsymbol{k}}z)\rho(\boldsymbol{b}+\hat{\boldsymbol{k}}z)$$

the scattering amplitude becomes

$$f(\theta, \phi) = \frac{k}{2\pi i}\int_0^{2\pi} d\eta\int_0^\infty b\,db\,e^{i2kb\sin\theta/2\cos\eta}\int_{-\infty}^{\infty} dz\frac{d\rho}{dz}$$

$$= \frac{k}{2\pi i}\int_0^{2\pi} d\eta\int_0^\infty b\,db\,e^{i2kb\sin\theta/2\cos\eta}\left[e^{i\chi(b)}-1\right] \tag{3.135}$$

where
$$\chi(b)=\frac{-1}{\hbar v}\int_{-\infty}^{\infty}V(\boldsymbol{b}+\hat{\boldsymbol{k}}z)\,dz\,. \tag{3.136}$$

and we have used equations (3.131) and (3.133) to evaluate the integral over z. If the potential is axially symmetric we can carry out the η integration using the formula [17]

$$\int_0^{2\pi} e^{i\lambda\cos\eta}\,d\eta=2\pi J_0(\lambda) \tag{3.137}$$

to obtain

$$f(\theta)=-ik\int_0^{\infty} J_0(2kb\sin\tfrac{1}{2}\theta)\left[e^{i\chi(b)}-1\right)b\,db\,. \tag{3.138}$$

In the WKB approximation the formula for the phase shift is

$$\delta_l=\int_{r_0}^{\infty}\left[k^2-\frac{2\mu V}{\hbar^2}-\frac{l(l+1)}{r^2}\right]^{\frac{1}{2}}dr-\int_{r_0}^{\infty}\left[k^2-\frac{l(l+1)}{r^2}\right]^{\frac{1}{2}}dr \tag{3.139}$$

where r_0 represents the zero of the integrand, and if $V/E\ll 1$ the integrands may be expanded to give

$$\delta_l=\frac{-1}{\hbar v}\int_0^{\infty}V(xyz')\,dz'\,.$$

Comparing this result with equation (3.136) we see that in the semi-classical limit $kb=l+\frac{1}{2}$ we have the correspondence

$$\chi\left(\frac{l+\frac{1}{2}}{k}\right)=2\delta_l\,. \tag{3.140}$$

Some care must be exercised in using equations (3.136) or (3.139) if the potential includes a Coulomb term. In this case the complete phase shift is $\delta_l+\sigma_l$ where σ_l is Coulomb phase shift due to a point nucleus which must be calculated from equation (3.23). (We can not use equation (3.136) to calculate σ_l as the integral diverges for a $1/r$ potential.) We write the potential as the sum of a nuclear term V_N and a Coulomb part V_c where V_c has the form due to a finite nucleus inside a distance R_c and the form due to a point charge beyond R_c, i.e.

$$\begin{aligned} V_c &= V_c^{\text{finite}} \qquad r<R_c\\ &= V_c^{\text{point}} \qquad r\geqslant R_c\,, \end{aligned}$$

so that the phase shift δ_l is given by

$$\hbar v \delta_l = -\int_0^\infty V_N dz - \int_0^{Z_c} (V_c^{\text{finite}} - V_c^{\text{point}}) dz \tag{3.141}$$

where $Z_c^2 = R_c^2 - b^2$. Comparison of this procedure with the discussion leading to equation (3.26) shows that the semi-classical method allows only an approximate treatment of the effect of the Coulomb potential.

A number of important approximations for the scattering amplitude have been developed for application to scattering processes in which there is strong absorption of the projectiles. These approximations are described in Chapter 8.

3.11 Approximate methods for inelastic scattering and reactions

We now consider approximations for non-elastic scattering, using the two potential formulae (3.125) and (3.126). We assume that the potential U is independent of the internal co-ordinates of the projectile and target and depends only on the channel co-ordinates and possibly the channel spin; it therefore describes elastic scattering and does not connect different states of the target nucleus. In contrast, the potential W is assumed to depend on the channel co-ordinates and some, though not necessarily all, of the internal co-ordinates. In the most general situation the potentials U and W will not be the same in the initial and final channels; for example, consider the reaction $a+A \rightarrow b+B$. In the entrance channel it is convenient to take the potential U to be the potential which describes elastic scattering of projectile a from nucleus A and similarly in the exit channel it is convenient to take the potential U to be the potential which describes the elastic scattering of b from B. Thus the total Hamiltonian can be written as

$$\begin{aligned} H &= H_a + H_A + T_{aA} + V_{aA} \\ &= H_a + H_A + T_{aA} + U_{aA} + (V_{aA} - U_{aA}) \\ &= H_b + H_B + T_{bB} + U_{bB} + (V_{bB} - U_{bB}), \end{aligned}$$

i.e. $W_{aA} = V_{aA} - U_{aA}$, $W_{bB} = V_{bB} - U_{bB}$. The transition matrix element is given by (see below)

$$T_{fi} = \langle \Psi_{bB}^- | V_{aA} - U_{aA} | \Phi_A \xi_a \chi_a^+ \rangle \tag{3.142a}$$

$$= \langle \chi_b^- \xi_b \Phi_B | V_{bB} - U_{bB} | \Psi_{aA}^+ \rangle \tag{3.142b}$$

where ξ_a, ξ_b, Φ_A, Φ_B are eigenstates of H_a, H_b, H_A and H_B respectively, and $\Psi^\pm$ is a solution of the Schrödinger equation with the complete Hamiltonian H. These expressions for T_{fi} are known as the 'prior' and 'post' forms for the transition matrix element. The expressions (3.142a, b) are derived by considering any two Hamiltonians H_1 and H_2 for a given system differing only through potential terms which tend to zero more rapidly than $1/r$. The stationary solutions of H_1 and H_2 at energy E are Ψ_1 and Ψ_2, and the transition operators corresponding to the two Hamiltonians are T_1 and T_2. By considering the asymptotic behaviour of the solutions in open and closed channels it can be shown that [31]

$$\langle f|T_1|i\rangle = \langle f|T_2|i\rangle + \langle \Psi^-_{2f}|H_1 - H_2|\Psi^+_{1i}\rangle$$

where f represents the final system (b, B, $\mathbf{k}_b$) and i the initial system (a, A, $\mathbf{k}_a$). If we now set $H_1 = H_a + H_A + U_{aA}$ and $H_2 = H_1 + (V_{aA} - U_{aA})$ we see that H_1 does not connect the initial and final states so that $\langle f|T_1|i\rangle \equiv 0$, and that $H_1\Psi^+_1 = \xi_a \Phi_A \chi^+_a$. Hence

$$T_{fi} = \langle f|T_2|i\rangle = \langle \Psi^-_{bB}|V_{aA} - U_{aA}|\xi_a \Phi_A \chi^+_a\rangle\,,$$

which verifies equation (3.142a). Equation (3.142b) can be obtained by setting $H_2 = H_b + H_B + U_{bB}$ and $H_1 = H + (V_{bB} - U_{bB})$.

In the rest of this section we use the 'post' form of the transition matrix element, but simplify the formalism and the notation by neglecting the internal structure of the projectile, labelling the initial and final states by i and f and neglecting the labels on the potential W. Thus the matrix element (3.142b) can be written as

$$T_{fi} = \langle \chi^-_f \Phi_f|W|\Psi^+_i\rangle\,.$$

The simplest possible approximation we can make for this problem is to replace χ^-_f and Ψ^+ by plane waves states, i.e. $\chi^-_f \simeq \phi_f$, $\Psi^+ \simeq \phi_i \Phi_i$, which yields the *plane wave Born approximation* (PWBA) for the transition matrix element. A much better approximation is achieved by using the Born series to express Ψ^+ in terms of χ^+,

$$\Psi^+ = \sum_\alpha \psi^+_\alpha \Phi_\alpha = \sum_\alpha \left[\sum_{n=0}^{\infty} (G^+_1 W)^n \chi^+\right]_\alpha \Phi_a,$$

and by assuming that terms with $\alpha \neq i$ may be neglected. This latter assumption allows us to replace Ψ^+ by $\psi^+_i \Phi_i$ and leads to the matrix element

$$T_{fi} = \langle \chi^-_f \Phi_f|W|\psi^+_i \Phi_i\rangle \tag{3.143}$$
$$= \langle \chi^-_f \Phi_f|W|\chi^+_i \Phi_i\rangle + \langle \chi^-_f \Phi_f|WG^+_1 W|\chi^+_i \Phi_i\rangle + \ldots \tag{3.144}$$

The first term in this series gives the *distorted wave Born approximation*

(DWBA) for the transition matrix element. It is a first Born approximation in W, but the *distorted wavefunctions* $\chi^{\pm}$ take account of the distortion of the plane waves by the potential U. The second and higher terms in the expansion (3.144) describe *multiple scattering* processes in which the potential acts more than once and the target nucleus passes through an intermediate state which may be the ground state or an excited state of the nucleus. Second-order processes of this type are discussed in Chapter 7. As noted in section 3.7, the explicit inclusion of more than two reaction channels leads to coupled equations, and a particular example is also described in Chapter 7.

It should be noted that the operator G_1 in equation (3.144) contains the nuclear Hamiltonian, i.e. we now have

$$G_1^+ = \frac{1}{E - H_0 - H_N - U + i\varepsilon} \tag{3.145}$$

$$G^+ = \frac{1}{E - H_0 - H_N - U - W + i\varepsilon}, \tag{3.146}$$

so that equation (3.143) can be written as

$$T_{fi} = \langle \chi_f^- \Phi_f | t^+ | \chi_i^+ \Phi_i \rangle \tag{3.147}$$

where

$$t^+ = W + WG^+W = W + WG_1^+ t^+ . \tag{3.148}$$

This transition operator describes scattering in the medium represented by the potential U. We can also define a similar transition operator $t(j)$ which describes the scattering of the projectile by the jth nucleon in the target nucleus as

$$t^+(j) = W(j) + W(j) G_1^+ t^+(j) \tag{3.149}$$

where

$$W = \sum_j W(j) . \tag{3.150}$$

We also define a wavefunction $\psi^+(j)$ such that

$$W(j)\psi^+ = t(j)\psi^+(j) , \tag{3.151}$$

but from equation (3.124)

$$W(k)\psi^+ = W(k)\chi^+ + W(k)G_1^+ \sum_j W(j)\psi^+$$

so that

$$t(k)\psi^+(k) = W(k)\chi^+ + W(k)G_1^+ \sum_j t(j)\psi^+(j)$$

and substituting for $t(k)$ from equation (3.149) we have

$$W(k)\psi^+(k) + W(k)G_1^+ t^+(k)\psi^+(k) = W(k)\chi^+ + W(k)G_1^+ \sum_j t(j)\psi^+(j)$$

$$\psi^+(k) = \chi^+ + G_1^+ \sum_{j \neq k} t(j)\psi^+(j).$$

Thus substituting for $\psi(k)$ in equation (3.143) and using equations (3.150) and (3.151) we have

$$T_{fi} = \langle \chi_f^- \Phi_f | \sum_j W(j) | \psi_i^+ \Phi_i \rangle$$

$$= \langle \chi_f^- \Phi_f | \sum_j t(j) | \chi^+ \Phi_i \rangle + \langle \chi_f^- \Phi_f | \sum_j t(j) G_1^+ \sum_{k \neq j} t(k) | \chi_i^+ \Phi_i \rangle + \dots \quad (3.152)$$

The approximation obtained by terminating equation (3.152) after the first term is called the *multiple scattering approximation*, since the operator $t(j)$ sums to all orders the interaction of the projectile with the jth nucleon in the presence of the potential U. The higher terms correspond to successive scatterings from different nucleons in the nucleus.

The transition operator for the scattering of free particles is given by

$$\tau^+(j) = W(j) + W(j) G_{\text{free}}^+ \tau^+(j) \quad (3.153)$$

with

$$G_{\text{free}} = \frac{1}{E - H_0 + i\varepsilon}. \quad (3.154)$$

If it can be assumed that the incident particle interacts with one nucleon at a time, that the amplitude of the incident wave reaching each nucleon in the target nucleus is unaffected by the presence of the other nucleons, and that the binding energies of the target nucleons are very much less than the incident energy, then it is permissible to make the *impulse approximation* and replace $t(j)$ by $\tau(j)$. If we continue to take account of the potential U in the distorted waves we have the *distorted wave impulse approximation* (DWIA),

$$T_{fi} = \langle \chi_f^- \Phi_f | \sum_j \tau(j) | \chi_i^+ \Phi_i \rangle, \qquad f \neq i, \quad (3.155)$$

but if we set $U = 0$ we have the *plane wave impulse approximation* (PWIA)

$$T_{fi} = \langle \phi_f \Phi_f | \sum_j \tau(j) | \phi_i \Phi_i \rangle, \qquad \text{all } f, \quad (3.156)$$

which applies to both elastic and non-elastic scattering. Since the energy-momentum conditions for nucleon–nucleus scattering are different from free nucleon–nucleon scattering these matrix elements are off the energy shell for free scattering.

The validity and use of these approximations are discussed in different contexts in later chapters. An important feature of the matrix elements for non-elastic processes is the presence of the distorted wavefunctions $\chi^{\pm}$. The wavefunction χ^{+} with the outgoing boundary condition can be obtained in the form of a partial wave expansion, using the methods described in sections 3.2 and 3.3, or in an approximate form using the method described at the beginning of this section. In the final state, the wavefunction χ^{-} with the incoming boundary condition is required, and since the complex conjugate of χ^{-} is always required the simplest way to obtain this function is through the condition of time-reversal invariance [31]

$$\chi^{-*}(\boldsymbol{k}, \boldsymbol{r}) = \chi^{+}(-\boldsymbol{k}, \boldsymbol{r}) . \tag{3.157}$$

Hence from equation (3.132) the high energy approximation for χ^{-*} is

$$\chi^{-*}(\boldsymbol{k}_f, \boldsymbol{r}) = \exp\left[-i\boldsymbol{k}_f . \boldsymbol{r} - \frac{i}{\hbar v_f}\int_0^{\infty} U_f(\boldsymbol{r}+\hat{\boldsymbol{k}}_f s)\, ds\right]. \tag{3.158}$$

For the partial wave expansion we have

$$\chi^{+}(\boldsymbol{k}, \boldsymbol{r}) = \sum_l i^l (2l+1) f_l(kr)\, P_l(\cos\theta) \tag{3.159}$$

where $\cos\theta = \boldsymbol{k}.\boldsymbol{r}/kr$. Hence, using the relation $P_l(-\cos\theta) = (-1)^l P_l(\cos\theta)$ we have

$$\chi^{-*}(\boldsymbol{k}_f, \boldsymbol{r}) = \sum_l (-i)^l (2l+1) f_l(k_f r)\, P_l(\cos\theta) \tag{3.160}$$

so that the radial functions appearing in the expansion for χ^{-*} have the same form and asymptotic behaviour as those appearing in the expansion for χ^{+}, the only differences arising from differences in the wave numbers and potentials in the initial and final channels. For charged particles, the expansions given in equations (3.158)–(3.160) must be multiplied by $e^{i\sigma_l}$, where σ_l is the Coulomb phase shift defined by equation (3.23), and the asymptotic behaviour of the radial function is given by equation (3.22). The expansions (3.158)–(3.160) are all referred to an axis of quantization along the momentum vector. In many cases it is convenient to take the axis of quantization along the direction of the incident beam so that the momentum vector for the scattered beam is at some angle to the z-axis. The appropriate forms for the partial wave expansions are then

$$\chi^{+}(\boldsymbol{k}_i, \boldsymbol{r}) = \sum_l i^l [4\pi(2l+1)]^{\frac{1}{2}} e^{i\sigma_l} f_l(k, r)\, Y_l^0(\hat{\boldsymbol{r}}) \tag{3.161}$$

$$\chi^{-*}(\boldsymbol{k}_f, \boldsymbol{r}) = 4\pi \sum_{lm} i^{-l} e^{i\sigma_l} f_l(k_f r)\, Y_l^m(\hat{\boldsymbol{r}})\, Y_l^{m*}(\hat{\boldsymbol{k}}_f) \tag{3.162}$$

where we have used the expansion formula [33]

$$P_l(\cos\theta) = \frac{4\pi}{2l+1} \sum_m Y_l^m(\hat{\boldsymbol{r}})\, Y_l^{m*}(\hat{\boldsymbol{k}})\,.$$

4 Electromagnetic interactions

4.1 Elastic electron scattering

The electromagnetic interaction of charged particles with nuclei is the most important source of information about the charge distribution in nuclei, and of the many processes falling in this category the most clear and unambiguous results are obtained from the elastic scattering of electrons. For electrons of very low energy the scattering by the nucleus can be attributed to a point charge, but at energies above about 50 MeV the detailed behaviour of the nuclear charge distribution has a substantial effect on the scattering process. In this interesting energy region we have a kinetic energy $T \leq 100\,mc^2$ so that the scattering must be described by the Dirac equation but the rest mass energy of the electron may be neglected in comparison with the total energy.

The Dirac equation is [32, 40]

$$[\boldsymbol{\alpha}\,.\,\boldsymbol{p}c+\beta mc^2+V]\Psi = E\Psi \tag{4.1}$$

where $\boldsymbol{\alpha}$, β are the Dirac matrices, V is the potential and E is the total energy. The four-component wavefunction Ψ may be represented by two two-component wavefunctions ϕ and ψ, and with neglect of the term involving the mass, and the representation [40]

$$\boldsymbol{\alpha} = \begin{pmatrix} \boldsymbol{\sigma} & 0 \\ 0 & -\boldsymbol{\sigma} \end{pmatrix} \qquad \beta = \begin{pmatrix} 0 & 1 \\ 1 & 0 \end{pmatrix}$$

for the matrices $\boldsymbol{\alpha}$, β, equation (4.1) reduces to two sets of uncoupled equations

$$[\boldsymbol{\sigma}\,.\,\boldsymbol{p}c+V-E]\phi = 0 \tag{4.2}$$

$$[-\boldsymbol{\sigma}\,.\,\boldsymbol{p}c+V-E]\psi = 0 \tag{4.3}$$

where $\boldsymbol{p}=\hbar\boldsymbol{k}$ and $E=\hbar ck$. The plane wave solutions of equations (4.2) and (4.3) obtained when $V=0$ are

$$\phi = u e^{i\boldsymbol{k}.\boldsymbol{r}}, \quad \psi = v e^{i\boldsymbol{k}.\boldsymbol{r}}$$

where
$$\boldsymbol{\sigma}.(\hbar c\boldsymbol{k})u = Eu, \quad \boldsymbol{\sigma}.(\hbar c\boldsymbol{k})v = -Ev.$$

Thus these solutions represent spin states parallel and anti-parallel to the momentum. In the presence of a spherically symmetric potential of finite extent the asymptotic form of the wavefunctions ϕ and ψ can be represented as in equation (3.27) and the cross-section is given by equation (3.29).

In order to bring the notation used in Chapter 3 into line with that used in the analysis of electron scattering, we rewrite equations (3.35) and (3.6) in terms of the total angular momentum j. This gives

$$\begin{aligned}\phi = \psi_{\frac{1}{2}\frac{1}{2}} &= \sum_{j-\frac{1}{2}=0}^{\infty} i^{j-\frac{1}{2}} F_j^+ \left[(j+\tfrac{1}{2})P_{j-\frac{1}{2}}\chi_{\frac{1}{2}}^{+\frac{1}{2}} - P_{j-\frac{1}{2}}^1 e^{i\phi}\chi_{\frac{1}{2}}^{-\frac{1}{2}}\right]\\ &+ \sum_{j-\frac{1}{2}=0}^{\infty} i^{j+\frac{1}{2}} F_j^- \left[(j+\tfrac{1}{2})P_{j+\frac{1}{2}}\chi_{\frac{1}{2}}^{+\frac{1}{2}} + P_{j+\frac{1}{2}}^1 e^{i\phi}\chi_{\frac{1}{2}}^{-\frac{1}{2}}\right]\\ &= \sum_j i^{j-\frac{1}{2}}\left[F_j^+ X_j^1 + iF_j^- X_j^2\right],\end{aligned} \tag{4.4}$$

$$\begin{aligned}e^{ikz}\chi_{\frac{1}{2}}^{+\frac{1}{2}} &= \sum_{j-\frac{1}{2}=0}^{\infty} i^{j-\frac{1}{2}} j_{j-\frac{1}{2}}(kr)\,2j\,P_{j-\frac{1}{2}}\chi_{\frac{1}{2}}^{+\frac{1}{2}}\\ &= \sum_j i^{j-\frac{1}{2}} j_{j-\frac{1}{2}}(kr)\left[X_j^1 + X_{j-\frac{1}{2}}^2\right]\\ &= \sum_j i^{j-\frac{1}{2}}\left[j_{j-\frac{1}{2}}(kr)X_j^1 + i j_{j+\frac{1}{2}}(kr)X_j^2\right],\end{aligned} \tag{4.5}$$

where
$$X_j^1 = (j+\tfrac{1}{2})P_{j-\frac{1}{2}}\chi_{\frac{1}{2}}^{+\frac{1}{2}} - P_{j-\frac{1}{2}}e^{i\phi}\chi_{\frac{1}{2}}^{-\frac{1}{2}} \tag{4.6}$$
$$X_j^2 = (j+\tfrac{1}{2})P_{j+\frac{1}{2}}\chi_{\frac{1}{2}}^{+\frac{1}{2}} + P_{j+\frac{1}{2}}e^{i\phi}\chi_{\frac{1}{2}}^{-\frac{1}{2}}, \tag{4.7}$$

and $F_j^\pm$ are the radial solutions for $j=l+\frac{1}{2}$, $j=l-\frac{1}{2}$ respectively. Because we are now solving the Dirac equation instead of the Schrodinger equation the $F_j^\pm$ satisfy different radial equations from the f_{lj} discussed in Chapter 3. In fact, the $F_j^\pm$ satisfy coupled radial equations of the form [32]

$$\frac{dF_j^+}{dr} - \frac{j+\frac{1}{2}}{r}F_j^+ + \frac{E-V}{\hbar c}F_j^- = 0 \tag{4.8}$$

$$\frac{dF_j^-}{dr} + \frac{j+\frac{1}{2}}{r}F_j^- - \frac{E-V}{\hbar c}F_j^+ = 0. \tag{4.9}$$

Using the asymptotic expansions (3.8) and (3.36) we find

$$f_{\frac{1}{2}\frac{1}{2}} = \frac{1}{2ik}\sum_j (j+\tfrac{1}{2})(e^{2i\delta_j}-1)[P_{j-\frac{1}{2}}+P_{j+\frac{1}{2}}] \tag{4.10}$$

$$f_{-\frac{1}{2}+\frac{1}{2}} = \frac{1}{2ik}\sum_j (e^{2i\delta_j}-1)[P^1_{j+\frac{1}{2}}e^{i\phi}-P^1_{j-\frac{1}{2}}e^{i\phi}], \tag{4.11}$$

and using the recurrence relations [17]

$$\begin{aligned}\sin\theta\, P^1_l &= -l\cos\theta\, P_l + l\,P_{l-1}\\ &= (l+1)\cos\theta\, P_l-(l+1)P_{l+1}\end{aligned}$$

we have
$$f_{-\frac{1}{2}+\frac{1}{2}} = [\tan\tfrac{1}{2}\theta\, e^{i\phi}]f_{\frac{1}{2}\frac{1}{2}}\,. \tag{4.12}$$

In deriving equations (4.10)–(4.12) we have assumed that the potential is spin-independent so that $\delta^+_j=\delta^-_j$. The spin-dependence of the formulae then arises from the use of the Dirac equation. Proceeding in the same way for the other wavefunction ψ we find

$$f_{+\frac{1}{2}-\frac{1}{2}} = [-\tan\tfrac{1}{2}\theta\, e^{i\phi}]f_{\frac{1}{2}\frac{1}{2}},\quad f_{-\frac{1}{2}-\frac{1}{2}} = f_{\frac{1}{2}\frac{1}{2}}, \tag{4.13}$$

and hence the cross-section is given by

$$\frac{d\sigma}{d\Omega} = [1+\tan^2\tfrac{1}{2}\theta]\,|f_{\frac{1}{2}\frac{1}{2}}|^2\,. \tag{4.14}$$

For a static spherically symmetric charge distribution $\rho_{\text{ch}}(r)$ the Coulomb potential is given by

$$V(r) = -Ze^2\int\frac{\rho(\boldsymbol{r}')\,d\boldsymbol{r}'}{|\boldsymbol{r}-\boldsymbol{r}'|} \tag{4.15}$$

$$= -4\pi Ze^2\left\{\frac{1}{r}\int_0^r \rho_{\text{ch}}(r')r'^2\,dr' + \int_r^\infty \rho_{\text{ch}}(r')\,r'\,dr'\right\} \tag{4.16}$$

which tends at large distances to a pure Coulomb potential. The incident plane wavefunction must now be replaced by a Coulomb wavefunction whose partial wave expansion has the asymptotic behaviour

$$L_j(kr) \to \frac{1}{kr}\sin[kr-\tfrac{1}{2}(j-\tfrac{1}{2})\pi+\alpha\log 2kr+\eta^c_j] \tag{4.17}$$

where $\alpha = Ze^2/\hbar c$ and η^c_j is the Coulomb phase shift. (Compare equations

(3.18)–(3.25) for the non-relativistic case.) Similarly, the scattered wave has the asymptotic behaviour

$$F_j^{\pm} \rightarrow \frac{1}{kr}\sin\left[kr - \tfrac{1}{2}(j \mp \tfrac{1}{2})\pi + \alpha \log 2kr + \eta_j\right] \tag{4.18}$$

where η_j is the phase shift due to the potential $V(r)$. The scattering amplitude is usually calculated in the form

$$f(\theta) = f_c(\theta) + \left[f(\theta) - f_c(\theta)\right] \tag{4.19}$$

with

$$f(\theta) - f_c(\theta) = \frac{1}{2ik}\sum_j (e^{2i\eta_j^c} - e^{2i\eta_j})(j+\tfrac{1}{2})\left[P_{j-\frac{1}{2}}(\cos\theta) + P_{j+\frac{1}{2}}(\cos\theta)\right]. \tag{4.20}$$

Since the difference between η_j and η_j^c tends to zero for large j, the series (4.20) converges, even though η_j and η_j^c do not separately tend to zero. However, since $f(\theta)$ is small at large angles the two terms of equation (4.19) are approximately equal but of opposite sign and must be calculated very accurately. Methods for doing this are described by Yennie, Ravenhall and Wilson [41].

In Born approximation the cross-section for elastic scattering is given by [32]

$$\frac{d\sigma}{d\Omega} = \left[\frac{q\cot\frac{1}{2}\theta}{4\pi\hbar c}\right]^2 \left|\int V(\boldsymbol{r})e^{i\boldsymbol{q}\cdot\boldsymbol{r}}d\boldsymbol{r}\right|^2 \tag{4.21}$$

where q is the momentum transfer which for elastic scattering is given by

$$q = 2k\sin\tfrac{1}{2}\theta\,. \tag{4.22}$$

Using the result $\displaystyle\int \frac{e^{i\boldsymbol{q}\cdot\boldsymbol{r}}}{|\boldsymbol{r}-\boldsymbol{r}'|}d\boldsymbol{r} = \frac{4\pi}{q^2}e^{i\boldsymbol{q}\cdot\boldsymbol{r}'}$ we have

$$\int V(\boldsymbol{r})e^{i\boldsymbol{q}\cdot\boldsymbol{r}}d\boldsymbol{r} = -\frac{4\pi Ze^2}{q^2}\int e^{i\boldsymbol{q}\cdot\boldsymbol{r}'}\rho(\boldsymbol{r}')d\boldsymbol{r}'\,, \tag{4.23}$$

where we have used equation (4.15) for the potential but have allowed for a non-symmetric charge distribution. Finally, the cross-section becomes

$$\frac{d\sigma}{d\Omega} = \frac{Z^2e^4\cos^2\frac{1}{2}\theta}{\hbar^2c^2q^2\sin^2\frac{1}{2}\theta}|F_{\text{ch}}(q)|^2 = \frac{Z^2e^4\cos^2\frac{1}{2}\theta}{4E^2\sin^4\frac{1}{2}\theta}|F_{\text{ch}}(q)|^2\,, \tag{4.24}$$

where F_{ch} is the *charge form factor* defined by equation (2.41). The first factor is the cross-section for scattering from a point charge* so that equation (4.24) can be written as

$$\frac{d\sigma}{d\Omega} = \frac{d\sigma}{d\Omega_{\text{point}}} |F_{\text{ch}}(q)|^2 . \tag{4.25}$$

It is found that the Born approximation is reasonably satisfactory for electron scattering from light nuclei ($A < 16$) at energies above about 100 MeV. It has the advantage that the form factor is obtained in a very direct way from the experimental data, and also that it can easily be used to calculate scattering from a non-symmetric distribution. For example, a nucleus with a non-zero electric quadruple moment has a charge distribution of the form

$$\rho(\mathbf{r}) = \rho_0(r) P_0(\cos\theta) + \rho_2(r) P_2(\cos\theta) , \tag{4.26}$$

which presents no problem in the Born approximation, but would lead to coupled equations in the exact phase shift analysis (see section 3.5). The deficiencies in the Born approximation show up in the region of the minima in the cross-sections since the Born approximation predicts deep minima whereas the experimental data show that in medium and heavy nuclei these minima are almost completely filled in.

The phase shift analysis leads directly to the theoretical cross-section. It is possible to define a form factor by dividing the cross-section by the point cross-section but the resulting form factor is no longer a function of momentum transfer only. Unlike the case of nucleon scattering it is not customary to take into account the loss of intensity in the elastically scattered beam, due to inelastic scattering and reactions, as the cross-sections for the latter processes are generally much smaller than for elastic scattering. Some estimates of the effect due to the excitation of intermediate nuclear states have been made [42] which indicate that the effect is small but not entirely negligible at large momentum transfer. The procedure adopted in the analysis of the experimental data is either to take a phenomenological form for the charge distribution and to determine the parameters appropriate to a given nucleus by a least squares fit to the data, or to take the form for the distribution predicted by a suitable nuclear model using equation (2.38) and to vary the model parameters to obtain a fit. In the latter case it is essential to include the finite size of the proton using equation (2.40) and for light nuclei it is desirable to take account of the centre-of-mass motion as described in sec-

* The cross-section for scattering from a point charge is also written as $\sigma_{\text{Mott}}(\theta)$ and referred to as the cross-section for Mott scattering.

tion 2.5. The information obtained from these calculations is discussed in the next section.

A more complete description of electron scattering can be given [43, 44] which includes not only the Coulomb interaction with the nuclear charge distribution, but also the interaction with the nuclear current and magnetization densities. In the usual semi-classical approach the nucleus is treated non-relativistically and the static densities are given by the matrix elements of the appropriate operators taken between the wavefunctions for the initial and final nuclear states, and the electromagnetic interaction is represented by a multipole expansion. The general expression for the cross-section is then given by

$$\frac{d\sigma}{d\Omega} = \sum_{\lambda=0}^{\infty} \frac{d\sigma}{d\Omega} C\lambda + \sum_{\lambda=1}^{\infty} \frac{d\sigma}{d\Omega} E\lambda + \sum_{\lambda=1}^{\infty} \frac{d\sigma}{d\Omega} M\lambda \tag{4.27}$$

where the three terms correspond to the Coulomb or longitudinal part of the interaction, the transverse electric part and the transverse magnetic part respectively. For a transition between nuclear states J_i and J_f the contributing multipoles are determined by the selection rule

$$|J_f - J_i| \leqslant \lambda \leqslant J_i + J_f , \tag{4.28}$$

and hence for elastic scattering from spin zero nuclei we have $\lambda = 0$ only so that the scattering is given by the monopole part of the Coulomb interaction, i.e. the cross-section is exactly that derived above for a spherically symmetric charge distribution. For elastic scattering from a nucleus with non-zero spin other contributions are possible. For example, for a nucleus with a 1^+ ground state the possible values of λ are 0, 1, 2, and since the parity of the Coulomb and transverse electric multipole operators is $(-1)^{\lambda}$ while that of the transverse magnetic multipole operators is $(-1)^{\lambda+1}$, this means that in this case the cross-section has $C0$, $C2$, and $M1$ contributions. (Time reversal invariance prevents a contribution from $E\lambda$ terms to the elastic scattering.) The $C0$ and $C2$ terms arise from the two parts of the charge distribution given in equation (4.26), and the results from a calculation of these two terms for ^{11}B $(J_i = \frac{3}{2}^-)$ using Born approximation and the shell model are shown in figure 4.1(a).

The transverse contributions to the cross-section have a dependence on the scattering angle θ which is different from that of the Coulomb contributions, and for elastic scattering the factor $\cos^2 \frac{1}{2}\theta / \sin^4 \frac{1}{2}\theta$ (see equation (4.24)) is replaced in the transverse terms by $(1 + \sin^2 \frac{1}{2}\theta)/2 \sin^4 \frac{1}{2}\theta$. Thus for scatter-

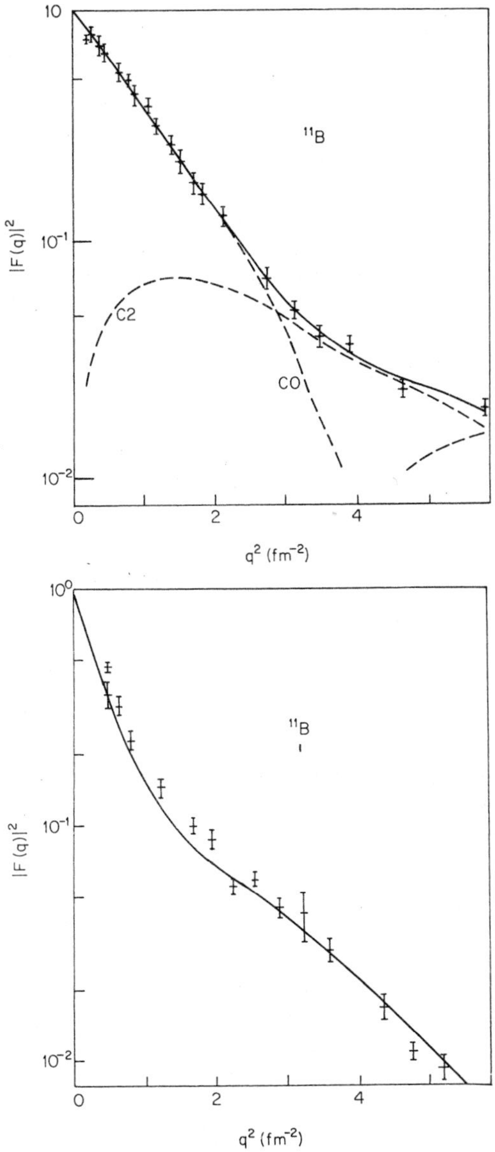

Figure 4.1 (**a**) The form factor for Coulomb elastic electron scattering from ^{11}B. [From I. S. Towner, Ph.D. Thesis, University of London, 1966.] (**b**) The form factor for magnetic elastic scattering from ^{11}B. [From work by D. U. L. Yu, reported in *Advances in Physics* **17**, 481 (1968).] The calculated curves shown in (a) and (b) are obtained using the same wave-functions.

ing in the region of $\theta \sim 180^\circ$ the Coulomb scattering disappears and hence, by varying the incident momentum to vary q, the behaviour of the form factor of the magnetic dipole distribution and possibly the magnetic octupole distribution can be investigated. Results for 180° elastic scattering from ^{11}B are shown in figure 4.1(b), together with the theoretical prediction obtained using Born approximation and the same nuclear wavefunctions that were used for the curves shown in figure 4.1 (a).

4.2 Nuclear structure studies with electron scattering and reactions

The analysis of elastic electron scattering data in terms of a phenomenological charge distribution has provided valuable information on the macroscopic properties of nuclei. At low momentum transfer the form factor can be expanded to give

$$
\begin{aligned}
F_{\text{ch}}(q) &= 4\pi \int_0^\infty \rho_{\text{ch}}(r) j_0(qr) r^2\, dr \\
&= 4\pi \int_0^\infty \rho_{\text{ch}}(r)(qr)^{-1}[(qr) - \tfrac{1}{6}(qr)^3 + \ldots] r^2\, dr \\
&= 1 - \tfrac{1}{6} q^2 \langle r^2 \rangle + \ldots . \qquad (4.29)
\end{aligned}
$$

Thus measurements at very low momentum transfer yield information about the mean square radius $\langle r^2 \rangle$ of the charge distribution, while measurements at larger momentum transfer are more sensitive to the shape of the charge distribution.

Many functional forms for the charge distribution have been used, but the most commonly used are the Fermi distribution

$$\rho_{\text{F}}(r) = \frac{\rho_0}{1 + e^{(r-c)/a}} \qquad (4.30a)$$

where ρ_0 is a constant determined by the normalization, c is the halfway radius, and the diffuseness a is related to the surface thickness as shown in figure 4.2, and the parabolic Fermi distribution

$$\rho_{\text{PF}}(r) = \left(1 + w\frac{r^2}{c^2}\right)\frac{\rho_0}{1 + e^{(r-c)/a}} \qquad (4.30b)$$

where w is a constant which may be positive or negative and which can cause a hump or depression near the origin. (If w is negative, ρ_{PF} must be modified so that it does not go negative but goes smoothly to zero.) These forms for the charge distribution have not been deduced from first principles but are phenomenological distributions which have been found by experience to lead to agreement with a wide selection of data.

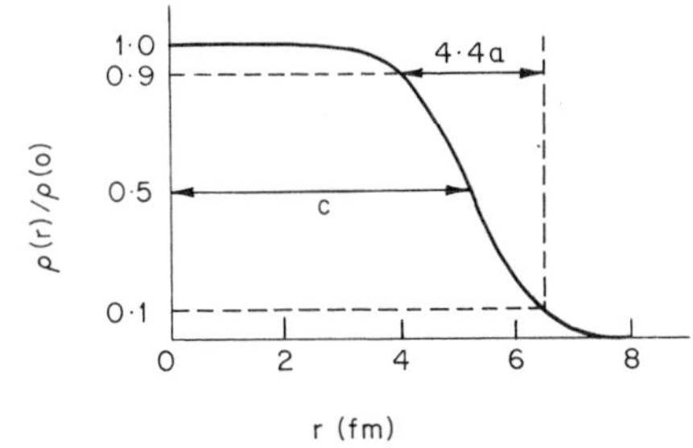

Figure 4.2 The fermi parametrization of the charge distribution of a fairly heavy nucleus.

For the Fermi distribution, the r.m.s. radius $\langle r^2\rangle^{\frac{1}{2}}$ and the radius R_{EQ} of the equivalent uniform distribution are related to the parameters c and a by the expression

$$R_{\text{EQ}}^2 = \tfrac{5}{3}\langle r^2\rangle = c^2 + \tfrac{7}{3}\pi a^2 \tag{4.31}$$

and hence the measurements at low q yield pairs of values of c and a which correspond to the same $\langle r^2\rangle$. Such measurements are usually made with electron beams of about 40–60 MeV. At higher energies the scattering is sensitive to the detailed behaviour of the charge distribution. For energies up to about 500 MeV, the electron scattering is sensitive primarily to the transition or surface region and fits to the cross-sections determine c and a, but at still higher energies the electrons penetrate into the interior region and to describe this region a model with more parameters or a more fundamental approach is required.

These analyses of data for a wide range of nuclei yield information on the variation of the central density, the r.m.s. radius, the halfway radius, and the surface thickness as functions of the mass number A [45]. From these results we obtain the information used in Chapter 2 as the basis for the construction of macroscopic models of nuclei, i.e. that the mean central density is constant except for very light nuclei and that to a good approximation c and R_{EQ} increase as $A^{\frac{1}{3}}$. It may be noted that the radii do not increase exactly as $A^{\frac{1}{3}}$ even for the most stable nuclei and the variation for nuclei away from the maximum stability line, e.g. through a set of isotopes, can depart significantly from this relationship. Thus the quantity γ defined by the relation

$$\gamma = \frac{3A}{R_{\mathrm{EQ}}} \frac{dR_{\mathrm{EQ}}}{dA} \tag{4.32}$$

is unity if $R_{\mathrm{EQ}} \propto A^{\frac{1}{3}}$, but in practice it is found that for reasonably spherical nuclei $\gamma \simeq 0.65$ for isotopes (same Z) and $\gamma \simeq 1.5$ for isotones (same N) [45].

Information on the microscopic structure of nuclei can also be obtained from elastic electron scattering. This is done by calculating the charge distribution and the current and magnetization densities using the nuclear wavefunctions given by a suitable microscopic model [46]. In the independent particle model the monopole part of the charge distribution involves all the single-particle states of the shell model configuration whereas the quadrupole part of the charge distribution and the current and magnetization densities depend only on the single-particle states in the unfilled shell or shells. For light nuclei the calculation of these quantities is often carried out using oscillator wavefunctions so that it is possible to make the correction for centre-of-mass motion exactly, and these wavefunctions yield satisfactory agreement with the data at low momentum transfer for most nuclei up to $A \simeq 16$. However, since the Fourier transform of e^{-r^2/a^2} yields $e^{-\frac{1}{4}q^2a^2}$, the form factor calculated from oscillator wavefunctions falls very rapidly at large momentum transfer and cannot be regarded as realistic. It is perfectly possible to calculate the single-particle wavefunctions in a realistic potential and the theoretical curves shown in figure 4.1 were calculated in this way. The agreement with the data in the two cases ensures that not only is the charge distribution of the Z protons in the $1s$ and $1p$ shells reasonably correctly given by the model but also that the wavefunction for the protons in the $1p$ shell separately yields agreement with the data.

In general, inelastic electron scattering to particular final states provides a more searching test of the model chosen for the nuclear states since it depends on the transition density (2.46). For example, in many nuclei there are low-lying levels which are strongly excited in $C2$ transitions and attempts have been made to fit the data using microscopic models of the collective excitation. Some results for the excitation of 2^+ states in ^{58}Ni and ^{60}Ni are shown in figure 4.3. (The excitation of the same states in proton scattering is discussed in section 7.3.) It is interesting that the effect of folding in the finite proton size is to damp the oscillation of the transition density but the effect on the cross-section is not significant except at large angles. The same damping effect occurs for the monopole density $\rho_0(r)$. For these calculations of inelastic scattering, plane wave Born approximation is not adequate but a distorted wave Born approximation for electron scattering has recently been developed [47] and the results shown in figure 4.3 were calculated by this method. Figure 4.4 shows the magnitude of the distortion effect due to elastic Coulomb scattering by the nucleus.

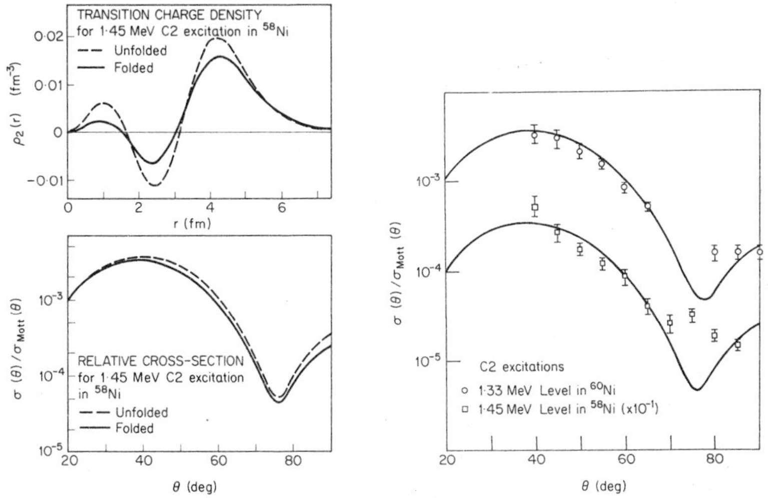

Figure 4.3 The transition density and cross-section for a one-phonon excitation in ^{58}Ni, and a similar cross-section for ^{60}Ni. For the full line curves the effect of the finite size of the proton has been folded in. [Adapted from J. T. Reynolds and D. S. Onley, *Nucl. Phys.* **66,** 1 (1965.)

There has been considerable interest in the possibility of using electron scattering to investigate short-range proton–proton correlations in nuclei [48]. The simplest approach is to use correlated nuclear wavefunctions to generate a many-particle charge distribution with which to fit the electron scattering data. This method has been used to study both elastic scattering and also inelastic scattering to particular low-lying states. In some calculations the scattering amplitude has been calculated in first Born approximation and, since in this case differences between the correlated and uncorrelated wavefunctions modify the effective one-particle density and show up in the region of the Born minima, the conclusions obtained from these calculations are open to doubt except for very light nuclei. More usually, the scattering amplitude is calculated from a multiple scattering expansion based on Born approximation, the non-relativistic form of which is given by equation (3.127), or based on the high energy approximation. The results obtained from a

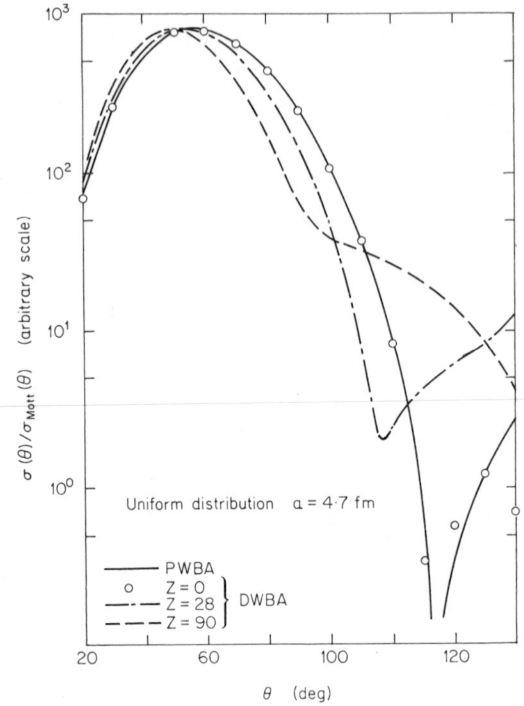

Figure 4.4 The form factor for a C2 transition induced by 183 MeV electrons obtained using the analytic plane wave Born approximation and the distorted wave Born approximation. [Adapted from T. A. Griffy, D. S. Onley, J. T. Reynolds and L. C. Biedenharn, *Phys. Rev.* **128,** 833 (1962).]

number of independent calculations are in qualitative agreement, but there are as yet few data suitable for an analysis of this type. It has been shown that the two-particle density function contributes in a very direct way to inelastic scattering summed over all the possible final nuclear states. In Born approximation the summed cross-section for excitation through the Coulomb part of the interaction is given by

$$\frac{d^2\sigma}{d\Omega_f dk_f} = \frac{d\sigma}{d\Omega}\mathrm{e-p} \sum_n \left| \langle n \right| \sum_{j=1}^{Z} e^{i\mathbf{q}.\mathbf{r}_j} \left| 0 \rangle \right|^2 \delta(E_i - E_f - q^2/2Am - E_n^A + E_0^A) \tag{4.33}$$

where $(d\sigma/d\Omega)$ e − p is the electron–proton cross-section and in the δ-function for energy conservation E_i, E_f are the initial and final electron energies, $q^2/2Am$ is the recoil energy of the nucleus and $E_n^A - E_0^A$ is the excitation energy of the state n. A sum rule can be obtained for this expression at constant q. Experimentally, this is done by adding the cross-sections for different scattering angles and final electron energies related by energy conservation and the requirement of constant q. Theoretically, it is done by using closure, assuming that the contribution from very highly excited states outside the experimentally accessible region is negligible. This gives

$$\frac{d^2\sigma}{d\Omega_f dk_f} = Z \frac{d\sigma}{d\Omega}_{\mathrm{e-p}}[1-F^2(q)+(Z-1)D(q)] \tag{4.34}$$

where $D(q)$ is the Fourier transform of the proton–proton correlation function (see section 2.6) and the finite size of the proton is included in free electron–proton cross-section. As $q \to 0$, $F^2(q) \to 1$ and $D(q) \to 0$ so that $(d^2\sigma/d\Omega_f dk_f) \to 0$, but for very large q, $F^2(q) \to 0$ and $D(q) \to 0$ so that

$$\frac{d^2\sigma}{d\Omega_f dk_f} \to Z \frac{d\sigma}{d\Omega}_{\mathrm{e-p}}$$

and the cross-section is given by the incoherent scattering from Z protons. Thus in principle it should be possible to determine the function $D(q)$ but suitable electron scattering data are difficult to obtain, although some data on high energy inelastic scattering from the lithium isotopes has been interpreted in terms of incoherent scattering [49]. Data on ^{12}C and ^{16}O have been quite successfully analysed using a single particle model and including correlations due to the exclusion principle [50].

Knock-out reactions with electrons also provide important information on the behaviour of nucleons in nuclei. So far almost all the experimental data refers to knock-out of protons, i.e. to the (e, e′, p) reaction. Conservation of energy and momentum for this process yields the equations

$$E_i = E_f + \frac{Q^2}{2(A-1)m} + E_{\mathrm{p}} + S_{\mathrm{pA}}$$

$$\boldsymbol{k}_i = \boldsymbol{k}_f + \boldsymbol{k}_{\mathrm{p}} + \boldsymbol{Q}$$

where $(E_i, \boldsymbol{k}_i)$, $(E_f, \boldsymbol{k}_f)$ are the initial and final total energy and momenta of the electron, $(E_{\mathrm{p}}, \boldsymbol{k}_{\mathrm{p}})$ are the energy and momentum of the ejected proton and $\boldsymbol{Q}$ is the recoil momentum of the residual nucleus. Thus a measurement of the energies and momenta of the outgoing proton and electron allows a determination of the separation energy S_{pA} of the struck proton. Such measurements have been carried out at Frascati with a 600 MeV electron beam [51]. In a poor resolution experiment a broad peak is observed which includes a range of excited states of the residual nucleus as shown in figure 4.5 and therefore location of the centre of this peak yields an experimentally averaged value for the separation energy of protons in a given shell, provided that the final nuclear states excited by knock-out from different shells do not overlap to any extent. On this assumption results have been obtained from the (e, e′, p) reaction for the mean separation energy of protons in inner shells of nuclei. The interpretation of these data is discussed in section 9.4. The

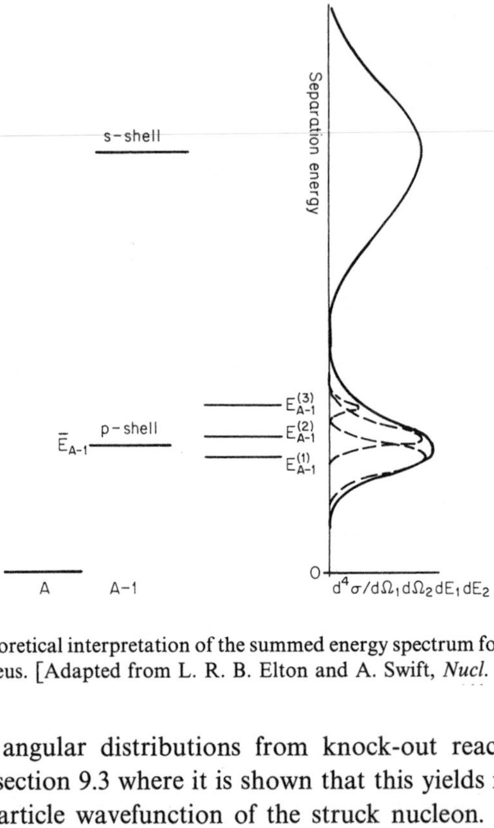

Figure 4.5 Theoretical interpretation of the summed energy spectrum for the (*p*,2*p*) reaction on a 1*p*-shell nucleus. [Adapted from L. R. B. Elton and A. Swift, *Nucl. Phys. A94*, **52** (1967).]

analysis of angular distributions from knock-out reactions is discussed in detail in section 9.3 where it is shown that this yields information about the single-particle wavefunction of the struck nucleon. In the case of the (e, e′, p) reaction, if the target has low Z the scattered electron can be represented to a good approximation by plane waves but the distortion of the outgoing proton by the nuclear potential due to the residual nucleons must not be neglected [52].

In the present context, the most important feature of the microscopic approach to electron scattering is that it provides a lowest order prediction for nucleon single particle wavefunctions, which may be used in a study of nuclear reactions. The simplest phenomenological procedure is to assume a suitable functional form for an effective local potential and then to determine the parameters for individual nuclei by fitting a selection of experimental data. For protons a suitable selection of data consists of the cross-section for elastic electron scattering, the mean separation energies and the spin-orbit splittings between the sub-shells. When these calculations were first started, it was frequently the case that only the separation energies for the least bound nucleons were known, and these data, together with the electron scattering were sufficient to determine, for each nucleus considered, one effective single-particle potential for the protons. The results shown in figure 4.1 (a) were obtained in this way, i.e. the parameters of the potential were varied until the binding energy and the charge distribution constructed from corresponding single-particle wavefunctions gave agreement with the electron scattering data. Although the calculation is in lowest order in the sense that it does not formally include residual interactions and is normally restricted to the lowest shell model configuration, some at least of the effects of residual interactions must be simulated by the wavefunctions since they have been obtained by fitting various pieces of experimental data. The degree to which this simulation is effective can be tested through analyses of nuclear reactions which give information on the validity of the assumption that the single-particle states are well separated in the manner indicated in figure 4.5, and the importance of short and long range correlations. The conclusions reached in these analyses are discussed in Chapter 9.

4.3 Coulomb excitation

Nuclear projectiles which are charged, such as protons, α-particles, heavy ions, interact with nuclei through both the Coulomb and nuclear interactions. For projectiles whose energies are well above the Coulomb barrier both interactions are effective and the scattering amplitudes add coherently and interfere, as shown in equation (3.25) for elastic scattering, but at very low energies the Coulomb barrier prevents the projectiles from penetrating the nucleus and experiencing the nuclear interactions. Even in the latter case inelastic scattering can occur, and this mechanism of inducing transitions in nuclei is known as Coulomb excitation [6, 43]. It is found that the nuclear states most strongly excited are the low-lying collective states, and this mode of excitation is therefore particularly useful for the study of rotational and vibrational excitations.

The Coulomb interaction between a projectile of charge $Z'e$ and a nucleus of charge Ze is given by

$$V^c = Z'e\sum_{i=1}^{Z}\frac{e}{|\boldsymbol{r}-\boldsymbol{r}_i|} = Z'e\sum_i\sum_{lm}\frac{4\pi e}{2l+1}\frac{r_<^l}{r_>^{l+1}}Y_l^m(\hat{\boldsymbol{r}}_i)\,Y_l^{m*}(\hat{\boldsymbol{r}})$$

where $r_>$, $r_<$ are the greater and lesser of r and r_i respectively, $\boldsymbol{r}_i$ is the co-ordinate of the ith nucleon in the nucleus and $\boldsymbol{r}$ is the co-ordinate of the centre-of-mass of the projectile relative to the centre-of-mass of the target. Since the projectile does not penetrate the nucleus we need only the potential for $r > r_i$ which can be written as

$$V^c = \frac{Z'Ze^2}{r} + Z'e\sum_{l=1}^{\infty}\sum_{m=l}^{-l}\frac{4\pi}{2l+1}r^{-(l+1)}Q_{lm}\,Y_l^{m*}(\hat{\boldsymbol{r}}) \tag{4.35}$$

where the Q_{lm} are the electric multipole operators

$$Q_{lm} = e\sum_i r_i^l\,Y_l^m(\hat{\boldsymbol{r}}_i)\,. \tag{4.36}$$

The first term in equation (4.35) yields the simple Rutherford elastic scattering but the higher terms give rise to excitations of order l.

The cross-section for Coulomb excitation may be calculated using a semi-classical approximation [43] or using DWBA [53]. In either case, the scattering states of the projectile are just the Rutherford scattering states. The dependence of the cross-section on nuclear structure arises through the matrix element of the Q_{lm} between the initial and final nuclear states. It is usual to define the reduced transition probability* associated with an excitation of order l as

$$\begin{aligned} B_l &= \sum_{M_iM_f}|\langle J_f M_f|\,Q_{lm}|J_i M_i\rangle|^2 \\ &= (2J_i+1)^{-1}|\langle J_f\|\,Q_l\|J_i\rangle|^2 \end{aligned} \tag{4.37}$$

where we have used the Wigner-Eckart theorem. If we transform the multipole operator to a body-fixed system

$$Q_{lm} = \sum_\nu Q'_{l\nu}\mathscr{D}^l_{m\nu}\,,$$

and use the nuclear wavefunctions given by the collective model we obtain

* The more usual notation for the reduced transition probability is $B(E\lambda)$, although the excitation is due to the longitudinal Coulomb part of the electromagnetic interaction, and not the transverse electric part.

$$B_l = (J_i K_i lv | J_f K_f)^2 | \langle \chi_{K_f} | Q'_{lv} | \chi_{K_i} \rangle |^2 \tag{4.38}$$

where χ represents the internal structure of the rotator. Thus measurement of B_l values provides information about the strength of transitions within rotational bands or between rotational bands, and hence about the intrinsic vibrational and rotational structure of the nucleus.

The measured values of B_l can be used to derive information about nuclear shapes if a suitable nuclear model is assumed. Thus for a permanently deformed nucleus the reduced quadrupole transition probability B_2 for a $0^+ \to 2^+$ transition is given, for small deformations, by

$$B_2(0^+ \to 2^+) = \left[\frac{3Z R_0^2 \beta_2}{4\pi} \right]^2$$

where β_2 is the deformation parameter defined in section 2.2. A compilation of B_2 and β_2 values has been given by Stelson and Grodzins [54]. The rotational model also provides a relation between B_2 values and quadrupole moments of excited states; for example, the quadrupole moment of the first 2^+ state in a $K=0$ band is given by

$$Q_2 = \pm \frac{32\pi}{35} [B_2(0^+ \to 2^+)]^{\frac{1}{2}} .$$

Since it is now experimentally possible to measure these quadrupole moments of excited states, comparison with the values predicted from B_2 values provides a check on the validity of the simple rotational model.

4.4 Muonic atoms

The muon is a particle whose rest mass energy is 105·7 MeV, i.e. it is approximately 207 times heavier than the electron. It has spin $\frac{1}{2}$, a small magnetic moment whose magnitude is very accurately given by quantum electrodynamics, and interacts through the electromagnetic and weak interactions. Thus the muon may be regarded essentially as a heavy electron, and the absence of any anomalous magnetic moment indicates that the muon should be described very accurately by the Dirac equation (4.1) although it will not be possible to neglect the terms involving the mass, unlike the case of electron scattering. Muons are formed by decay of kaons and pions, and in turn decay through the weak interaction with a half-life of $1{\cdot}5 \times 10^{-6}$ secs [3].

The negative muon can be captured by an atom and go into an extra-nuclear bound state, so forming a *muonic atom**. The muon is first captured into a state of high principal quantum number n and then makes transitions to lower shells with the emission of x-rays. In this discussion we denote the quantum numbers of the muon states using the customary atomic notation and not the nuclear notation introduced in section 2.2, so that, for example, the lowest d state is associated with $n=3$ and not $n=1$. Since the Bohr radii of the atomic states are inversely proportional to the mass, the muon wavefunctions penetrate much more deeply into the nuclear region than is the case for electrons. This means that the muon wavefunctions are much more sensitive to the shape and finite extent of the nuclear charge distribution, and that the screening effect of the electrons is a sufficiently small correction that the muonic atom may be treated to a good approximation as hydrogen-like. The properties of muonic atoms are usually discussed in terms of the energy shifts of each atomic level compared to the energies predicted for a point nuclear charge. Since the effective Coulomb potential inside the nuclear Coulomb radius is less than the potential due to a point charge the muon states are less strongly bound in a real nucleus than they would be for a point nucleus. This energy shift is greatest for the inner shells and decreases as the penetration of the muon wavefunction into the nucleus decreases. It can be detected by measuring the energies of x-rays emitted when the muon makes a radiative transition between two states.

The experimental study of muonic x-rays has recently received considerable impetus due to the introduction of the Li-drifted Ge detector whose high resolution has made possible accurate studies of the x-rays emitted in transitions between higher shells and of fine and hyperfine structure [55]. Prior to this work, precise measurements had been made primarily on the $2p \rightarrow 1s$ transition for a wide range of nuclei. Using perturbation theory it can be shown that the theoretical transition energy depends on a linear combination of even moments $\langle r^k \rangle$ of the nuclear charge distribution, with $\langle r^2 \rangle$ having by far the greatest weight. Thus analysis of the X-ray data essentially determines the mean square radius. The results are in good agreement with those obtained from low-energy electron scattering. With Ge (Li) detectors the fine structure of the K and L x-rays can be resolved so that the transitions

$$2p_{\frac{1}{2}} \rightarrow 1s_{\frac{1}{2}}\,, \quad 2p_{\frac{3}{2}} \rightarrow 1s_{\frac{1}{2}}\,,$$
$$3d_{\frac{5}{2}} \rightarrow 2p_{\frac{3}{2}}\,, \quad 3d_{\frac{3}{2}} \rightarrow 2p_{\frac{1}{2}}\,,$$

* Although muonic atoms are sometimes referred to as *μ-mesic atoms*, the muon belongs, with the electron and the neutrino, to the family of leptons, and not to the family of mesons which have spin zero and are strongly interacting (see Chapter 10).

etc., can be studied. In addition, transitions between the higher levels give information on higher moments ξ of the nuclear charge distribution. The data are usually analysed in terms of the Fermi distribution and pairs

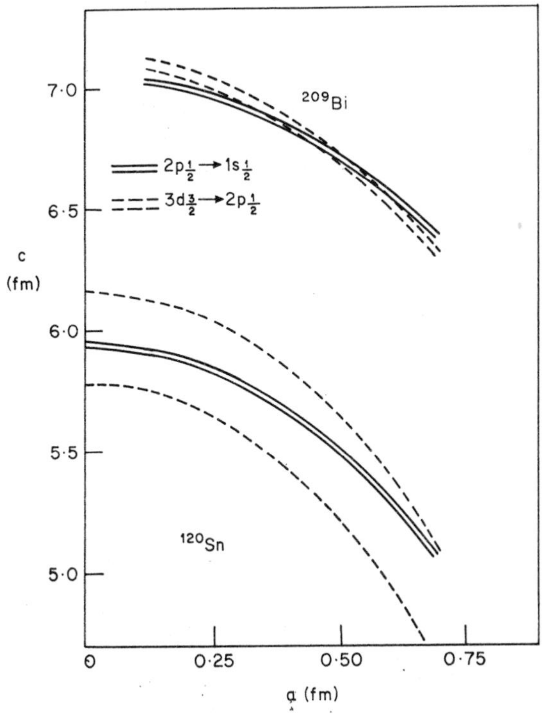

Figure 4.6 The *c-a* plots for muonic x-ray transitions in ^{120}Sn and ^{209}Bi. [From H. L. Acker, G. Backenstoss, C. Daum, J. C. Sens and S. A. de Wit, *Nucl. Phys.* **87,** 1 (1966).]

of values of *c* and *a* are found which give the measured energy for each transition. A *c–a* plot of these values yields a line for each transition and the intersection of these lines should in principle yield an exact determination of *c* and *a*. In practice, owing to the experimental uncertainties in the measured values, each line becomes a band and the point of interaction becomes an area, as shown in figure 4.6. It is found that the parameters are well determined only for heavy nuclei. For nuclei with low and medium *Z* the 3*d* and 2*p* muon wavefunctions do not penetrate very deeply into the nucleus and the experimental uncertainties in the x-ray measurements for transitions between these states are still such that the $c-a$ values are not well determined as can be seen from figure 4.6. Although the data are almost always analysed in terms of the Fermi distribution the results do not serve to establish this as the unique form for the charge distribution [56]. Figure 4.7 shows

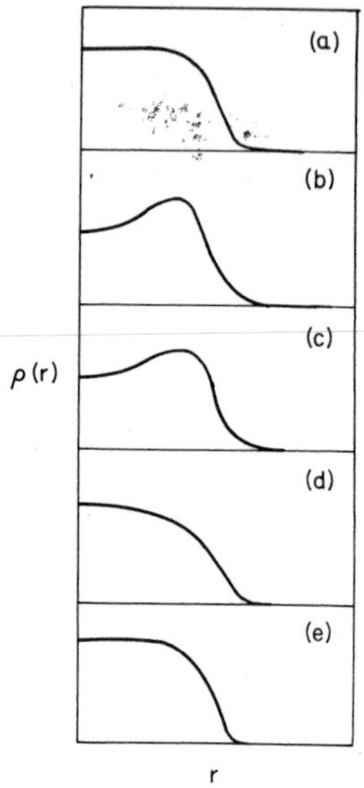

Figure 4.7 (**a**)–(**d**). Different shapes of charge distributions which have the same $\langle r^2 \rangle$ and $\langle r^4 \rangle$ moments. (**e**) A shape of the same type as (**d**) but having the same values of c and a as (**a**). [From R. C. Barrett, ref. 56.]

some distributions for ^{209}Bi which have the same $\langle r^2 \rangle$ and $\langle r^4 \rangle$ moments (curves a, b, c, d) or the same half-way radius and surface thickness (curves a, e). The distributions have almost identical moments up to $\langle r^{10} \rangle$, the $2p_{\frac{1}{2}}-1s_{\frac{1}{2}}$ energies agree to within 10 keV, and the $4f_{\frac{5}{2}}-3d_{\frac{3}{2}}$ energies agree to within 3 eV. This shows what high accuracy is needed in both theory and experiment in order to determine a third parameter of the charge distribution.

In some cases the x-ray spectra emitted by muonic atoms show additional structure due to the interaction of the muon with the magnetic dipole moment and the electric quadrupole moment of the nucleus. By analogy with the spectra of electronic atoms this is known as hyperfine structure, although the hyperfine splitting in muonic atoms may be of the same order as the fine structure splitting, namely ~100 keV. Because the magnetic moment of the muon is so small, the electric quadrupole effect is usually dominant. In addition to the static effect, which leaves the nucleus in its ground state, there is a dynamic effect in which the nucleus is excited to low-lying collective states through the quadrupole interaction. If the dynamic effect occurs, hyperfine structure can be observed even for spin zero nuclei for which the static effect would be absent. The rotational model provides a natural description of this effect which has been analysed in terms of a deformed charge distribution of the form

$$\rho = \rho_0 \left[1 + \exp\left\{\frac{r - c(1 + \beta Y_2^0)}{a(1 + \beta' Y_2^0)}\right\}\right]^{-1}. \tag{4.39}$$

(Compare, for example, the treatment of the excitation of collective states by inelastic scattering given in sections 7.3 and 7.4). This provides information on the magnitude and sign of β and β' and hence on the nature of the deformation, and on the ratio of the quadrupole moments of the excited and ground states which can be compared with results obtained from Coulomb excitation.

There are a number of corrections to the basic theory which must be taken into account when detailed calculations are made. We discuss here the two corrections due to vacuum polarization and nuclear polarization, and refer the reader to the paper by Barrett, Brodsky, Erickson and Goldhaber [57] for a discussion of the other, more sophisticated, corrections. The static charge distribution of the nucleus provides an external electromagnetic field and this field can cause polarization of the vacuum by producing a separation of charge in the form of virtual electron-positron pairs. This vacuum polarization produces a change in the external field independently of the interacting particle [40], and in muonic atoms this has the effect of increasing the binding energy of the muon particularly in the most tightly bound states. For example, the correction for the 1*s* muon state in ^{208}Pb is 67 keV. The Coulomb interaction between the nucleus and the muon should in fact be represented by a dynamical model which includes virtual excitation of intermediate states of the nucleus accompanied by virtual transitions of the muon. This distortion or polarization of the nuclear charge distribution causes an increase in binding of the muon but the most recent estimates indicate that the effect is small, of the order of a few keV.

5 The compound nucleus

5.1 Resonances

The excitation function for a given process always shows fluctuations [3, 6]. For low-energy projectiles these fluctuations consist of high, narrow, peaks which are called *resonances*. At very low energies charged particles do not cause nuclear reactions because of the large Coulomb barrier, but resonance

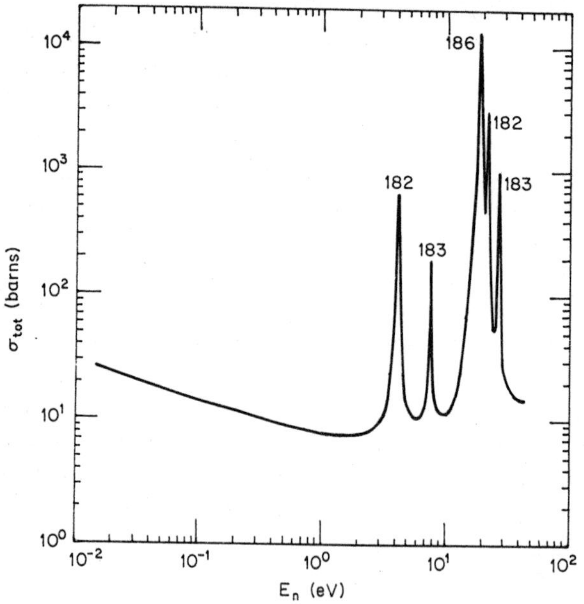

Figure 5.1 The cross-section for low energy neutrons on natural tungsten showing resonances in the isotopes ^{182}W, ^{183}W, ^{186}W. [From R. R. Spencer and K. T. Faler, Proc. Conf. on Slow Neutron-Capture Gamma-Ray Spectroscopy, 1966, Argonne National Laboratory Report ANL 7282 (1966).]

behaviour is observed in low-energy neutron scattering from medium mass nuclei and in neutron capture, or (n, γ), reactions on heavy nuclei. These resonances are exceedingly sharp. In medium mass nuclei, the widths are of the order of eV and the spacings of the order of keV, and in heavy nuclei these magnitudes are substiantially reduced. This resonance behaviour persists for incident energies of up to about 1 MeV for heavy nuclei and about 10 MeV for light nuclei, and this low energy region is therefore called the *resonance region.* At higher energies the widths of the resonances become comparable with the spacing, and at sufficiently high energies the widths are very much greater than the average spacing, so that in this region of overlapping resonances the excitation functions become much smoother functions of energy. This energy region is called the *continuum region.* In the continuum region the fine structure of the resonances is often not resolved and then the excitation functions show broad maxima whose positions vary smoothly and slowly with mass number. These *giant resonances* have widths up to an MeV and spacings of about 10–20 MeV. Typical resonance behaviour observed in total cross-sections is shown in figures 5.1 and 5.2.

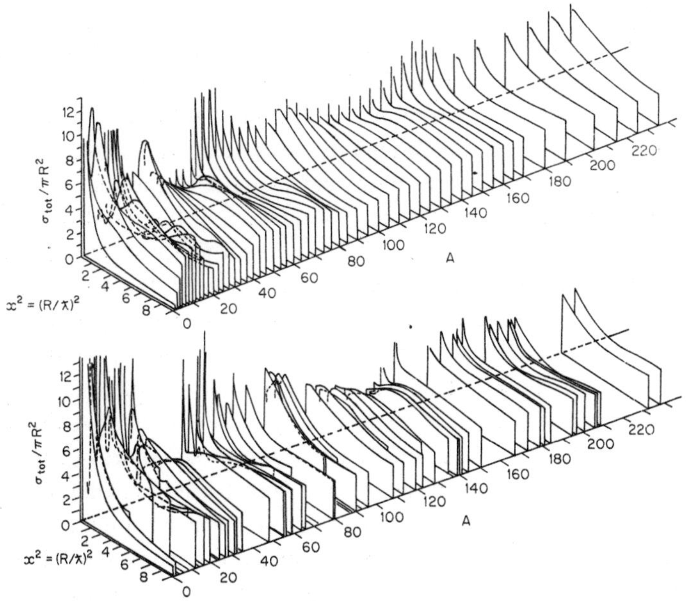

Figure 5.2 (**a**) Observed neutron total cross-sections averaged over the resonances as a function of energy and mass number. (**b**) Calculated neutron total cross-sections calculated for a complex square well potential of radius $R = 1{\cdot}45\,A^{1/3}$ fm and strength $-V_0(1+i\zeta)$. [From H. Feshbach, C. E. Porter and V. F. Weisskopf, *Phys. Rev.* **96,** 448 (1954).]

The formation of a resonance at a particular incident energy may be associated with the formation of a long-lived state of the compound system composed of the projectile and the target. Since the projectile carries kinetic energy the total energy of the compound system is positive and the state of the system is a virtual or quasi-stationary state and not a true stationary state. This means that the state of the compound system must eventually decay with emission of the original particle, a different particle, or a γ-ray. The width Γ of this virtual state is related to its lifetime τ by the usual uncertainty relation

$$\Gamma = \hbar/\tau \tag{5.1}$$

from which it follows that the narrowest resonances have the longest lifetimes. For level widths of $\sim 1{\cdot}0$ eV, the lifetime of the compound system is

$$\tau = \frac{\hbar}{\Gamma} \sim 10^{-15} \text{ secs}$$

which is very long compared with the nuclear transit time $2R/v$ which is $\sim 10^{-21}$ secs.

The earliest model for nuclear scattering, a single-particle potential model, was introduced by Bethe in 1935. The incident projectile was assumed to move in an attractive real potential so that the quasi-stationary levels are the single particle levels in this potential with positive energy, and a resonance will occur whenever the incident energy corresponds to the energy of one of these single-particle levels. For a simple square well potential of depth V_0 and range R, the s-wave resonances occur when $KR = (n+\frac{1}{2})\pi$ where $K^2 = 2\mu(E+V_0)/\hbar^2$, and the p-wave resonances occur when $KR \approx n\pi$. For a given nucleus the spacing of these levels is about 10 MeV, depending on the choice of potential parameters. If it is assumed that, for low energies, V_0 is independent of mass number A while $R \propto A^{\frac{1}{3}}$, the positions of the resonances can be located as functions of A for fixed incident energy. It is evident that this model contains many of the features associated with giant resonances but apparently contains none of the features associated with the fine resonances. If the fine resonances are also to be associated with quasi-stationary levels, the levels in question must be closely spaced levels of a many particle system (see figure 5.3), i.e. we must abandon the single-particle picture and assume that many nucleons participate in the formation of the compound system.

Historically it happened that the fine resonances were first investigated in 1935–36 by Moon, Szilard, Fermi and others, while the giant resonances were first observed by Barschall and collaborators in a series of experiments during 1948–52. For this reason the potential model languished for some years while the main emphasis was placed on the explanation of the fine resonances.

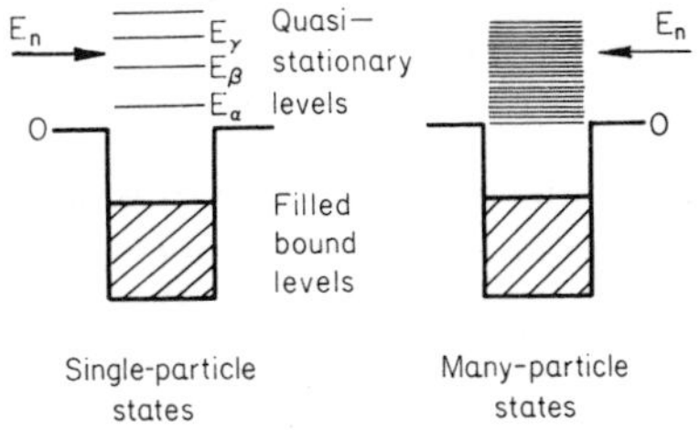

Figure 5.3 Neutron energy levels in a single-particle and compound system. [Adapted from W. E. Burcham, ref. 1.]

5.2 Elementary theory of the compound nucleus

The compound nucleus model was introduced by Bohr to explain the narrow resonances observed in experiments with low-energy neutrons. It is assumed that the incident neutron and the target form a compound nucleus in which many of the target nucleons participate collectively and the kinetic energy of the incident neutron is shared with these target nucleons [58, 59]. As we have seen, the participation of many target nucleons is necessary to produce the closely spaced levels of the compound system. On the basis of this many-particle picture, it must take some time for a single nucleon in the compound nucleus to acquire sufficient energy to be emitted and therefore the emission of the excess energy of the compound system in the form of a γ-ray is highly probable. This picture therefore predicts that the capture process (n, γ) will predominate compared with elastic scattering, and this is in agreement with experiment.

In the compound nucleus model it is assumed that the mode of decay of the compound nucleus is independent of its mode of formation. Thus we may represent a reaction as a two-stage process

$$A+a \rightarrow C^* \rightarrow b+B$$

where C is the compound nucleus, and the decay of C^* should not depend on the nature of the projectile a and the target A. Comparison of the decay of the same compound nucleus formed by different primary interactions have verified this *independence hypothesis* in the intermediate energy region. Such comparisons must of course be carried out for the same range of excitation energies. For example, the nucleus ^{64}Zn can be formed as the compound nucleus in a variety of reactions, including

$$^{60}\mathrm{Ni}+\alpha \rightarrow {}^{64}\mathrm{Zn}^*$$

$$\mathrm{Cu}^{63}+p \rightarrow {}^{64}\mathrm{Zn}^*$$

Figure 5.4 shows the results obtained by Ghoshal for the formation of ^{64}Zn* by these two reactions and the subsequent decay through the following modes

$$^{64}\mathrm{Zn}^* \rightarrow \mathrm{n}+{}^{63}\mathrm{Zn}$$

$$\rightarrow \mathrm{n}+\mathrm{n}+{}^{62}\mathrm{Zn}$$

$$\rightarrow \mathrm{n}+\mathrm{p}+{}^{62}\mathrm{Cu}\,.$$

These results may clearly be taken as a verification of the independence hypothesis in this energy region. It was also assumed by Bohr that because

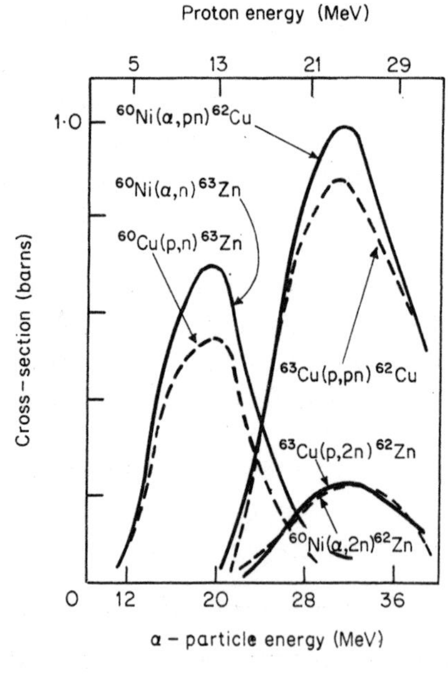

Figure 5.4 Yield of decay products from the compound nucleus ^{64}Zn. [From S. Ghoshal, *Phys. Rev.* **80,** 939 (1950).]

the nucleon–nucleon interaction is a strong short-range interaction the formation of a compound state occurs immediately and with a probability of unity. This asymption of *immediate formation* is in conflict with the potential model of scattering and with the shell model of the nucleus. Fortunately, this assumption is not essential and by allowing the probability of formation to be less than unity a reconciliation of these models has been achieved (see sections 6.1 and 6.2).

We now derive the cross-sections for resonant processes for the simple case of *s*-wave neutron scattering. The independence hypothesis implies that the cross-section for a particular process $\alpha \to \beta$ can be factorized into the cross-section $\sigma_c(\alpha)$ for the formation of the compound nucleus through the entrance channel α, and the probability that the compound nucleus will decay through channel β. Thus

$$\sigma_{\beta\alpha} = \sigma_{\mathrm{cn}}(\alpha)\frac{\Gamma_\beta}{\Gamma} \tag{5.2}$$

where Γ_β is the partial width for decay through channel β and Γ is the total width

$$\Gamma = \sum_\beta \Gamma_\beta \,. \tag{5.3}$$

We now make the assumption that the nucleus has a well-defined surface. We also assume that the incident neutron does not interact with the target nucleus at separation distances greater than the *channel radius R*, and define the quantity ρ_l as the logarithmic derivative of the lth partial wave at the surface multiplied by R, i.e.

$$\rho_l = \left(\frac{R}{u_l}\frac{du_l}{dr}\right)_{r=R}$$

where $u_l = r\psi_l$. Since u_l and its derivative are continuous at R, ρ_l is determined by the conditions in the interaction region $r < R$. Using equations (3.10), (3.11), and (3.65), we have

$$u_0(r) = \frac{i}{2k}\left[e^{-ikr} - \eta_0 e^{ikr}\right] \qquad r \geqslant R$$

and hence

$$\eta_0 = \frac{\rho_0 + ikR}{\rho_0 - ikR}e^{-2ikR} \tag{5.4}$$

from which we may deduce that if ρ_0 is real, $|\eta_0|$ is unity and there is no reaction, but if Im $\rho_0 < 0$ then $|\eta_0| < 1$. The cross-section for *s*-wave elastic scattering is obtained from equation (3.66) and using equation (5.4) becomes

$$\sigma_{\text{sc}} = \frac{\pi}{k^2}|1-\eta_0|^2 = \frac{\pi}{k^2}\left|e^{2ikR}-1-\frac{2ikR}{\rho_0-ikR}\right|^2 \tag{5.5}$$

and similarly the absorption cross-section is obtained from equations (3.67) and (5.4) as

$$\sigma_{\text{abs}} = \frac{\pi}{k^2}[1-|\eta_0|^2] = \frac{\pi}{k^2}\left[\frac{-4kR\,\text{Im}\,\rho_0}{(\text{Re}\,\rho_0)^2+(\text{Im}\,\rho_0-kR)^2}\right] \tag{5.6}$$

We have already deduced that Im $\rho_0 < 0$, and since σ_{abs} has its maximum value when Re$\rho_0 = 0$, we may expand ρ_0 around the resonance energy as

$$\rho_0 = -a(E-E_s)-ib+\ldots \tag{5.7}$$

where a is a real constant whose dimension is (energy)$^{-1}$, b is a real dimensionless constant, and E_s is the resonance energy. The scattering cross-section then becomes

$$\sigma_{\text{sc}} = \frac{\pi}{k^2}\left|e^{2ikR}-1+\frac{2ikR/a}{(E-E_s)+i(b+kR)/a}\right|^2. \tag{5.8}$$

Thus the scattering amplitude consists of a non-resonant part

$$A_{\text{pot}} = \frac{1}{2ik}(e^{2ikR}-1) \tag{5.9}$$

which is the amplitude for *potential* or *shape-elastic scattering*, and a resonant part

$$A_{\text{res}} = \frac{1}{2ik}\left[\frac{+2ikR/a}{(E-E_s)+i(b+kR)/a}\right]$$

which represents the scattering arising from re-emission of the absorbed neutron by the compound nucleus, or *compound-elastic scattering*. We now define the *partial width* for re-emission of the incident neutron through channel α as

$$\Gamma_\alpha = 2kR/a\,,$$

the *total width* as

$$\Gamma = 2(b+kR)/a\,,$$

and the *reaction width* as

$$\Gamma_r = \sum_{\beta \neq \alpha} \Gamma_\beta = \Gamma - \Gamma_\alpha = 2b/a\,.$$

These widths* should also be labelled by s to indicate that they apply to the resonance occurring at $E = E_s$. The cross-section for compound elastic scattering now becomes

$$\sigma_{\mathrm{ce},\alpha} = \frac{\pi}{k^2}\,\frac{\Gamma_\alpha^2}{(E-E_s)^2 + (\frac{1}{2}\Gamma)^2},$$

and the absorption cross-section becomes

$$\sigma_{\mathrm{abs}} = \frac{\pi}{k^2}\,\frac{\Gamma_r\Gamma_\alpha}{(E-E_s)^2 + (\frac{1}{2}\Gamma)^2}.$$

The cross-section for compound nucleus formation is obtained by adding the cross-sections for those processes which involve formation of the compound nucleus through channel α, i.e.

$$\sigma_{\mathrm{cn}}(\alpha) = \sigma_{\mathrm{abs}} + \sigma_{\mathrm{ce},\alpha} = \frac{\pi}{k^2}\,\frac{\Gamma\Gamma_\alpha}{(E-E_s)^2 + (\frac{1}{2}\Gamma)^2} \tag{5.10}$$

and from equation (5.2) the cross-section for the process $\alpha \rightarrow \beta$ is

$$\sigma_{\beta\alpha} = \frac{\pi}{k^2}\,\frac{\Gamma_\beta\Gamma_\alpha}{(E-E_s)^2 + (\frac{1}{2}\Gamma)^2} \tag{5.11}$$

These formulae are the one-level dispersion formulae introduced by Breit and Wigner [59]. They resemble closely the dispersion formulae of physical optics, the resonance formulae for electrical circuits, and many other resonance phenomena [60]. They can be generalized to take account of spin and to describe resonances in other partial waves [7, 61].

At low energies and far from resonance the scattering cross-section obtained from equation (5.9) is $4\pi R^2$ which is just the scattering from an im-

* It is also useful to define the *reduced width* γ_β^2 for decay in channel through the relation

$$\gamma_\beta^2 = \Gamma_\beta / 2P_\beta$$

where the *penetrability* P_β is the imaginary part of the logarithmic derivative of the outgoing wave in the channel β evaluated at the surface $r = R$ and multiplied by the channel radius R. For s-wave neutrons the outgoing wave is just e^{ikr} multiplied by the appropriate S-matrix element and hence $P = kR$. Thus the reduced width determines the probability that the components specified by β will appear at the surface so that the compound system can break-up through this mode.

penetrable sphere of radius R. For this reason the description of elastic scattering given by this model is often called *hard-sphere scattering*. Near resonance the potential scattering and shape-elastic scattering interfere to give the cross-section the distinctive shape shown in figure 5.5. Also near resonance, the cross-section for compound nucleus formation reduces to $4\pi\Gamma_\alpha/k^2\Gamma$. The resonance formulae (5.10) and (5.11) can be applied only near resonance because the expansion of ρ in equation (5.7) contains only sufficient terms to give validity close to the resonance, but by fitting these formulae to the experimental cross-sections it is possible to determine precisely the resonance energy, resonance width, and the angular momentum and parity of the compound state.

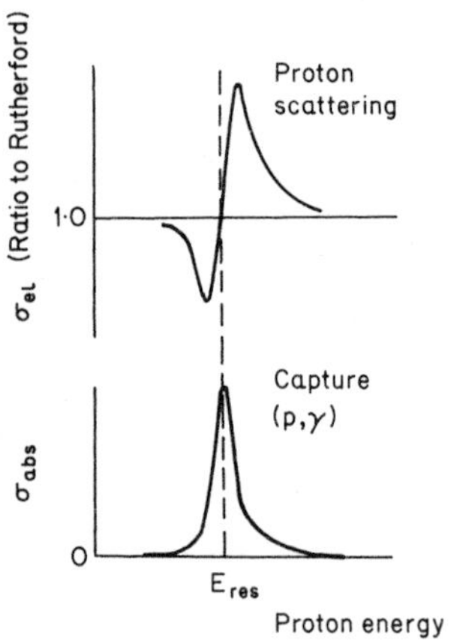

Figure 5.5 Resonances in the total cross-sections for proton scattering and capture, showing the effect of interference in the scattering cross-section. [From W. E. Burcham, ref. 1.]

The results obtained so far are valid if it is correct to assume that a long-lived compound state is formed. This requires that the widths of the resonances are much less than their spacing, and this requirement is satisfied in the resonance region. In the region where the widths of the resonances are of the order of their spacing an incident neutron can excite several compound states whose relative phases will depend on the mode of excitation. In this situation it seems very likely that this phase relationship will influence the modes of decay, and the validity of the independence hypothesis is, therefore, very doubtful [62]. On the other hand, if the energy of the incident particle is sufficient to excite the system high in the continuum region where the density of overlapping levels is very high and a very large number of states are excited simultaneously, the phase relations between these states may be regarded as random at least with regard to the effect on the modes of decay. If the compound nucleus is sufficiently long-lived, thermodynamic equilibrium is set up and the distribution of the available energy follows the laws of statistical mechanics. The assumption that this equilibrium situation is reached is the basis of the statistical model; it is assumed that the energy (and the total angular momentum) determines the properties of the compound nucleus so that, apart from the energy dependence, all possible decay modes occur with equal probability.

The simple model can be applied in this case also. We assume, however, that in the continuum region many channels are open for the decay of the compound state and the probability of re-emission through the entrance channel is very small and therefore compound elastic scattering is negligible. In this non-elastic region the wavefunction of the incoming neutron will be attenuated so that

$$u_0(r) = e^{-iKr}, \qquad r \leqslant R$$

where K is the wave number in the internal region, and the s-wave logarithmic derivative is given by

$$\rho_0 = -iKR\,.$$

The cross-section for the formation of the compound nucleus is then given by equations (5.4) and (5.6) as

$$\sigma_{cn}(\alpha) = \sigma_{abs} = \frac{\pi}{k^2}\,\frac{4kK}{(k+K)^2} \tag{5.12}$$

and the cross-section for a particular channel reaction is

$$\sigma_{\beta\alpha} = \frac{\pi}{k^2}\,\frac{4kK}{(k+K)^2}\,\frac{\Gamma_\beta}{\Gamma}\,. \tag{5.13}$$

At low energies equation (5.12) reduces to

$$\sigma_{abs} = \frac{4\pi}{kK} \propto \frac{1}{v}$$

yielding the $1/v$ *law* for low energy neutron capture. The quantity $4kK/(k+K)^2$ is called the transmission coefficient T_0 for s-wave neutrons. The cross-section σ_{cn} can also be obtained by averaging equation (5.10) over many overlapping levels. If $\bar{\Gamma}_\alpha$ is the mean width for resonances due to particles in channel α and D is the mean spacing of levels within an energy interval I we may write

$$\sigma_{\text{cn}}(\alpha) = \frac{1}{I}\int_{E-\frac{1}{2}I}^{E+\frac{1}{2}I} \frac{\pi}{k^2} \sum_s \frac{\Gamma^s \Gamma^s_\alpha}{(E-E_s)^2+(\frac{1}{2}\Gamma^s)^2}\, dE \tag{5.14}$$

$$= \frac{\pi}{k^2}\frac{2\pi}{I}\sum_s \Gamma^s_\alpha = \frac{\pi}{k^2}\, 2\pi \frac{\bar{\Gamma}_\alpha}{D}, \tag{5.15}$$

where the energy interval is chosen so that the variation of k^2 can be neglected, and combining this with equation (5.12) we have

$$\frac{\bar{\Gamma}_\alpha}{D} = \frac{1}{2\pi}\frac{4kK}{(k+K)^2} \approx \frac{2k}{\pi K}. \tag{5.16}$$

In order to compare data at different excitation energies it is convenient to define an energy independent quantity called the s-wave *strength function* as

$$\frac{\Gamma^0_\alpha}{D} = \left(\frac{E_0}{E}\right)^{\frac{1}{2}} \frac{\bar{\Gamma}_\alpha}{D} \tag{5.17}$$

where E_0 is an arbitrary energy taken to be 1 eV*.

The cross-sections $\sigma_{\beta\alpha}$ and σ_{abs} can be related to the reflection coefficients and the S-matrix elements defined in Chapter 3, but these coefficients must reproduce the rapid fluctuations in the cross-sections due to the resonances and are therefore very complicated functions of energy. If it is assumed that the reflection coefficients can be subjected to an energy averaging process similar to that used in equation (5.14) to yield a smoothly varying function of energy, averaged cross-sections can be defined in the energy interval I. Comparison of equation (5.12) and equation (3.67) then yields the expression for the *transmission coefficient*

$$T_0 = 1 - |\eta_0|^2 \tag{5.18}$$

where the energy averaged value of η_0 is to be understood.

5.3 Formal theory of compound reactions

In the region of configuration space outside the interaction region the wavefunction in a given channel may be expressed in the product form of equation

* The strength function is also defined as $\bar{\gamma}^2_\alpha/D$ where $\bar{\gamma}^2_\alpha$ is the average reduced width. For a black nucleus, the value for s-wave neutrons is $(\pi KR)^{-1}$, using equation (5.16) and the definition of γ^2. Thus the expression $\bar{\gamma}^2/D$ has the disadvantage compared with $\bar{\Gamma}/D$ that it depends on the arbitrary radius R.

(3.70), and the results of section 3.7 may be used to derive expressions for the cross-sections and transmission coefficients in terms of the S-matrix elements. We recall that in section 3.7 the symbol α represented the quantum numbers (x, j, m, l, Z) where x labels the pairs of particles and their relative energy, j is the total angular momentum, and Z is the channel spin composed of the intrinsic angular momenta I_1 and I_2 of the interacting particles. The absorption cross-section can be obtained as a generalization of equation (3.77) (or (3.80)). In this case it is necessary to include the two Clebsch-Gordan coefficients $(l0Z\mu|jm)$, $(l'\nu'Z'\mu'|jm)$ arising from the expansion (3.70) and on carrying out the sums over μ, μ', ν' this yields a factor $(2j+1)/(2l+1)$. The cross-section is then given by

$$\sigma_{\text{abs}} = \frac{\pi}{k_\alpha^2} \sum_{\beta \neq \alpha} \sum_{jll'ZZ'} \frac{1}{g_\alpha^j} |S_{\alpha\beta}^j|^2 \tag{5.19}$$

where $1/g_\alpha^j$ is the spin weighting factor

$$\frac{1}{g_\alpha^j} = \frac{2j+1}{(2I_1+1)(2I_2+1)}.$$

In the statistical model, it can be argued that because the S-matrix elements vary rapidly with energy the statistical assumption implies that there is a random phase relation between the different components of the S-matrix. The process of energy averaging then eliminates the cross terms and gives

$$\sigma_{\text{abs}} = \frac{\pi}{k_\alpha^2} \sum_{jlZ} \frac{1}{g_\alpha^j} (1-|S_{lZ}^j|^2) = \frac{\pi}{k_\alpha^2} \sum_{jlZ} \frac{1}{g_\alpha^j} T_{lZ}^j(\alpha) \tag{5.20}$$

where we have used the unitary and symmetry properties of the S-matrix in the form

$$\sum_\beta \sum_{l'Z'} S_{lZ,l'Z'}^j S_{lZ,l'Z'}^{j*} = 1$$

and have introduced the general definition of the transmission coefficient

$$T_{lZ}^j(\alpha) = 1-|S_{lZ}^j|^2 . \tag{5.21}$$

We now represent the cross-section for a transition from an initial pair of particles labelled by x to a final pair of particles labelled by x' as $\sigma_{xx'}$ which can be written as

$$\sigma_{xx'} = \sum_{jll'ZZ'} \sigma_{\beta\alpha}^j = \sum_{jll'ZZ'} \sigma_{\text{cn}}^j(\alpha) \frac{\Gamma_\beta^j}{\Gamma} \tag{5.22}$$

In the non-elastic region, we have $\Sigma\, \sigma^j_{\text{cn}}(\alpha) = \sigma_{\text{abs}}$. Also, using equation (3.84) the reciprocity theorem gives

$$k_\alpha^2\, g_\alpha^j\, \sigma^j_{\beta\alpha} = k_\beta^2\, g_\beta^j\, \sigma^j_{\alpha'\beta'} \tag{5.23}$$

and combining equations (5.22) and (5.23) we find, for a given channel α,

$$\Gamma^j_\alpha = k_\alpha^2\, g_\alpha^j\, \sigma^j_{\text{cn}}(\alpha)$$

so that using equations (5.20) and (5.21) we have

$$\frac{\Gamma^j_\beta}{\Gamma} = \frac{\sum\limits_{l'Z'} T^j_{l'Z'}(\beta)}{\sum\limits_{l''Z''\gamma} T^j_{l''Z''}(\gamma)}.$$

Finally, the cross-section $\sigma_{xx'}$ can be written as

$$\sigma_{xx'} = \frac{\pi}{k_\alpha^2} \sum_j \frac{(2j+1)}{(2I_1+1)(2I_2+1)} \sum_{lZ} T^j_{lZ}(\alpha) \frac{\sum\limits_{l'Z'} T^j_{l'Z'}(\beta)}{\sum\limits_{l''Z''\gamma} T^j_{l''Z''}(\gamma)}. \tag{5.24}$$

Equation (5.24) represents the basic cross-section formula of the statistical theory [63]. A similar expression can be derived for the differential cross-section. These formulae have been applied to a very wide range of processes proceeding through a compound nucleus with considerable success [64]. The transmission coefficients may be calculated using the black-nucleus or strong-coupling model which was used at the beginning of this section to describe *s*-wave neutrons, or they may be calculated using the complex potential of the optical model (see Chapter 6). The magnitude of the transmission coefficient determines the rate of decay of the compound nucleus through the corresponding channel, but even if the individual coefficients are small the compound nucleus may still decay very rapidly if the number of possible decay processes is large. When the excitation energy E^* of the compound nucleus is high, the number of final states becomes large and the average cross-sections must be modified by the inclusion of a *level density function* $\rho(E^*)$. If the final states are not discrete, the quantity of interest is the evaporation spectrum of the decay products, which is given by [63]

$$N(E)\,dE = \sum_{\Delta E} \sigma_{xx'}$$

where ΔE is the energy interval between E and $E+dE$, and $\sigma_{xx'}$ is now the modified cross-section. The level density function and evaporation spectrum are often characterized by a *nuclear temperature* [65], denoted by θ, such that at high excitation energy $\rho(E^*) \sim e^{E^*/k\theta}$. The evaporation spectrum is produced by the 'boiling off' of nucleons after the nucleus has reached thermal equilibrium at temperature θ.

The evaluation of the cross-section formulae given in this section requires a knowledge of the S-matrix. We have so far used only the simplest model, namely that of a black-nucleus, to examine the s-wave phase shifts. The construction of a more physically realistic S-matrix involves a description of the nuclear system in the interior or interaction region; this in turn should provide a more fundamental basis for the description of resonances and a link between the parameters of the resonance and the nuclear structure. A critical examination of the methods used for tackling this problem has recently been given by Lane and Robson [66]. The methods fall essentially into three groups: (i) those which impose boundary conditions at the channel radii in order to define the resonance behaviour, (ii) those which use a Hamiltonian modified by projection operators to produce resonances but avoid the use of channel radii, and (iii) those which are based on the analytical properties of the S-matrix and define resonances as poles in the S-matrix. The last two groups are attractive from the point of view of formal theory and one of the forms of method (ii) is used in section 6.2 to examine the formal properties of the optical potential and the connection between single-particle and compound processes. However, to complete this chapter we consider the application of method (i) to the scattering of s-wave neutrons*, following the method of Vogt [64].

The eigenstates of the nuclear Hamiltonian in the interior region are denoted by X_λ with energy E_λ. These states are required to satisfy the boundary condition

$$r\frac{dX_\lambda}{dr} + bX_\lambda = 0 \tag{5.25}$$

at the channel radius $r = R$, where the constant b is a real number. The true nuclear wavefunction Φ for the compound system is not stationary, but since the X_λ form a complete set it is possible to expand Φ in terms of the X_λ, i.e.

$$\Phi = \sum_\lambda A_\lambda X_\lambda, \tag{5.26}$$

* It should be noted that this example is included as an illustration of the method, and not as a practical numerical method.

where A_λ are the expansion coefficients

$$A_\lambda = \int_0^R X_\lambda \Phi \, dr \,. \tag{5.27}$$

The differential equations for Φ and X_λ are (remembering that we are dealing with s-wave neutrons)

$$-\frac{\hbar^2}{2m}\frac{d^2\Phi}{dr^2} + V\Phi = E\Phi \tag{5.28}$$

$$-\frac{\hbar^2}{2m}\frac{d^2 X_\lambda}{dr^2} + VX_\lambda = E_\lambda X_\lambda \,, \qquad r \leqslant R \,, \tag{5.29}$$

and multiplying equation (5.28) by X_λ and equation (5.29) by Φ, subtracting and integrating, we have

$$\frac{\hbar^2}{2m}\left(\Phi \frac{dX_\lambda}{dr} - X_\lambda \frac{d\Phi}{dr}\right)\bigg|_{r=R} = (E - E_\lambda)\int_0^R \Phi X_\lambda \, dr \,.$$

Hence, using equations (5.25) and (5.27) we have

$$A_\lambda = (E_\lambda - E)^{-1} \frac{\hbar^2}{2mR} X_\lambda(R)\{R\Phi'(R) + b\Phi(R)\}$$

where the prime indicates the differential with respect to r, and substituting in equation (5.26) gives

$$\Phi(R) = \mathscr{R}\{R\Phi'(R) + b\Phi(R)\} \tag{5.30}$$

where the function $\mathscr{R}$ which relates the value of Φ at the surface to its derivative at the surface is given by

$$\mathscr{R} = \frac{\hbar^2}{2mR}\sum_\lambda \frac{X_\lambda(R)\,X_\lambda(R)}{E_\lambda - E} \,. \tag{5.31}$$

Now, rearranging equation (5.30) we have

$$R\frac{\Phi'(R)}{\Phi(R)} = \frac{1 - b\mathscr{R}}{\mathscr{R}} \,,$$

but this is just the logarithmic derivative ρ_0 which can be inserted into equation (5.4) to determine the S-matrix element η_0 in terms of the $\mathscr{R}$-function. This gives

$$S_0 = \eta_0 = \left\{1 + \frac{2ikR\mathscr{R}}{1-(b+ikR)\mathscr{R}}\right\} e^{-2ikR}. \tag{5.32}$$

Finally, we assume that E is near to a particular E_λ, say E_α, neglect all terms $\lambda \neq \alpha$ in equation (5.31), and define

$$\Gamma_\alpha = \hbar^2 k X_\alpha^2(R)/m$$

$$\Delta_\alpha = -b\Gamma_\alpha/2kR,$$

so that the S-matrix element becomes

$$S_0 = \left[1 + \frac{i\Gamma_\alpha}{(E_\alpha + \Delta_\alpha - E) - \frac{1}{2}i\Gamma_\alpha}\right] e^{-2ikR} \tag{5.33}$$

and the scattering cross-section is

$$\sigma_{sc} = \frac{\pi}{k^2}\left| e^{2ikR} - 1 + \frac{i\Gamma_\alpha}{(E - E_\alpha - \Delta_\alpha) + \frac{1}{2}i\Gamma_\alpha}\right|^2. \tag{5.34}$$

From this analysis we see that the procedure of imposing the boundary conditions at the channel radius leads to isolated s-wave resonances of Breit-Wigner form. If the constant b is non-zero the position of the maximum in the cross-section is shifted* to $E_\alpha + \Delta_\alpha$. In general, a nucleus can decay through many channels and when the formalism is extended to take this into account the $\mathscr{R}$-function becomes a matrix. In this *$\mathscr{R}$-matrix theory* [6, 67] the constant b is real and the X_λ and E_λ can be chosen to be real so that the eigenvalue problem is Hermitian. In the *Kapur-Peierls theory* [68] the constant b is chosen to be $-ikR$ in each channel, so that the denominator $1-(b+ikR)\mathscr{R}$ in equation (5.32) is unity. This has the effect of simplifying the matrix manipulation but causes the X_λ and E_λ to be complex and energy-dependent. The usefulness of these methods depends on the complexity of the problem and particularly on the extent to which the resonances are well separated.

* The level shift does not appear in the simple form of the Breit-Wigner formula because $E_\alpha + \Delta_\alpha$ is defined as the resonance energy.

6 The optical model

6.1 The theory of average cross-sections

The compound nucleus model predicts that the average cross-sections decrease smoothly with energy and the strength function should be essentially independent of mass number A (see equations (5.15) and (5.16)). Thus the existence of the giant resonances, shown in figure 5.2, and the marked A-dependence of the strength function, shown in figure 6.1, cannot be explained

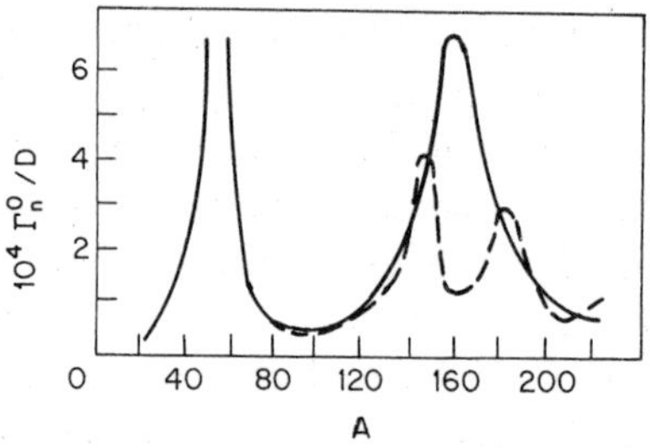

Figure 6.1 Calculated values for the *s*-wave neutron strengths functions obtained using an optical potential. The full and dashed curves refer to a spherical and spheriodal potential, respectively. The data follow the dashed curve.

in terms of the compound nucleus model. Further, the width of the giant resonances implies a lifetime τ of the same order as the nuclear transit time and a mean free path $\lambda = 1/\tau$ for the incident neutron of the same order of magnitude as the nuclear diameter, and suggests that the formation of the compound state does not take place immediately or with a probability of unity. This long mean free path is consistent with the shell model of the nucleus, and the spacings of the giant resonances are consistent with a simple potential model. All these observations suggest that an incident neutron can penetrate the target nucleus and maintain its independent-particle motion without necessarily or immediately forming a compound state. The neutron may subsequently emerge from the nucleus or it may participate in the formation of a compound state, so that the formation of the compound state appears as an absorption effect on the incident beam. This process may be described by a complex potential whose real part describes the average potential energy of the neutron inside the nucleus and may be expected to resemble the shell model potential for bound states, and whose imaginary part describes the absorption effect due to compound nucleus formation. The strength of the imaginary part determines the magnitude of the mean free path (see section 3.6). These ideas form the basis of the *optical model* and the complex potential is called the *optical potential.* The model is not restricted to the interaction of neutrons with nuclei, but for the sake of simplicity in the formulae we continue to discuss this particular application.

In the optical model, the target nucleus is represented by a potential which is a function of a limited number of variables such as the relative co-ordinate $\boldsymbol{r}$, the spin co-ordinates and the energy of the incident particle. This model can therefore not reproduce the detailed resonance behaviour of the many-particle system, and in order to understand exactly what the cross-sections calculated from the optical potential do represent we return to the theory of average cross-sections. We assume, as before, that the reflection coefficients and the cross-sections can be averaged over an energy interval I such that $D \ll I \ll E$ where D is the mean spacing of the resonances, and denote such an energy averaged quantity by $\langle\ \rangle$. We apply the averaging procedure to the cross-sections for elastic scattering and for absorption, defined by equations (3.66) and (3.67). This gives

$$\begin{aligned}
\langle\sigma_{\text{sc}}\rangle &= \frac{\pi}{k^2}\sum_l (2l+1)\langle|1-\eta_l|^2\rangle \\
&= \frac{\pi}{k^2}\sum_l (2l+1)|1-\langle\eta_l\rangle|^2 + \frac{\pi}{k^2}\sum_l (2l+1)[\langle|\eta_l|^2\rangle - |\langle\eta_l\rangle|^2] \\
\langle\sigma_{\text{abs}}\rangle &= \frac{\pi}{k^2}\sum_l (2l+1)\langle(1-|\eta_l|^2)\rangle = \frac{\pi}{k^2}\sum_l (2l+1)(1-\langle|\eta_l|^2\rangle)\,. \\
&= \frac{\pi}{k^2}\sum_l (2l+1)(1-|\langle\eta_l\rangle|^2) - \frac{\pi}{k^2}\sum_l (2l+1)[\langle|\eta_l|^2\rangle - |\langle\eta_l\rangle|^2]\,.
\end{aligned}$$

Now from section 5.2 we know that the total cross-section for elastic scattering is the sum of the contributions from shape-elastic and compound-elastic scattering, while the cross-section for absorption is the difference between the cross-section for compound nucleus formation and compound-elastic scattering. In the continuum region as the widths become very much greater than the spacing and the excitation functions vary smoothly with energy, the averaging procedure is no longer a mathematical device but represents the real physical process. Thus we expect that as the incident energy increases

$$\left.\begin{aligned} \langle|\eta_l|^2\rangle - |\langle\eta_l\rangle|^2 &\to 0 \\ \sigma_{\mathrm{ce},\alpha} &\to 0 \\ \sigma_{\mathrm{cn}}(\alpha) &\to \sigma_{\mathrm{abs}}\,. \end{aligned}\right\} \tag{6.1}$$

Thus if we associate the averaged reflection coefficient $\langle\eta_l\rangle$ with the quantity calculated from the complex potential of the optical model, we see that the elastic cross-section calculated from the optical potential is the shape-elastic cross-section

$$\sigma_{\mathrm{se},\alpha} = \frac{\pi}{k^2}\sum_l (2l+1)|1-\langle\eta_l\rangle|^2 \tag{6.2}$$

and the non-elastic cross-section calculated from the optical potential is the cross-section for compound nucleus formation

$$\sigma_{\mathrm{cn}}(\alpha) = \frac{\pi}{k^2}\sum_l (2l+1)(1-|\langle\eta_l\rangle|^2)\,. \tag{6.3}$$

In the high-energy limit these cross-sections correspond to the measured cross-sections for elastic scattering and absorption. This limit occurs at about 10 MeV for nucleon scattering from medium and heavy nuclei and at about 20–30 MeV for light nuclei. Similarly, the differential cross-section calculated from the optical potential is that for shape-elastic scattering and can be compared directly with experiment only if the compound-elastic scattering is negligible. Methods have been developed which allow estimates of the compound elastic scattering to be made [69], and it is found that in many cases, though by no means all, the compound elastic scattering provides a small isotropic background contribution to the measured differential cross-section.

We now consider the relation between the single-particle description of the potential model and the many-particle description of the compound nucleus model. If the independent-particle picture were an exact description of the nuclear system, all nuclear states would be single-particle states and there would be no coupling between these states. However, the sum of two-nucleon interactions can not be entirely represented by a single-particle potential and the additional residual interactions cause the nuclear states to have a more complicated structure. However, it is always possible in principle to expand the true nuclear state in terms of a complete set of single-particle states. This means that a given single-particle state may contribute to several nuclear states, i.e. the single-particle strength is distributed over several nuclear states, and the extent to which this occurs will depend on the strength of the residual interactions. In the strong-coupling model, the residual interactions are assumed to be so strong that all configurations of the same spin and parity have the same strengths. In the intermediate coupling model, the residual interactions are assumed to be weaker so that, although each single particle state is split into many compound states, these states are spread around the single-particle energy and the amount of mixing of different single-particle states into the same compound state is small.

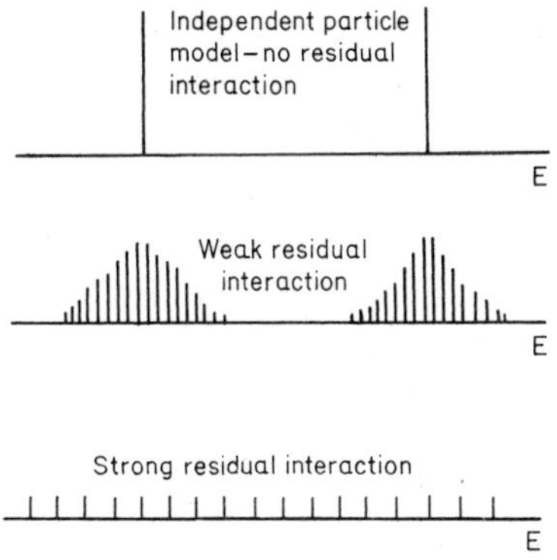

Figure 6.2 Distribution of compound states as a function of energy. The length of each line indicating the position of a compound state is proportional to the reduced width for excitation of that state.

These ideas [6, 70] are illustrated in figure 6.2. Thus we can understand why where are many narrow compound resonances without invoking the strong interactions and immediate formation necessary in Bohr's original model. If we seek to examine a single-particle process involving a particular single-particle state, the strength with which this process occurs at any energy E will depend on the strength with which this single-particle state is admixed into the nuclear states at the energy E, and since in the intermediate coupling model the single-particle strength is spread over a restricted region around the original single particle energy, we may understand why the low resolution experiments reveal the single-particle features of the system.

In the next section we derive formal expressions for the optical potential and the scattering solution. The purpose of this work is to examine the general properties of the optical potential and so to predict what properties may be expected for the phenomenological optical potential which has been used to analyse and correlate a vast amount of elastic scattering data. The form and use of the phenomenological potential are described in section 6.3 and 6.4. Finally, the description of elastic scattering in the high energy limit is discussed in section 6.5, using the impulse approximation.

6.2 Formal theory of the generalized optical potential

It is a general feature of models, such as the compound nucleus model and the optical model, that they allow the description of certain phenomena in terms of certain model parameters, and the purpose of more fundamental nuclear theory is to relate these model parameters to the properties of the many-body system. Although the many-body problem is intractable, the use of a formal theory can provide a connection between particular phenomena and the general properties of the many-body system. The construction of a formal theory of nuclear reactions has been approached in several different ways, and we follow here the unified theory due to Feshbach [71].

We consider the problem of a nucleon scattered by a nucleus, and assume that the incident energy is such that only the elastic channel is open. We have to solve the Schrödinger equation

$$(E-H)\Psi = 0$$

with the Hamiltonian H given by

$$H = H_0 + V + H(\xi) \tag{6.4}$$

where $H_0 = T$ is just the kinetic energy operator for the incident nucleon, V is the interaction potential, and $H(\xi)$ is the internal Hamiltonian of the target nucleus which has co-ordinates ξ. The eigenstates of $H(\xi)$ are $\Phi_\alpha(\xi)$ with energy ε_α, and for convenience we put the ground state energy ε_0 equal to zero. The total wavefunction Ψ may be written as an expansion in the complete set of nuclear states Φ_α

$$\Psi = \sum_\alpha \psi_\alpha \Phi_\alpha \,. \tag{6.5}$$

It is assumed here the incident nucleon is distinguishable, and the problem of antisymmetrization between the incident nucleon and the target nucleon will not be considered.

Because of our initial assumptions, the open channel part of Ψ is just $\psi_0 \Phi_0$, and in order to define the optical potential for this problem we must find the equation for ψ_0. In Feshbach's formalism this is done by defining projection operators P and Q which project on and off the open channels.

Thus
$$P\Psi = \psi_0 \Phi_0 , \quad Q\Psi = (1-P)\Psi . \tag{6.6}$$

Also
$$P^2 \Psi = P\Psi , \tag{6.7}$$
$$PQ\Psi = QP\Psi = 0 . \tag{6.8}$$

The construction of these projection operators depends on the problem under consideration, and in the present case it will be satisfactory to take

$$P = |\Phi_0)(\Phi_0| \tag{6.9}$$

where the round brackets imply summations and integration over the target co-ordinates. The Schrödinger equation is now

$$(E-H)(P+Q)\Psi = 0$$

and premultiplication by P yields

$$(E-PHP)P\Psi = (PHQ)Q\Psi \tag{6.10}$$

while premultiplication by Q yields

$$(E-QHQ)Q\Psi = (QHP)P\Psi \tag{6.11}$$

where we have used equations (6.7) and (6.8). Equation (6.11) may be inverted to give

$$Q\Psi = \frac{1}{E-H_{QQ}} H_{QP} P\Psi \tag{6.12}$$

where $H_{QQ} = QHQ$, $H_{PQ} = PHQ$, etc. (If there are open non-elastic channels, it is necessary to replace $E-H_{QQ}$ in equation (6.12) by $E-H_{QQ}+i\varepsilon$ in order to ensure that $Q\Psi$ has only outgoing waves in these channels.) The equation for $P\Psi$ can be obtained by combining equations (6.12) and (6.10) to give

$$\left(E-H_{PP}-H_{PQ}\frac{1}{E-H_{QQ}}H_{QP}\right)P\Psi = 0 . \tag{6.13}$$

If we now use the form for the projection operator P given in equation (6.9), we have

$$H_{PP} = |\Phi_0)[H_0 + (\Phi_0|V|\Phi_0)](\Phi_0|$$
$$H_{PQ} = |\Phi_0)(\Phi_0|VQ$$

and equation (6.13) becomes

$$\left[E - H_0 - (\Phi_0|V|\Phi_0) - \left(\Phi_0\left|VQ\frac{1}{E-H_{QQ}}QV\right|\Phi_0\right)\right]\psi_0 = 0\,. \quad (6.14)$$

Examination of equation (6.14) leads to the definition of the *generalized optical potential* as

$$V_{\text{opt}} = (\Phi_0|V|\Phi_0) + \left(\Phi_0\left|VQ\frac{1}{E-H_{QQ}}QV\right|\Phi_0\right). \quad (6.15)$$

If V is a local interaction and exchange effects are neglected, the term $(\Phi_0|V|\Phi_0)$ is a single-particle potential which is *local* in configuration space (i.e. it has the form $V(\boldsymbol{r})$) and represents the interaction between the incident nucleon and the target nucleus in its ground state (potential scattering). The second term represents scattering which proceeds through an intermediate excited state of the nucleus, the operator $(E-H_{QQ})^{-1}$ being the propagator for the nucleon within the excited nucleus (see section 3.9). The presence of the operator $(E-H_{QQ})^{-1}$ also causes the potential to be *non-local* in configuration space. This follows from equation (3.103) and means that the representation of V_{opt} in configuration space has the form

$$\langle \boldsymbol{r}'|V_{\text{opt}}|\boldsymbol{r}\rangle = V(\boldsymbol{r})\,\delta(\boldsymbol{r}-\boldsymbol{r}') + K(\boldsymbol{r}, \boldsymbol{r}') \quad (6.16)$$

so that equation (6.14) becomes

$$[E - H_0 - V(\boldsymbol{r})]\,\psi(\boldsymbol{r}) = \int K(\boldsymbol{r}, \boldsymbol{r}')\psi(\boldsymbol{r}')d\boldsymbol{r}'\,. \quad (6.17)$$

It also follows from equation (3.103) that in this special case the optical potential is real, but in general the presence of the operator $(E-H_{QQ}+i\varepsilon)^{-1}$ will cause the potential to be complex.

The second term in equation (6.15) will vary rapidly with energy whenever E is in the vicinity of an eigenvalue of H_{QQ} and will therefore give rise to a marked fluctuation in the cross section in the energy region. This suggests that the occurrence of an eigenvalue of H_{QQ} should be associated with resonance scattering. If the eigenstates of H_{QQ} are given by

$$(E_t - H_{QQ})\Psi_t = 0$$

the propagator can be written, using equation (3.103), as

$$\frac{1}{E-H_{QQ}} = \sum_t \frac{|\Psi_t\rangle\langle\Psi_t|}{E-E_t}$$

where the angular bracket indicates integration and summation over all co-ordinates. For scattering in the vicinity of an isolated resonance at $E = E_s$ the generalized optical potential can be rewritten in the form

$$\begin{aligned} V_{\text{opt}} &= (\Phi_0|V|\Phi_0) + \sum_{t \neq s} \frac{(\Phi_0|VQ|\Psi_t\rangle\langle\Psi_t|QV|\Phi_0)}{E-E_t} \\ &\quad + \frac{(\Phi_0|VQ|\Psi_s\rangle\langle\Psi_s|QV|\Phi_0)}{E-E_s} \\ &= U + \frac{(\Phi_0|VQ|\Psi_s\rangle\langle\Psi_s|QV|\Phi_0)}{E-E_s} \end{aligned} \tag{6.18}$$

The term defined as U will vary slowly with energy and includes the effects of distant resonances. In this way the distant compound resonances combine to give the same effect as a single-particle resonance.

We may define scattering solutions $\chi^{\pm}$ which satisfy the equation

$$(E - H_0 - U)\chi^{\pm} = 0 \tag{6.19}$$

and, using equation (3.124), the wavefunction ψ_0 is given by

$$\psi_0^+ = \chi_0^+ + \frac{1}{E-H_0-U+i\varepsilon} \frac{(\Phi_0|VQ|\Psi_s\rangle\langle\Psi_s|QV|\Phi_0)\psi_0^+}{E-E_s}. \tag{6.20}$$

The transition matrix element is given by equation (3.125) as

$$T_{fi} = \langle\phi|U|\chi_0^+\rangle + \frac{(\Phi_0|VQ|\Psi_s\rangle\langle\Psi_s|QV|\Phi_0)\psi_0^+}{E-E_s} \tag{6.21}$$

where the first term is the transition matrix element for elastic potential scattering. We now manipulate equation (6.20) in an algebraic manner by multiplying on the left with $(E-E_s)\langle\Psi_s|QV|\Phi_0)$ and integrating over all co-ordinates. This gives

$$\left[E - E_s - \langle\Psi_s|QV|\Phi_0) \frac{1}{E-H_0-U+i\varepsilon} (\Phi_0|VQ|\Psi_s\rangle\right] \langle\Psi_s|QV|\psi_0^+\Phi_0\rangle$$
$$= (E-E_s)\langle\Psi_s|QV|\Phi_0\chi_0^+\rangle. \tag{6.22}$$

The coefficient of $\langle \Psi_s | QV | \psi_0^+ \Phi_0 \rangle$ in this expression may be rewritten using the relation

$$\lim_{\varepsilon \to +0} \frac{1}{Z - Z_0 + i\varepsilon} = \frac{\mathscr{P}}{Z - Z_0} - i\pi\delta(Z - Z_0) \tag{6.23}$$

where $\mathscr{P}$ indicates the principal value, and defining the quantities Δ_s and Γ_s as

$$\Delta_s = \langle \Psi_s | QV | \Phi_0) \frac{\mathscr{P}}{E - H_0 - U} (\Phi_0 | VQ | \Psi_s \rangle \tag{6.24}$$

$$\Gamma_s = 2\pi \langle \Psi_s | QV \delta(E - H_0 - U) VQ | \Psi_s \rangle \,, \tag{6.25}$$

so that equation (6.22) becomes

$$[E - E_s - \Delta_s + \tfrac{1}{2} i \Gamma_s] \langle \Psi_s | QV | \psi_0^+ \Phi_0 \rangle = (E - E_s) \langle \Psi_s | QV | \Phi_0 \chi_0^+ \rangle \,,$$

and finally, substituting for $\langle \Psi_s | QV | \psi_0^+ \Phi_0 \rangle$ into equation (6.21) we obtain

$$T_{fi} = \langle \phi | U | \chi_0^+ \rangle + \frac{\langle \chi_0^- \Phi_0 | VQ | \Psi_s \rangle \langle \Psi_s | QV | \Phi_0 \chi_0^+ \rangle}{E - E_s - \Delta_s + \tfrac{1}{2} i F_s} \,.$$

Comparison with equation (5.8) shows that this is just the Breit-Wigner one-level formula, and hence the existence of an eigenvalue of H_{QQ} at E_s gives rise to a resonance at $E_s + \Delta_s$ with width Γ_s.

The energy shift Δ_s and the width Γ_s arise as a result of the coupling of the compound state Ψ_s with the product state for the entrance channel. This follows from the structure of equations (6.24) and (6.25), but can be seen rather more clearly if we examine the structure of the expression for Γ_s within the framework of the shell model. We first expand the scattering function χ^+ defined by equation (6.19) in partial waves in the form

$$\chi^+ = \sum_{lm} \chi_{lm}^+ \,.$$

The χ_{lm} provide a complete set of scattering states so that we may write

$$|\Phi_0) \delta(E - H_0 - U) (\Phi_0| = \sum_{lm} |\Phi_0 \chi_{lm}^+ \rangle \langle \chi_{lm}^+ \Phi_0| \,,$$

and the width Γ_{sl} for the resonance s in the lth partial wave becomes

$$\Gamma_{sl} = 2\pi |\langle \Psi_s | QV | \Phi_0 \chi_{lm}^+ \rangle|^2 \,.$$

Now if the ground state of target nucleus represented by Φ_0 is a closed shell, the product state in the entrance channel, represented by $\chi\Phi_0$ is a one-particle, no-hole state. In contrast the compound state consists of linear combinations of shell model states in which at least one nucleon has been excited. In the framework of the shell model we associate QV with the residual interaction and if this is a two-body interaction the only components of Ψ_s which will connect with the entrance channel are those that differ by a single particle-hole interaction, i.e. the components with two-particle one-hole character. The importance of this result is that it allows us to picture the formation of the compound nucleus as occurring through a succession of collisions of the incident nucleon with the target nucleons. In the first collision a two-particle one-hole state is formed through the action of the residual interactions; such a state is called a *doorway state* since it alone is connected with the incident channel and it is coupled to more complicated particle-hole configurations by successive collisions. Thus we obtain the picture of a nuclear reaction illustrated in figure 6.3 and originally due to Weisskopf.

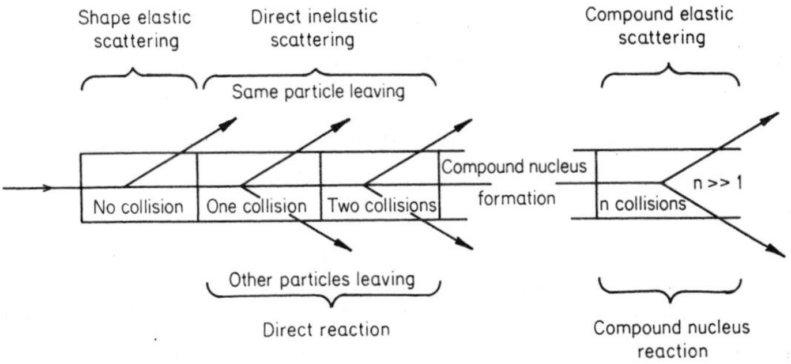

Figure 6.3 Representation, due to Weisskopf, of nuclear reactions in terms of successive collision. [This version from *Physics Education* **4,** 82 (1969).]

The first collision gives rise to a *direct reaction* for which there is still a strong overlap between the excited state and the ground state of the system, while the successive collisions create the compound nucleus through a hierarchy of (n + 1) particle n-hole states [72]. There is now no conflict with the shell model, as the formation of the compound nucleus can be brought about by the weak residual interactions and, in accordance with the optical model, the formation of the compound state does not occur immediately or with a probability of unity. In this picture there is a finite probability that a state can decay back to the preceding state, e.g. the doorway state can decay back to the entrance channel, and this process gives rise to additional resonances known as *intermediate structure* [73] with widths and spacings intermediate between the fine structure of the compound resonances and the gross structure of the giant resonances.

6.3 Properties of the optical potential for nucleons

In analyses of experimental data on elastic scattering of nucleons by nuclei at incident energies below about 300 MeV it is customary to assume a rather general functional form for the optical potential and to determine the parameters as functions of mass number A of the target and incident energy E. It is found that the optical model is more successful for medium and heavy nuclei than for light nuclei. For light nuclei, isolated resonances can affect the scattering process for energies up to 30 MeV, and for very light nuclei the importance of exchange effects renders the optical model of doubtful validity. We summarize in this section the essential features of the phenomenological optical potential for nucleons. The optical potentials for composite particles, such as α-particles and deuterons, are discussed in Chapter 8, and the use of the optical model at high energies is discussed in Chapter 10. Further details are given in several books and review articles [11, 37, 74, 75].

The optical potential for nucleons is usually taken to be local and to consist of the sum of a spin-independent term, a spin-dependent term, a Coulomb term and a symmetry term. The spin-independent term is usually written in the form

$$V(r) = -[Vf(r) + iWg(r)]\,. \tag{6.26}$$

The radial form factors $f(r)$ and $g(r)$ are chosen partly for mathematical convenience and partly on general physical arguments. A satisfactory choice for

$$f(r) = \left[1 + \exp\left(\frac{r-R}{a}\right)\right]^{-1}$$

with radius R and diffuseness a. This functional form is the same as the Fermi distribution for nuclear charge distributions (see Chapter 4) but the radius of the potential in general exceeds the radius of the charge or matter distributions. Since the imaginary part of the optical potential takes account of reactions which remove particles from the elastically scattered beam it is not obvious that it should have the same radial behaviour as the real part. At low energies the Pauli principle severely inhibits nucleon-nucleon collisions inside the nucleus so that the imaginary part of the optical potential may be expected to peak in the nuclear surface but this effect decreases with increasing incident energy. It is common, therefore, to use volume absorption represented by the Saxon-Woods or some similar shape at medium and high energies and surface absorption represented by a surface peaked function at low energies. A general functional form which combines this behaviour is

$$Wg(r) = \left[W - 4W_D \frac{d}{dx'}\right](1+e^{x'})^{-1}, \tag{6.27}$$

where $$x' = (r - R')/a'.$$

In order to fit the polarization observed in elastic nucleon scattering it is necessary to include a spin-dependent term in the optical potential. By analogy with the shell model potential this is usually taken to be a spin-orbit coupling potential of the form

$$V_{so}(r) = +(V_{so} + iW_{so})\frac{b}{r}\frac{d}{dr}h(r)\boldsymbol{l}.\boldsymbol{\sigma} = -(V_{so} + iW_{so})H(r)\boldsymbol{l}.\boldsymbol{\sigma} \tag{6.28}$$

where $h(r)$ is usually taken to be of Saxon-Woods form, i.e.

$$h(r) = (1+e^{x_s})^{-1}$$

with $$x_s = (r - R_s)/a_s. \tag{6.29}$$

The constant b is introduced to give the correct dimensions, and various choices are given in the literature. For the calculation described in section 6.4 the value $b = (\hbar/m_\pi c)^2$ has been used. The spin-orbit potential defined in equation (6.28) is appropriate for the scattering of spin $\frac{1}{2}$ projectiles (i.e. nucleons) from spin zero nuclei. For targets with non-zero spin I the possibility arises of spin-spin interactions involving $\boldsymbol{I}.\boldsymbol{\sigma}$ and of more complicated interactions. At present there is no evidence that such interactions are required to fit the data and they are not normally included in optical model calculations.

For proton scattering it is necessary to include a Coulomb term in the optical potential to take account of the interaction of the incident proton with the charge distribution due to the protons in the nucleus. In almost all cases it is adequate to treat the nucleus as a uniformly charged sphere of radius R_c, so that the Coulomb term is given by

$$\begin{aligned} V_c(r) &= \frac{Z_1 Z_2 e^2}{2R_c}\left(3 - \frac{r^2}{R_c^2}\right) \qquad r < R_c \\ &= \frac{Z_1 Z_2 e^2}{r} \qquad r \geqslant R_c \end{aligned} \tag{6.30}$$

where $Z_1 e$ is the charge of the incident particle and $Z_2 e$ is the charge of the target. A further difference between the optical potentials for protons and neutrons arises through I-spin effects [76] since, if the two-nucleon interaction contains a term of the form $\boldsymbol{t}_i . \boldsymbol{t}_j$, the averaging process implied in the definition of the generalized optical potential will give rise to a term of the form $\boldsymbol{t} . \boldsymbol{T}$ in the real part of the optical potential. The total I-spin of the system can take the values $T \pm \frac{1}{2}$ where T is the I-spin of the target, so that the eigenvalues of $\boldsymbol{t} . \boldsymbol{T}$ are $+\frac{1}{2}T$ and $-\frac{1}{2}(T+1)$. Now for incident neutrons the total spin can take the value $T+\frac{1}{2}$ only (since $t_z = +\frac{1}{2}$ using convention B defined in section 2.2), so that the real part of the potential can be written as

$$\begin{aligned} V_N &= V_0 + V_1 (\boldsymbol{t} . \boldsymbol{T})_N / A \\ &= V_0 + \tfrac{1}{2} T V_1 / A \end{aligned}$$

where V_1 is the strength of the I-spin dependent term and the factor $1/A$ is included for convenience. For protons, both total I-spins values can occur and the coupling coefficients for the two cases yield the weighting factors $(2T+1)^{-1}$ and $2T(2T+1)^{-1}$. Thus the mean potential for the protons is given by

$$\begin{aligned} V_p &= \frac{1}{2T+1}\left[V_0 + \frac{TV_1}{2A}\right] + \frac{2T}{2T+1}\left[V_0 - \frac{(T+1)V_1}{2A}\right] \\ &= V_0 - \tfrac{1}{2} T V_1 / A . \end{aligned}$$

Since the I-spin can be replaced by the neutron excess $(N-Z)$ through the relation $T = \frac{1}{2}(N-Z)$ the real part of the optical potential can be written as

$$V = V_0 \pm \tfrac{1}{4}\left(\frac{N-Z}{A}\right) V_1 \tag{6.31}$$

where the plus sign refers to neutrons and the minus sign to protons. The constant V_1 is found to be ~ 100 MeV and opposite in sign to V_0, and this means that the real potential is stronger for protons than for neutrons. The second term in equation (6.31) is called the symmetry term. It is not yet clear whether the imaginary part of the optical potential also contains a symmetry term (see section 7.5).

For a precise determination of the parameters of the optical potential for a given energy and target a complete range of experimental data is required, e.g. differential cross-section and polarization over a wide range of angles and the absorption cross-section. Very often such a wide range of data is not available, and then certain ambiguities arise. The most important of these is the VR^n-ambiguity, where V and R are the depth and radius, respectively, of the real part of the potential and n is a constant which is about 2; if V and R are varied in such a way that VR^n remains constant the calculated differential cross-section is insensitive to the variation.

From the formal theory of the optical potential we expect the parameters of the phenomenological potential to be energy dependent. It is found that the depth of the real part decreases smoothly as a function of energy and eventually changes sign. The energy at which the real part becomes repulsive is not yet well-defined but seems to be in the region of 200–400 MeV. It is not yet clear whether the shape of the potential also changes in this region. The imaginary part increases with energy up to about 100 MeV, mainly due to the diminishing effect of the exclusion principle, and then remains fairly constant up to 300 MeV. The formal theory also shows that the optical potential is in general non-local. In calculations with a phenomenological potential the effects of this non-locality are studied through the introduction of the range of the non-locality β as a parameter. It is customary to assume that the non-locality occurs only in the spin-independent term and can be described by the same form factors for the real and imaginary parts, although these assumptions are not entirely justified [37]. In an extensive study of neutron elastic scattering, Perey and Buck [77] have shown that data covering the energy range 0·4–24 meV can be fitted with an energy-independent non-local potential of the form

$$V(\boldsymbol{r}, \boldsymbol{r}') = U(\tfrac{1}{2}|\boldsymbol{r}+\boldsymbol{r}'|)\,H(|\boldsymbol{r}-\boldsymbol{r}'|/\beta)$$

where U and H are taken to be of Saxon-Woods and Gaussian form respectively. They have also shown that it is possible to find an equivalent local, but energy-dependent, potential. Thus it appears that a part, at least, of the observed energy dependence of the local optical potential is due to neglect of the non-locality. The local optical potentials for complex projectiles are also energy dependent, and in a study of elastic deuteron scattering Rawitscher [78] has shown that explicit coupling of the stripping cannels (see Chapter 9) can give a non-locality in the deuteron optical potential and an equivalent local potential whose energy dependence is in agreement with experiment.

Many attempts have been made to calculate optical potentials from the fundamental theory. A realistic calculation of this type is an exercise in many-body theory since, as we have seen in section 6.2, the formalism demands a knowledge of the particle-hole configurations involved and also, as discussed in section 2.2, the blocking effect of the Pauli principle must be taken into account at low and medium energies. These complications particularly effect the calculation of the imaginary part of the potential. For the real part, some success has been achieved using the simple formula

$$V_{\text{opt}} = \int V_{\text{eff}}(|\boldsymbol{r}-\boldsymbol{r}'|)\,\rho(\boldsymbol{r}')\,d\boldsymbol{r}'$$

where $\rho(\boldsymbol{r}')$ is the nuclear matter distribution. This expression is obtained from the first term of equation (6.15) by writing the interaction potential as a sum of two-body interactions between the projectile and the target nucleons, and has been applied to the scattering of nucleons [79] and α-particles [80] from nuclei at medium energies. At higher energies, above 100 MeV, the use of impulse approximation leads to a more useful approximation for nucleon optical potentials, and the relevant formulae are derived in section 6.5.

6.4 An example of an optical model analysis

From equations (3.21), (3.30) and (3.31) the optical model wavefunction can be written as

$$\psi = \sum_{lj\lambda} i^l [4\pi(2l+1)]^{\frac{1}{2}} (l0s\mu|jm)(l\lambda s\nu|jm) f_{lj}(kr)\, Y_l^\lambda(\theta,\phi)\,\chi_s^\mu\, e^{i\sigma_l}\,. \tag{6.32}$$

It is convenient to define a radial function $u_{lj}(kr)$ such that

$$u_{lj}(kr) = krf_{lj}(kr)\,. \tag{6.33}$$

If we consider nucleon scattering, so that $j=l\pm\frac{1}{2}$ and the eigenvalues of $\boldsymbol{l}.\boldsymbol{\sigma}$ for these two values of j are l and $-(l+1)$, respectively, the radial equations for the u_{lj} become

$$\left[\frac{d^2}{dr^2} + k^2 - \frac{l(l+1)}{r^2} + Vf(r) + iWg(r) - V_c(r) + (V_{\text{so}} + iW_{\text{so}})h(r)\begin{Bmatrix} l \\ -(l+1)\end{Bmatrix}\right] u_{lj}^{\pm}(kr) = 0 \tag{6.34}$$

where u_{lj}^{+} corresponds to $j=l+\frac{1}{2}$, and u_{lj}^{-} to $j=l-\frac{1}{2}$. It may be noted that the sign of the central imaginary term has been taken to be negative so this term acts in the physically correct manner as a sink of particles. In contrast the imaginary part of the spin-orbit potential acts as a source for one j value and as a sink for the other; some care must be exercised therefore if this term is included, but in most optical model analyses W_{so} is taken to be zero.

At the origin, u_{lj} must be zero since f_{lj} is finite. At asymptotically large distances, u_{lj} behaves as

$$u_{lj}(kr) \to \tfrac{1}{2}kr\left[H_l^{(2)}(kr) + \eta_{lj}^{+} H_l^{(1)}(kr)\right] \tag{6.35}$$

where $H^{(1)}$ and $H^{(2)}$ are the solutions of equation (6.34) without the nuclear potential which have outgoing and incoming character. For neutrons, these are the usual spherical Hankel functions

$$h_l^{(1)} = j_l + i\eta_l\,, \quad h_l^{(2)} = j_l - i n_l\,,$$

while for protons the spherical Bessel functions j_l and n_l are replaced by the regular and irregular Coulomb functions F_l and G_l. The reflection coefficient is determined by numerically integrating the differential equation (6.34) outwards until the nuclear potential is negligible and then matching the function u_{lj} and its derivative to the asymptotic form. Elaborate computer programmes are avaliable which perform these integrations, tabulate the reflection coefficients, and compute, print and plot the differential cross-section, polarization and absorption cross-section corresponding to a given set of potential parameters. Many of these programmes include a search routine which allows a pre-selected group of parameters to be varied in order to minimise the discrepancy between the calculated values and the experimental data. Full details of the computational methods are given in Hodgson's book [74] and in the program reports.

An unusually complete set of data exists for proton scattering from a range of nuclei at around 30 MeV. Figure 6.4 shows the data for the differential cross-sections measured at 30·3 MeV and for the polarization measured at 28·5 MeV. Some of the potential parameters obtained by fitting these data on ^{58}Ni are given in table 6.1. Potential 2 represents a very simple potential in which the radial form factors were fixed and were the same for each part of the potential, there was no surface absorption, and a search was made on only two parameters V and W. For potential 1, all these restraints were relaxed except that the radial parameters of the real central term and the spin-orbit term were kept equal, and for potential 5 all constraints were relaxed and a search made on all parameters. The results calculated from potentials 1 and 5 are shown in figure 6.4 by the full and dashed curves respectively. (It should be noted that many more sets of parameters are given in the papers quoted in table 6.1, and we have not necessarily chosen the ones which give the best fits to the data.) Potentials 3 and 4 do not yield fits

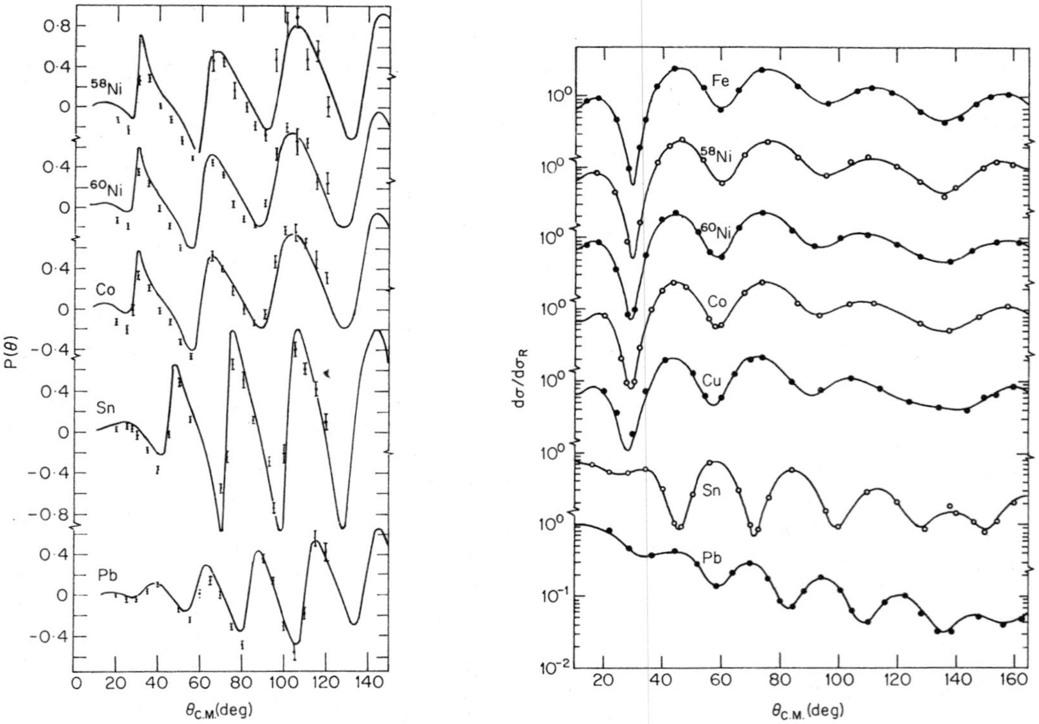

Figure 6.4 The differential cross-sections and polarization for 29 MeV proton scattering from a range of nuclei. The curves through the points are optical model fits to the data. [From G. R. Satchler, *Nucl. Phys.* **A92,** 273 (1967).]

Table 6.1 Parameters of the optical potential for 29 MeV proton scattering from ^{58}Ni.

Potential	V (MeV)	r_0 (fm)	a_0 (fm)	W (MeV)	W_D (MeV)	r' (fm)	a' (fm)	V_{so} (MeV)	r_s (fm)	a_s (fm)
1 (a)	56·6	1·072	0·801	4·11	4·20	1·381	0·497	6·36	1·072	0·801
2 (b)	40·61	1·25	0·65	11·69	0	1·25	0·65	6·0	1·25	0·65
3	40·61	1·25	0·65	0	0	1·25	0·65	6·0	1·25	0·65
4	40·61	1·25	0·65	11·69	0	1·25	0·65	0	1·25	0·65
5 (c)	53·9	1·099	0·764	3·86	4·21	1·364	0·529	7·08	0·952	0·950

The potential is

$$V(r) = -V(1+e^{x})^{-1} - i\left(W - 4W_D\frac{d}{dx'}\right)(1+e^{x'})^{-1}$$

$$+\left(\frac{\hbar}{m_\pi c}\right)^2 V_{so}\frac{1}{r}\frac{d}{dr}(1+e^{x_s})^{-1}\boldsymbol{\sigma}\cdot\boldsymbol{l} + V_c(r)$$

$$x = (r - r_0A^{\frac{1}{3}})/a, \quad x' = (r - r'A^{\frac{1}{3}})/a', \quad x_s = (r - r_sA^{\frac{1}{3}})/a_s$$

REFERENCES

(a) G. R. Satchler, *Nuclear Physics* **A92** 273 (1967), table 2.
(b) R. C. Barrett, A. D. Hill, and P. E. Hodgson, *Nuclear Physics* **62** 133 (1965), table 3.
(c) G. R. Satchler, *Nuclear Physics* **A92** 273 (1967), table 3.

to the data and are used here only to illustrate the effects of different parts of the potential.

The reflection coefficients calculated from potentials 1 and 2 are shown in figure 6.5. Potential 5 yields reflection coefficients which are very close to those given by potential 1, but it can be seen from the figure that there is a marked difference between the values obtained from potentials 1 and 2 especially for low partial waves. These differences show up even more markedly when we consider the behaviour of the optical model wavefunction. Using equations (3.32) to (3.35) the wavefunction can be written in the form

$$|\psi|^2 = |\sum_{lj} A_{lj}(kr)P_l^0(\cos\theta)|^2 + |\sum_{lj} B_{lj}(kr)P_l^1(\cos\theta)|^2 \qquad (6.36)$$

where the first term gives the no-spin-flip component and the second gives the spin-flip component which is zero in the absence of a spin-orbit potential. From this expression we can obtain the behaviour of $|\psi|$ as a function of r (i.e. of distance from the centre of the potential) for various values of the angle θ between the radius vector $\boldsymbol{r}$ and the incident beam direction $\boldsymbol{k}$. Remembering that we have taken the axis of quantization along the incident beam direction, calculation of $|\psi|$ at $\theta = 0°$ and $\theta = 180°$ gives a picture of the

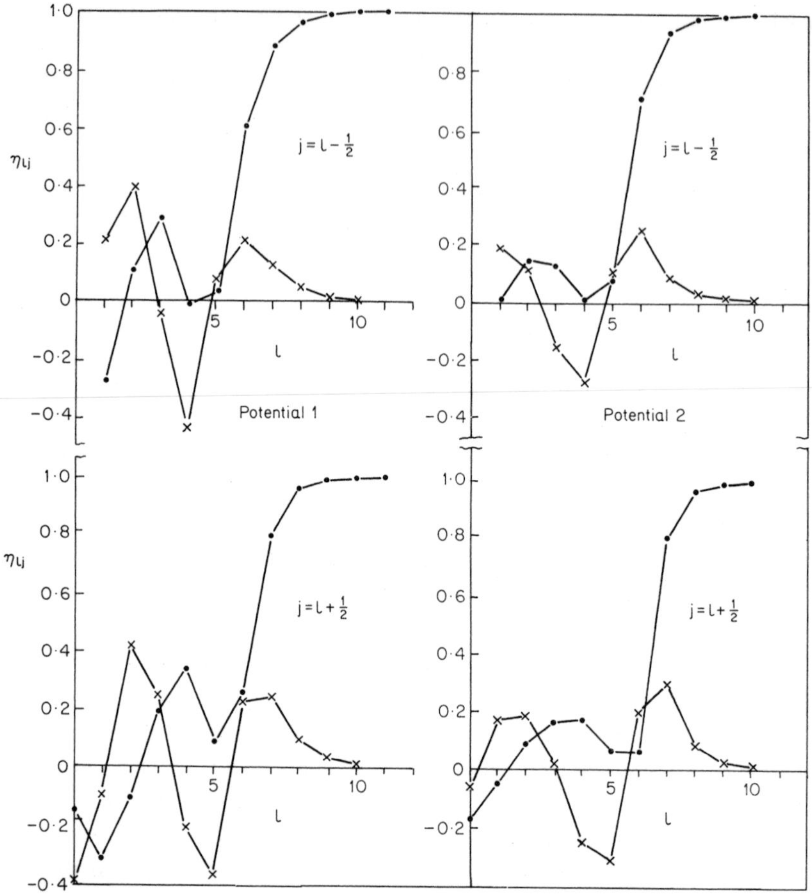

Figure 6.5 Reflection coefficients calculated using the potentials listed in Table 6.1.

optical model wavefunction along the z-axis through the centre of the potential. Results are shown in figure 6.6. The beam is incident from the left so that the left-hand-side of the figure represents the 'bright' side of the nucleus and the right-hand-side represents the 'dark' side, where these terms are used in the sense of geometrical optics. In fact, it can be seen that the refraction and diffraction caused by the potential gives rise to a region of high intensity in the geometrical shadow region. The height of this peak or *focus* [81] depends on the relative strengths of the real and imaginary parts of the potential, and in the case of potential 3 which has no imaginary part the focus rises to a value of $|\psi| = 4.5$. (The normalization is such that $|\psi| = 1$

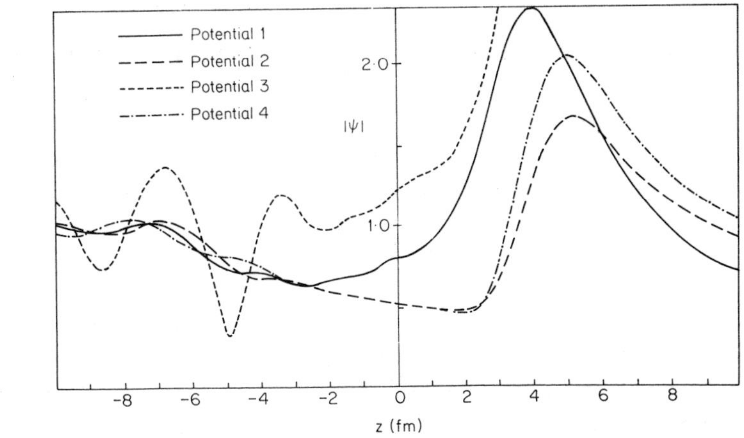

Figure 6.6 The modulus of the wavefunctions predicted by the potentials listed in Table 6.1 calculated along the *z*-axis through the centre of the nucleus.

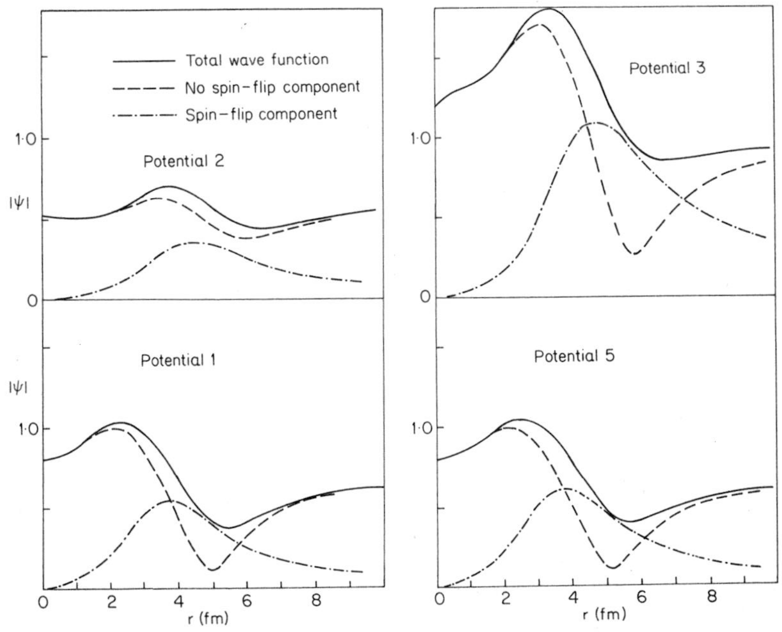

Figure 6.7 The modulus of the wavefunctions predicted by the potentials listed in Table 6.1 calculated at 30° to the positive *z*-axis. The full line represents the total wavefunction while the dashed lines show the no-spin-flip and spin-flip contributions.

for a plane wave.) On the dark side, the wavefunctions calculated from potentials 1 and 2 are substantially different, but on the bright side the behaviour is rather similar, while the wavefunction given by potential 4 differs from that given by potential 2 only in the surface region where the spin-orbit potential is effective. Comparison with the wavefunction given by potential 3 on the bright side illustrates the damping effect of the imaginary part of the potential. Figure 6.7 shows the behaviour of $|\psi|$ (full line) at 30° to the positive z-axis, for several potentials. Comparison of the results for potentials 1, 2 and 3 with those given on the right hand side of figure 6.6 shows that the focussing effect has almost completely disappeared, and in general the focus occurs on the axis and persists for only a few degrees on either side. The dashed curves in figure 6.7 show the modulus of the no-spin-flip and spin-flip terms respectively. The spin-flip term is zero at the origin, as it must be from the basic equation for $|\psi|$, and rises to a peak in the surface region, while the no-spin-flip term has a minimum rather further out. The results given by potentials 1 and 5 are very similar, but the change in the radial behaviour of the spin-orbit potential has caused a very slight shift in the positions of the maxima and minima.

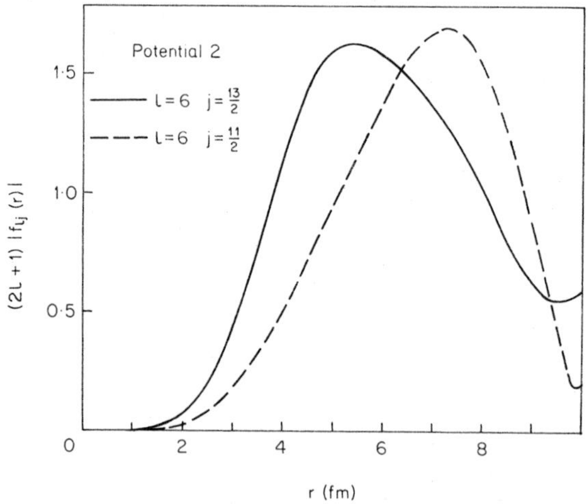

Figure 6.8 Comparison of the radial parts of the wavefunction for $j=l-\frac{1}{2}$ and $j=l+\frac{1}{2}$.

In figure 6.8 we have plotted the radial part of the wavefunction for a particular partial wave in the form $(2l+1)f_l(kr)$. From equation (6.34) it can be seen that for $j=l+\frac{1}{2}$ the spin-orbit potential has the same sign as the real central potential so that the effective potential is increased, whereas for $j=l-\frac{1}{2}$ the reverse is true and the potential is decreased. We would, therefore, expect that the partial wave with $j=l+\frac{1}{2}$ is pulled in towards the nucleus more than the partial wave with $j=l-\frac{1}{2}$ because of the stronger attractive potential. This behaviour is clearly displayed in figure 6.8. Also, because the partial wave with $j=l+\frac{1}{2}$ is pulled in to the nucleus we would expect a strong absorptive effect and a smaller reflection coefficient than for the partial wave with $j=l-\frac{1}{2}$, and this is confirmed by the results shown in figure 6.5.

Some pictures of the optical model wavefunctions for α-particles are given in Chapter 8. These pictures are included to give some insight into the scattering phenomena as described by the optical model. It must be remembered, however, that the quantities calculated in an optical model analysis depend on the asymptotic behaviour of the wavefunction and it is quite possible that none of the wavefunctions plotted in these figures represent the true variation of intensity in the interior of the nucleus. The question whether other processes, apart from elastic scattering, can be used to investigate the interior region and thereby to determine a preferred set of optical potential parameters is discussed in later chapters.

6.5 Impulse approximation for elastic scattering and the optical potential

At energies above 100 MeV the lifetime of the compound states is of the same order as the transit time and the compound states are therefore not expected to play any role in the scattering process. At these energies, the energy of the projectile is considerably in excess of the kinetic energy or the binding energy of the target nucleons, at least in light nuclei, so that the requirements for the validity of impulse approximation may be satisfied.

In impulse approximation the transition matrix element for elastic scattering is given by

$$T_{fi} = \langle \phi | T^{+} | \phi \rangle \tag{6.37}$$

where

$$T^{+} = (\Phi_0 | \sum_j \tau(j) | \Phi_0) \tag{6.38}$$

and we have used equations (3.115) and (3.156). In order to evaluate this matrix element we use the density function $\rho(\boldsymbol{r}, \boldsymbol{r}')$, defined as

$$\rho(\boldsymbol{r}, \boldsymbol{r}') = \int \Phi_0^*(\boldsymbol{r}', \boldsymbol{r}_2, \ldots \boldsymbol{r}_A)\, \Phi_0(\boldsymbol{r}, \boldsymbol{r}_2, \ldots \boldsymbol{r}_A)\, d\boldsymbol{r}_2 \ldots d\boldsymbol{r}_A\,, \tag{6.39}$$

so that $\rho(\boldsymbol{r})$ is the nuclear density in the ground state, and the form factor is

$$F(q) = \int e^{i\boldsymbol{q}.\boldsymbol{r}} \rho(\boldsymbol{r}) d\boldsymbol{r} \,, \tag{6.40}$$

where $\boldsymbol{q}$ is the momentum transferred by the projectile in scattering from momentum $\boldsymbol{k}$ to momentum $\boldsymbol{k}'$, i.e.

$$\boldsymbol{q} = \boldsymbol{k} - \boldsymbol{k}' \,. \tag{6.41}$$

We take the momenta of the jth nucleon in the target to be $\boldsymbol{k}_j$, $\boldsymbol{k}'_j$ before and after collision so that the transition matrix element becomes

$$\begin{aligned} T_{fi}(\boldsymbol{k}', \boldsymbol{k}) &= \sum \int d\boldsymbol{k}_j d\boldsymbol{k}'_j \delta(\boldsymbol{k} + \boldsymbol{k}_j - \boldsymbol{k}' - \boldsymbol{k}'_j) \\ &\quad \times \int d\boldsymbol{r}_j d\boldsymbol{r}'_j \exp(-i\boldsymbol{k}_j . \boldsymbol{r}_j) \exp(i\boldsymbol{k}'_j . \boldsymbol{r}'_j) \rho(\boldsymbol{r}_j, \boldsymbol{r}'_j) \langle \boldsymbol{k}', \boldsymbol{k}'_j | \tau(j) | \boldsymbol{k}, \boldsymbol{k}_j \rangle \\ &= A \int d\boldsymbol{k}_1 \int d\boldsymbol{r}_1 d\boldsymbol{r}'_1 \exp[-i\boldsymbol{k}_1 . (\boldsymbol{r} - \boldsymbol{r}'_1)] \exp[i(\boldsymbol{k} - \boldsymbol{k}') . \boldsymbol{r}'_1] \rho(\boldsymbol{r}_1, \boldsymbol{r}'_1) \\ &\quad \times \langle \boldsymbol{k}', \boldsymbol{q} + \boldsymbol{k}_1 | \tau(1) | \boldsymbol{k}, \boldsymbol{k}_1 \rangle \end{aligned} \tag{6.42}$$

where the δ-function expresses momentum conservation in the two-nucleon collision and we have, for the moment, ignored the dependence of $\tau(j)$ on spin and I-spin. We now make the crucial assumption in this development, namely that the dependence of the two-nucleon scattering amplitude on $\boldsymbol{k}_1$, the momentum of the struck nucleon, can be neglected [36]. This is equivalent to the assumption that the kinetic energy of the target nucleon is negligible compared with that of the incident nucleon. The momentum $\boldsymbol{k}_1$ then appears only in the factor $\exp[-i\boldsymbol{k}_1 . (\boldsymbol{r}_1 - \boldsymbol{r}'_1]$ so that integration over $\boldsymbol{k}_1$ yields $(2\pi)^3 \delta(\boldsymbol{r}_1 - \boldsymbol{r}'_1)$, and the transition matrix element reduces to

$$\begin{aligned} T_{fi}(\boldsymbol{k}', \boldsymbol{k}) &= A(2\pi)^3 \langle \boldsymbol{k}', \boldsymbol{q} | \tau | \boldsymbol{k}, 0 \rangle \int d\boldsymbol{r}\, e^{i\boldsymbol{q}.\boldsymbol{r}} \rho(\boldsymbol{r}) \\ &= A(2\pi)^3 \langle \boldsymbol{k}', \boldsymbol{q} | \tau | \boldsymbol{k}, 0 \rangle F(q) \,. \end{aligned} \tag{6.43}$$

Thus the assumption made above brings about a factorization of the transition matrix element into the product of the matrix element for two-nucleon scattering and the form factor which describes the overlap of the nuclear wavefunctions.

In nucleon–nucleon scattering the relation between $|\boldsymbol{k}|$ and $|\boldsymbol{k}'|$ is determined by the kinematics of the nucleon–nucleus system which are different from those of the two-nucleon system, i.e. the matrix element for two-nucleon scattering required in equation (6.43) is off the energy shell for two-nucleon scattering. However, at the energies in question, the two-nucleon scattering amplitude varies rapidly with q^2 but only slowly with energy, it is therefore assumed that the matrix element $\langle k', q|\tau|k, 0\rangle$ can be treated as a function of momentum transfer alone [82]. We write this function as $\tau(E, q^2)$, and it can be connected directly to the two-nucleon scattering amplitude M by the relation

$$M(E, q^2) = \frac{-\mu_0}{2\pi\hbar^2}(2\pi)^3\,\tau(E, q^2) \tag{6.44}$$

where μ_0 is the reduced mass for two nucleon scattering. The scattering amplitude for elastic nucleon-nucleus scattering in the impulse approximation then becomes

$$\begin{aligned} f(\theta, \phi) &= \frac{-\mu_A}{2\pi\hbar^2}\,T_{fi}(\mathbf{k}', \mathbf{k}) \\ &= A\frac{\mu_A}{\mu_0}M(E, q^2)F(q) = \frac{2A^2}{A+1}M(E, q^2)F(q)\,, \end{aligned} \tag{6.45}$$

where we have used the usual relations for the reduced masses in terms of the nucleon mass m,

$$\mu_0 = \tfrac{1}{2}m\,, \quad \mu_A = Am/(A+1)\,.$$

The scattering amplitude M is given by equation (3.56) which we rewrite using the suffices o and j, to indicate the incident and target nucleon respectively. Thus

$$\begin{aligned} M_j(E, q^2) = {} & A(q) + B(q)(\boldsymbol{\sigma}_o.\hat{\mathbf{n}})(\boldsymbol{\sigma}_j.\hat{\mathbf{n}}) + C(q)(\boldsymbol{\sigma}_o + \boldsymbol{\sigma}_j).\hat{\mathbf{n}} \\ & + E(q)(\boldsymbol{\sigma}_o.\hat{\mathbf{q}})(\boldsymbol{\sigma}_j.\hat{\mathbf{q}}) + F(q)(\boldsymbol{\sigma}_o.\hat{\mathbf{P}})(\boldsymbol{\sigma}_j.\hat{\mathbf{P}}) \end{aligned} \tag{6.46}$$

where $\hat{\mathbf{n}}$, $\hat{\mathbf{q}}$, $\hat{\mathbf{P}}$ are as defined in equations (3.47), (3.54), and (3.55). The spin-dependence of M is already explicit, and to make the I-spin dependence similarly explicit it is customary to express each coefficient in the form

$$A(q) = A_\alpha(q) + A_\beta(q)\mathbf{t}_o.\mathbf{t}_j \tag{6.47}$$

where $\mathbf{t}_o$, $\mathbf{t}_j$ are the I-spin operators for nucleons o and j. The matrix element of $A(q)$ taken between $T=0$ states of the colliding nucleons is

$$A_0 = \langle T=0|A_\alpha + A_\beta\mathbf{t}_o.\mathbf{t}_j|\,T=0\rangle = A_\alpha - 3A_\beta$$

and similarly the matrix element for $T=1$ states is

$$A_1 = A_\alpha + A_\beta\,.$$

Since p–p and n–n scattering occurs only in relative $T=1$ states we can set

$$A_1 = A_{\rm pp} = A_{\rm nn}$$

where $A_{\rm pp}$, $A_{\rm nn}$ are the relevant coefficients for p–p and n–n scattering. (It is assumed that Coulomb scattering is treated separately and that the nucleon–nucleon interaction is charge independent.) The n–p system may exist in $T=0$ and $T=1$ states with equal weight, so that

$$A_{\rm np} = \tfrac{1}{2}(A_1 + A_0)\,.$$

Combining these expressions we have

$$A_\alpha = \tfrac{1}{4}(3A_1 + A_0) = \tfrac{1}{2}(A_{\rm pp} + A_{\rm np}) \tag{6.48}$$

$$A_\beta = \tfrac{1}{4}(A_1 - A_0) \;= \tfrac{1}{2}(A_{\rm pp} - A_{\rm np})\,. \tag{6.49}$$

The factorization of the transition matrix element does not now occur until the matrix elements of the spin and I-spin operators have been evaluated. Thus the scattering amplitude (6.45) must be replaced by

$$f(\theta,\phi) = \frac{\mu_A}{\mu_0}\,\overline{M}(E,q^2) = \frac{2A}{A+1}\,\overline{M}(E,q^2) \tag{6.50}$$

where

$$\overline{M}(E,q^2) = \langle \Phi_0 | \sum_0 e^{i\boldsymbol{q}.\boldsymbol{r}_j} M_j(E,q^2) | \Phi_0 \rangle\,. \tag{6.51}$$

Finally, using equations (3.42) and (3.50) the cross-section for elastic scattering is given by

$$\frac{d\sigma}{d\Omega} = \frac{1}{2(2J+1)}\left(\frac{2A}{A+1}\right)^2 \text{Trace}\,(\overline{M}^\dagger\,\overline{M}) \tag{6.52}$$

and the polarization normal to the plane of scattering is given by

$$P = \text{Trace}(\overline{M}^\dagger \boldsymbol{\sigma}_0 . \hat{\boldsymbol{n}} \overline{M})/\text{Trace}(\overline{M}^\dagger\,\overline{M})\,. \tag{6.53}$$

For nuclei with $A=2Z$, $T=0$ and $J=0$, the sum over the terms in M_j which are linear in $\boldsymbol{\sigma}_j$ and $\boldsymbol{t}_j$ is zero, and the cross-section and polarization are given by the general formulae

$$\frac{d\sigma}{d\Omega} = \left(\frac{2A}{A+1}\right)^2 A^2[|A_\alpha(q)|^2 + |C_\alpha(q)|^2]F^2(q)$$

$$P = 2\,\mathrm{Re}(A_\alpha C_\alpha^*)/[|A_\alpha|^2 + |C_\alpha|^2]\,.$$

Thus the polarization is independent of the form factor and comparison with the experimental data provides a test of the parameters of the two-nucleon scattering amplitude. In general the agreement is satisfactory for momentum transfer up to about $\boldsymbol{q} \sim 1\ \mathrm{fm}^{-1}$ (which for light nuclei means scattering angles up to about 25° at 150 MeV). For light nuclei in which the proton and neutron distributions are probably identical, the form factor can be obtained from fits to elastic electron scattering (see Chapter 4). Comparison with the data for elastic proton scattering then provides a test of impulse approximation, and shows that the approximation has reasonable validity at low momentum transfer and incident energies above 100 MeV. For nuclei with $J \neq 0$ or $T \neq 0$, the formulae become more complicated and the polarization is no longer independent of the form factor. A general formalism with which the matrix element (6.51) can be evaluated has been given by Kerman, McManus and Thaler [36].

We may use the formal theory of section 3.10 to derive an optical potential from the transition matrix of the impulse approximation. We consider here only the Born approximation $T \sim V$, and using equations (3.128) and (6.50) we have

$$V_{\mathrm{opt}}(\boldsymbol{r}) = \frac{-2\pi\hbar^2}{\mu_0}\frac{1}{(2\pi)^3}\int e^{-i\boldsymbol{q}.\boldsymbol{r}}\,\overline{M}(E, q^2)\,d\boldsymbol{q}\,,$$

and hence for a nucleus with $J = 0$ and $T = 0$ the potential is

$$V_{\mathrm{opt}}(\boldsymbol{r}) = \frac{-\hbar^2 A}{(2\pi)^2\mu_0}\int e^{-i\boldsymbol{q}.\boldsymbol{r}}[A_\alpha(q) + C_\alpha(q)\boldsymbol{\sigma}_0 . \boldsymbol{n}]F(q)\,d\boldsymbol{q}\,. \qquad (6.54)$$

The spin-independent part of the equation (6.54) can be compared directly with general form for the optical potential given by equation (6.26). This gives

$$Vf(r) = \frac{+\hbar^2 A}{(2\pi)^2\mu_0}\int e^{-i\boldsymbol{q}.\boldsymbol{r}}\,\mathrm{Re}\,A_\alpha(q)F(q)\,d\boldsymbol{q} \qquad (6.55)$$

$$Wg(r) = \frac{\hbar^2 A}{(2\pi)^2\mu_0}\int e^{-i\boldsymbol{q}.\boldsymbol{r}}\,\mathrm{Im}\,A_\alpha(q)F(q)\,d\boldsymbol{q} \qquad (6.56)$$

For heavy nuclei the quantity $A_\alpha(q)$ varies slowly compared with the form factor $F(q)$ so that the optical potential is given approximately by

$$Vf(r) \approx \frac{A\hbar^2}{(2\pi)^2\mu_0} \text{Re } A_\alpha(0) \int e^{-i\boldsymbol{q}.\boldsymbol{r}} F(q)\, d\boldsymbol{q} \approx A \frac{2\pi\hbar^2}{\mu_0} \text{Re } A_\alpha(0)\, \rho(r) \quad (6.57)$$

$$Wg(r) \approx A \frac{2\pi\hbar^2}{\mu_0} \text{Im } A_\alpha(0)\, \rho(r) . \quad (6.58)$$

In order to make a similar comparison for the spin-dependent term we first manipulate the Born amplitude given by the phenomenological spin-orbit potential (6.28). The scattering amplitude is

$$f^{\text{B}}_{\text{so}} = \frac{-\mu_A}{2\pi\hbar^2}(+b)(V_{\text{so}}+iW_{\text{so}}) \int e^{i(\boldsymbol{k}-\boldsymbol{k}').\boldsymbol{r}} \frac{1}{r}\frac{dh(r)}{dr} \boldsymbol{\sigma}.(\boldsymbol{r} \wedge \hbar\boldsymbol{k})d\boldsymbol{r}$$

$$= \frac{-\mu_A}{2\pi\hbar}(+b)(V_{\text{so}}+iW_{\text{so}}) \int e^{i(\boldsymbol{k}-\boldsymbol{k}').\boldsymbol{r}} \boldsymbol{\sigma}.(\nabla h(r) \wedge \boldsymbol{k})d\boldsymbol{r} .$$

But replacing $\boldsymbol{\sigma}.\nabla h \wedge k$ by $\boldsymbol{k}.\boldsymbol{\sigma} \wedge \nabla h$ and integrating by parts we have

$$f^{\text{B}}_{\text{so}} = \frac{\mu_A}{2\pi\hbar}(+b)(V_{\text{so}}+iW_{\text{so}})\boldsymbol{k}.\boldsymbol{\sigma} \wedge \int \nabla [e^{i(\boldsymbol{k}-\boldsymbol{k}').\boldsymbol{r}}]h(r)d\boldsymbol{r}$$

$$= \frac{\mu_A}{2\pi\hbar}(+b)(V_{\text{so}}+iW_{\text{so}})\boldsymbol{k}.\boldsymbol{\sigma} \wedge i(\boldsymbol{k}-\boldsymbol{k}') \int e^{i\alpha.\boldsymbol{r}} h(r)d\boldsymbol{r} .$$

Now $$\boldsymbol{k}.\boldsymbol{\sigma} \wedge (\boldsymbol{k}-\boldsymbol{k}') = k^2 \sin\theta\, \boldsymbol{\sigma}.\hat{\boldsymbol{n}}$$

and hence $$f^{\text{B}}_{\text{so}} = \frac{-\mu_A}{2\pi\hbar^2} \int e^{i\boldsymbol{q}.\boldsymbol{r}} [-ib(V_{\text{so}}+iW_{\text{so}})k^2 \sin\theta\, h(r)\boldsymbol{\sigma}.\hat{\boldsymbol{n}}]d\boldsymbol{r} . \quad (6.59)$$

Comparing the quantity in square brackets in equation (6.59) with the spin-dependent part of V_{opt} given by equation (6.54) we have

$$bW_{\text{so}}k^2 \sin\theta\, h(r) = \frac{-\hbar^2 A}{(2\pi)^2\mu_0} \int e^{-i\boldsymbol{q}.\boldsymbol{r}} F(q) \text{Re } C_\alpha(\boldsymbol{q})d\boldsymbol{q} \quad (6.60)$$

$$-bV_{\text{so}}k^2 \sin\theta\, h(r) = \frac{-\hbar^2 A}{(2\pi)^2\mu_0} \int e^{-i\boldsymbol{q}.\boldsymbol{r}} F(q) \text{Im } C_\alpha(q)d\boldsymbol{q} \quad (6.61)$$

and for heavy nuclei

$$bW_{\text{so}}k^2 \sin\theta\, h(r) \approx -A \frac{2\pi\hbar^2}{\mu_0} \text{Re } C_\alpha(0)\, \rho(r) \quad (6.62)$$

$$bV_{so}k^2 \sin\theta\, h(r) \approx A\frac{2\pi\hbar^2}{\mu_0} \operatorname{Im} C_\alpha(0)\,\rho(r)\,. \tag{6.63}$$

Thus we obtain, through the impulse approximation, an optical potential of the same general form as the phenomenological potential. For heavy nuclei the radial behaviour of this potential approaches that of the nuclear density distribution, but for light nuclei the variation of the two-nucleon scattering amplitude with momentum transfer is as important as the variation of the form factor (see section 10.1) and, since the real and imaginary part of $A(q)$ and of $C(q)$ do not necessarily have the same q-dependence, this means that the radial behaviour of the real and imaginary parts of the central and spin-orbit parts of the potential may differ from each other and from the density distribution. The presence of the $\boldsymbol{t}_0.\boldsymbol{t}_j$ term in equation (6.47) ensures that the potential contains an I-spin dependent term.

7 Direct inelastic scattering

7.1 Inelastic scattering and nuclear models

The study of direct inelastic scattering from nuclei has proved to be a major source of information on nuclear structure and nuclear spectroscopy. The energy spectra reveal the existence and location of excited states of the nucleus, while the shape and magnitude of the differential cross-section for the excitation of a particular state gives information about the properties of this state and its mode of excitation. In this chapter we are concerned with the interpretation and analysis of these differential cross-sections.

In order to make theoretical predictions which may be compared with the experimental data on direct reactions, it is necessary to introduce a number of simplifying assumptions to reduce the many-body problem to a tractable form. In section 3.11 we introduced formally the approximations which can be made in the description of the interaction between the projectile and target nucleus. In the present chapter, it will also be necessary to assume definite models for the target nucleus in order to construct wavefunctions for particular initial and final states of the target nucleus. In a sense, there are now two uncertainties in the calculations, the choice of the effective interaction and the choice of the nuclear model, and in some cases it may be convenient to discuss these points separately, but they are certainly not independent of each other as they arise from the same many-body problem. If, in the higher energy range, it is permissible to use the impulse approximation there is then no uncertainty in the interaction which is determined by the parameters of the free interaction. A series of experiments on inelastic proton scattering in the energy range 150–185 MeV and the interpretation of these data in terms of impulse approximation have yielded a substantial and valuable amount of information but the problem of obtaining good energy resolution in this energy region has so far placed a limit on the information which can be obtained. In using the impulse approximation we effectively make two successive approximations to the many-body transition operator, i.e.

$$T \rightarrow \sum_j t(j) \rightarrow \sum_j \tau(j), \tag{7.1}$$

where $\tau(j)$ is the free two-nucleon transition operator and $t(j)$ is the effective two-nucleon operator inside the nucleus (see section 3.11). For incident energies below 100 MeV it is too inaccurate to use $\tau(j)$ and it is essential to take account of the effect of the nuclear medium in which the two nucleons scatter and the effect of the exclusion principle in blocking intermediate states which are occupied by other nucleons. It is, however, extremely convenient to describe the interaction in terms of a sum of single scatterings involving one target nucleon at a time, and the essence of the *microscopic* description of inelastic scattering at medium energies is the assumption that the transition operator $t(j)$ can be approximated by a local effective interaction. Both the impulse approximation and microscopic approximation are appropriate for use in conjunction with nuclear wavefunctions which are based on microscopic models such as the shell model or the Nilsson model, and the detailed application of these approximations is described in sections 7.2 and 7.3.

Although a direct reaction is considered to involve a rather limited disturbance of the nucleus so that there remains a good overlap between the initial and final nuclear wavefunctions, the reaction need not necessarily be thought of in single-particle terms. The collective model of the nucleus allows for a very simple excitation of the nucleus through a change in the collective state of the system. In many nuclei there are vibrational states or rotational states (see Chapter 2) which lie closer to the ground state than those states involving single-particle excitations and which are very strongly excited in inelastic scattering. The excitation of these states is most commonly represented by means of a *macroscopic* model, the *extended optical model*, which is described in section 7.3. The strong excitation of these collective states implies that they are strongly coupled to the ground state of the nucleus, and in this case it may not be sufficiently accurate to use distorted wave Born approximation to calculate the cross-section for excitation of one of these states. The more accurate *strong coupling approximation* is described in section 7.4.

Throughout this chapter, general methods for the analysis of direct inelastic scattering are described, which are applicable to all types of projectiles. The special methods and approximations appropriate for strongly absorbed projectiles are described in Chapter 8. Many of the methods used in the description of inelastic nucleon scattering can be very naturally extended to cover the charge-exchange (p, n) reaction and for this reason this reaction is included in this chapter and discussed in section 7.5.

7.2 Impulse approximation for inelastic scattering

In impulse approximation the transition matrix element for inelastic scattering from the initial state i to a definite final state f is given by equation (3.155) as

$$T_{fi}(\boldsymbol{k}', \boldsymbol{k}) = \langle \Phi_f \chi_f^- | \sum_j \tau(j) | \Phi_i \chi_i^+ \rangle \tag{7.2}$$

where $\chi^\pm$ are the distorted wavefunctions for the projectile. In order to express this matrix element in terms of the two-nucleon scattering amplitude, as was done in section 6.5 for elastic scattering, we expand the distorted wavefunctions in terms of plane waves states,

$$\chi^+(\boldsymbol{k}, \boldsymbol{r}) = \int a(\boldsymbol{p}) e^{i\boldsymbol{p}.\boldsymbol{r}} d\boldsymbol{p}$$

$$\chi^-(\boldsymbol{k}', \boldsymbol{r}) = \int a(\boldsymbol{p}') e^{-i\boldsymbol{p}'.\boldsymbol{r}} d\boldsymbol{p}'. \tag{7.3}$$

Equation (7.2) now becomes

$$\begin{aligned} T_{if}(\boldsymbol{k}', \boldsymbol{k}) = \sum_j &\int d\boldsymbol{k}_j d\boldsymbol{k}'_j \delta(\boldsymbol{p} + \boldsymbol{k}_j - \boldsymbol{p}' - \boldsymbol{k}'_j) \\ &\times \int d\boldsymbol{r}_j d\boldsymbol{r}'_j \exp(-i\boldsymbol{k}_j . \boldsymbol{r}_j) \exp(i\boldsymbol{k}'_j . \boldsymbol{r}'_j) \rho_{if}(\boldsymbol{r}_j, \boldsymbol{r}'_j) \\ &\times \int d\boldsymbol{p}\, d\boldsymbol{p}'\, a(\boldsymbol{p}) a(\boldsymbol{p}') \langle \boldsymbol{p}', \boldsymbol{k}'_j | \tau(j) | \boldsymbol{p}, \boldsymbol{k}_j \rangle \end{aligned} \tag{7.4}$$

where ρ_{if} is the transition density. In order to simplify this expression we must assume, as before, that the dependence of the matrix element $\langle \boldsymbol{p}', \boldsymbol{p} - \boldsymbol{p}' + \boldsymbol{k}_j | \tau(j) | \boldsymbol{p}, \boldsymbol{k}_j \rangle$ on $\boldsymbol{k}_j$ can be neglected so that integration over $\boldsymbol{k}_j$ again yields $(2\pi)^3 \delta(\boldsymbol{r}_j - \boldsymbol{r}'_j)$, but *in addition* we must assume that the variation of the matrix element for the two-nucleon scattering with the momenta $\boldsymbol{p}$ and $\boldsymbol{p}'$ contained in the distorted waves is sufficiently small so that the matrix element can be replaced by a mean value. We take this mean value to be the value corresponding to the free momenta $\boldsymbol{k}$ and $\boldsymbol{k}'$. With these assumptions, equation (7.4) reduces to

$$T_{fi}(\boldsymbol{k}', \boldsymbol{k}) = \frac{-2\pi\hbar^2}{\mu_0} \sum_j \int d\boldsymbol{r}_j \chi_f^{-*}(\boldsymbol{k}', \boldsymbol{r}_j) M_j(E, q^2) \rho_{if}(\boldsymbol{r}_j) \chi_i^+(\boldsymbol{k}, \boldsymbol{r}_j) \quad (7.5)$$

where we have used equation (6.44). It may be noted that in order to achieve this result for inelastic scattering it is necessary to make a more powerful assumption than was required in the case of elastic scattering [83]. The structure of the final form of the matrix element, given by equation (7.5), appears as if we had replaced the interaction $\tau(j)$ in equation (7.3) by $\tau_j(E, q^2)\,\delta(\boldsymbol{r}-\boldsymbol{r}_j)$. For this reason the matrix element (7.5) is sometimes referred to as the zero-range form of DWIA for inelastic scattering, although the q^2 dependence of $\tau_j(E, q^2)$ means that a large part of finite range behaviour of the two-nucleon interaction has been taken into account. The corrections to this approach due to full finite-range effects have been discussed by Haybron [84].

The scattering amplitude for inelastic scattering can be written in the form

$$f(\theta, \phi) = \frac{\mu_A}{\mu_0} \langle \Phi_f \chi_f^- | \sum_j M_j(E, q^2) | \chi_i^+ \Phi_i \rangle \quad (7.6)$$

$$= \left(\frac{2A}{A+1} \right) G(q), \quad (7.7)$$

where comparison of these two equations defines $G(q)$ and we have assumed that the energy loss is negligible compared with the incident energy. The cross-section for inelastic scattering in distorted wave impulse approximation (DWIA) is then given by

$$\left(\frac{d\sigma}{d\Omega} \right)_{fi} = \frac{1}{2(2J_i+1)} \left(\frac{2A}{A+1} \right)^2 \operatorname{Trace}(G^\dagger G) \quad (7.8)$$

and the polarization normal to the scattering plane is given by

$$P = \operatorname{Trace}(G^\dagger \boldsymbol{\sigma}_0 . \hat{\boldsymbol{n}}\, G)/\operatorname{Trace}(G^\dagger G) . \quad (7.9)$$

Plane wave impulse approximation (PWIA) is obtained by replacing the distorted wavefunctions in equation (7.6) by plane waves, i.e. by replacing $\chi_f^{-*}\chi^+$ by $e^{i\boldsymbol{q}.\boldsymbol{r}_j}$. In PWIA the expressions for the polarization have a particularly simple form if only one value of angular momentum transfer is allowed and if one state has $T=0$. For example, for transitions in ^{6}Li we obtain the following formulae:

$$1^+ \to 3^+,\ \Delta T=0:\ P = \frac{2[\mathrm{Re}(A_\alpha C_\alpha^*) + \frac{5}{7}\mathrm{Re}(B_\alpha C_\alpha^*)]}{|A_\alpha|^2 + |C_\alpha|^2 + \frac{4}{7}|E_\alpha|^2 + \frac{5}{7}(|B_\alpha|^2 + |C_\alpha|^2 + |F_\alpha|^2)}$$

$$1^+ \to 0^+,\ \Delta T=1:\ P = \frac{2\,\mathrm{Re}(B_\beta C_\beta^*)}{|B_\beta|^2 + |C_\beta|^2 + |E_\beta|^2 + |F_\beta|^2}.$$

where the coefficients A_α, etc., are defined in equations (6.46)–(6.49). These formulae reflect the general result that if one state has $T=0$, then for transitions with $\Delta T=0$ only the coefficients labelled α contribute while for transitions with $\Delta T=1$ only the coefficients labelled β contribute. This leads to a substantial difference in the predicted values of the polarization for the two types of transition. The numerical coefficients associated with the combinations of coefficients $|B|^2+|C|^2+|F|^2$ and with $|E|^2$ depend on the

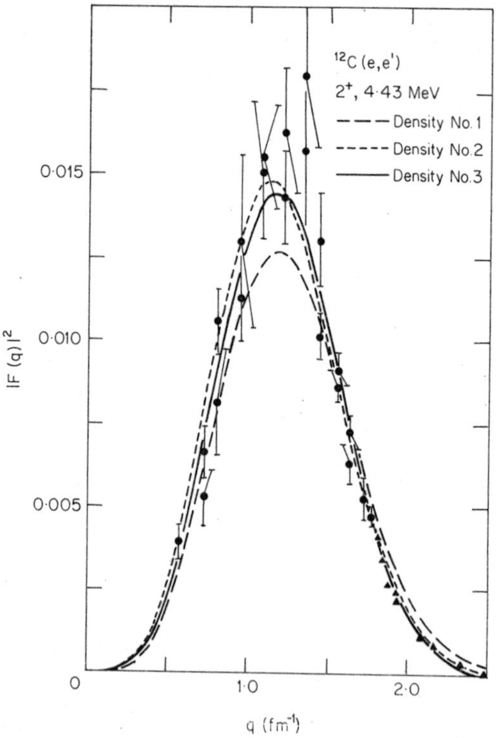

Figure 7.1 (**a**) Fits to the form factor for excitation of the lowest 2^+ state in ^{12}C using various parametrizations of the transition density.

coupling scheme used to construct the nuclear wavefunctions [36]. The values given above arise from the use of LS coupled wavefunctions for ^{6}Li.

Many calculations using DWIA have been carried out for light nuclei and have been compared with experimental data at incident energies of 150–185 MeV. As in the case of elastic scattering, it is possible to connect inelastic nucleon scattering and inelastic electron scattering from nuclei with $A=2Z$, $J_i=0$ and $T_i=0$, through the transition density $\rho_{if}(\hat{r})$. Combined analyses of inelastic proton scattering and inelastic electron scattering in a model independent way have demonstrated the validity of impulse approximation for light nuclei and low momentum transfer. Some results from a recent analysis of this type are shown in figure 7.1. In most calculations using DWIA the transition densities are obtained from a suitable nuclear structure calculation with the object of testing the nuclear model, and have shown that it is possible to differentiate between models in this way.

In the treatment of DWIA given above we have ignored the spin-depen-

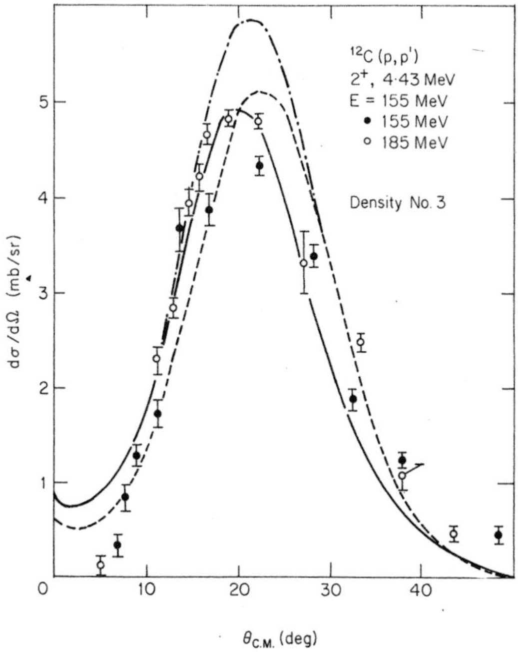

Figure 7.1 (b) Fits to the cross-section for excitation of the same state by proton inelastic scattering. The three curves refer to different forms for the nucleon–nucleon interaction. [From R. M. Haybron, M. B. Johnson and R. J. Metzger, *Phys. Rev.* **156**, 1136 (1967).]

dence of the distorted waves which arises from the spin-orbit term in the optical potential. In many cases it is adequate to omit the spin-orbit term when calculating the distorted waves, but there are also cases for which this is not true especially in calculations of the polarization. A description of the full treatment of spin dependence has been given by Haybron and McManus [85].

7.3 Distorted wave theory for inelastic scattering

In the previous section we have seen how DWIA can be applied to inelastic scattering at energies above 100 MeV. At lower energies the approximation most frequently used is the distorted wave Born approximation (DWBA). In this approximation the transition matrix element can be written as

$$T_{fi}(\boldsymbol{k}', \boldsymbol{k}) = \int d\boldsymbol{r}\, \chi_f^{-*}(\boldsymbol{k}', \boldsymbol{r})\, P(\boldsymbol{r})\, \chi_i^{+}(\boldsymbol{k}, \boldsymbol{r})\,. \tag{7.10}$$

The function $P(\boldsymbol{r})$ is given by

$$P(\boldsymbol{r}) = (\Phi_f | V(\boldsymbol{r}, \xi) | \Phi_i) \tag{7.11}$$

$$= \int d\xi\, \rho_{if}(\xi)\, V(\boldsymbol{r}, \xi) \tag{7.12}$$

where ξ represents all the relevant internal co-ordinates of the target, and $P(\boldsymbol{r})$ may be regarded as the effective interaction for nucleon–nucleus scattering. It is convenient to make a multipole expansion of the interaction potential,

$$V(\boldsymbol{r}, \xi) = \sum_{LM} V_{LM}(r, \xi)[i^L Y_L^M(\hat{\boldsymbol{r}})]^*\,, \tag{7.13}$$

so that, using the Wigner-Eckart theorem, the effective interaction for a transition between nuclear states J_i and J_f becomes

$$(J_f M_f | V | J_i M_i) = \sum_L (J_i M_i L M | J_f M_f) \langle J_f \| V_L \| J_i \rangle [i^L Y_L^M(\hat{\boldsymbol{r}})]^*\,. \tag{7.14}$$

The reduced matrix element is now a function of r only and is usually written as the product of a strength A_L and a form factor $F_L(r)$, i.e.

$$\langle J_f \| V \| J_i \rangle = A_L F_L(r) \tag{7.15}$$

where the magnitude and form of the factors depend on the nuclear model chosen. The transition matrix element now becomes

$$T_{fi} = \sum_{LM} (J_i M_i L M | J_f M_f)\, T_{LM} \tag{7.16}$$

where $$T_{LM} = i^{-L} A_L \int d\boldsymbol{r}\, \chi_f^{-*}(\boldsymbol{k}', \boldsymbol{r}) F_L(r) Y_L^{M*}(\hat{\boldsymbol{r}}) \chi_i^+(\boldsymbol{k}, \boldsymbol{r}) . \tag{7.17}$$

The differential cross-section is obtained by taking the sum over the final states M_f and the average over the initial states M_i. Thus

$$\frac{d\sigma}{d\Omega} = \left(\frac{\mu}{2\pi\hbar^2}\right)^2 \frac{k'}{k} \frac{1}{2J_i+1} \sum_{M_i M_f} \left| \sum_{LM} (J_i M_i LM | J_f M_f) T_{LM} \right|^2 \tag{7.18}$$

but since

$$\sum_{M_i M_f} (J_i M_i LM | J_f M_f)(J_i M_i L'M' | J_f M_f) = \frac{2J_f+1}{2L+1} \delta_{LL'} \delta_{MM'} \tag{7.19}$$

the cross-section reduces to

$$\frac{d\sigma}{d\Omega} = \left(\frac{\mu}{2\pi\hbar^2}\right)^2 \frac{k'}{k} \sum_{LM} \frac{2J_f+1}{(2J_i+1)(2L+1)} |T_{LM}|^2 . \tag{7.20}$$

In order to evaluate the matrix element T_{LM} we need partial wave expansions for the distorted wavefunctions. We again ignore the spin-orbit term in the optical potential and take the momenta $\boldsymbol{k}$, $\boldsymbol{k}'$ to lie in the x–z plane with the z-axis along $\boldsymbol{k}$. Then, using equations (3.161) and (3.162) the distorted wavefunctions are given by

$$\chi_i^+(\boldsymbol{k}, \boldsymbol{r}) = \sum_l i^l [4\pi(2l+1)]^{\frac{1}{2}} e^{i\sigma_l} f_l(kr) Y_l^0(\hat{\boldsymbol{r}})$$

$$\chi_f^{-*}(\boldsymbol{k}', \boldsymbol{r}) = 4\pi \sum_{l'm'} i^{-l'} \exp(i\sigma_{l'}) f_{l'}(k'r) Y_{l'}^{m'}(\hat{\boldsymbol{r}}) Y_{l'}^{m'*}(\hat{\boldsymbol{k}}')$$

where σ_l is the Coulomb phase shift, and the angular co-ordinates of $\boldsymbol{k}'$ are $(\theta, 0)$ where θ is the scattering angle. Substituting these expressions into equation (7.17) and integrating over the angular co-ordinates of $\boldsymbol{r}$ we have

$$T_{LM} = 4\pi A_L (2L+1)^{\frac{1}{2}} \sum_{ll'} i^{l-l'-L} e^{i(\sigma_l + \sigma_{l'})} U_{ll'L} Y_{l'}^{-M}(\theta, 0)$$
$$\times (2l'+1)^{\frac{1}{2}} (l' M L -M | l 0)(l' 0 L 0 | l 0) \tag{7.21}$$

where $U_{ll'L}$ is the radial integral,

$$U_{ll'L} = \int_0^\infty f_l(kr) F_L(r) f_{l'}(k'r) r^2 dr \tag{7.22}$$

$$= (kk')^{-1} \int_0^\infty u_{l'}(k'r) F_L(r) u_l(kr) dr . \tag{7.23}$$

The radial function $u_l(kr)$ is defined by equation (6.33) and can be computed numerically by the methods outlined in section 6.4. In a calculation of elastic scattering only the asymptotic behaviour of u_l is of interest, but in a distorted wave calculation the value for u_l obtained after each step in the integration must be normalized and stored. The values are then used for the evaluation of the radial integrals $U_{ll'L}$. The number of partial waves required in these calculations depends on the incident energy and on the range of the form factor F_L, since if the latter has a long range there can be a reasonable overlap with the high partial waves which are excluded from the interior region by the angular momentum barrier [86].

The expression (7.21) for the transition matrix element has a general and fairly simple form. It is evident that the characteristic features of the nuclear excitation enter through the reduced matrix element $A_L F_L(r)$. In the language of the collective model the nuclear excitation can be associated with a change in the state of vibration about a mean spherical shape or a change in the state of rotation of a permanently deformed shape (see Chapter 2). The macroscopic approach to the description of these excitations is based on the assumption that the interaction potential $V(\mathbf{r}, \xi)$ follows the deformation of the nuclear surface so that in this extended optical model the complex potential becomes a function of the collective co-ordinates. We take this potential to be $V[r-R(\theta', \phi')]$ where θ', ϕ' are referred to the body-fixed axes, and make a Taylor-series expansion about $R=R_0$, which gives

$$V(r-R) = V(r-R_0) - \delta R \frac{d}{dr} V(r-R_0) + \tfrac{1}{2}(\delta R)^2 \frac{d^2}{dr^2} V(r-R_0) + \ldots . \tag{7.24}$$

We can now associate the first term of this expansion with the usual spherically symmetric optical potential used to describe elastic scattering and to generate the distorted waves. The higher terms will provide the inelastic excitation and some corrections to elastic scattering.

In the rotational model it is assumed that the nucleus has a permanent deformation so that the nuclear surface can be defined in the body-fixed system by (see section 2.3)

$$R(\theta', \phi') = R_0[1 + \sum_{kq} \alpha'_{kq} Y_k^q(\theta', \phi')] .$$

In the space-fixed system the surface is defined by

$$R(\theta, \phi) = R_0[1+\sum_{k\nu} \alpha_{k\nu} Y_k^\nu(\theta, \phi)]$$
$$= R_0[1+\sum_{k\nu} Y_k^{\nu *}(\theta, \phi) \sum_q \alpha_{kq}'^* \mathscr{D}_{\nu q}^k(\hat{s})] \qquad (7.25)$$

where the collective co-ordinates are related by

$$\alpha_{k\nu} = \sum_q \alpha'_{kq} \mathscr{D}_{\nu q}^{k*}(\hat{s}), \quad \alpha'_{kq} = \sum_\nu \alpha_{k\nu} \mathscr{D}_{\nu q}^k(\hat{s}),$$

the spherical harmonics transform as

$$Y_k^\nu(\theta, \phi) = \sum_\lambda Y_k^\lambda(\theta', \phi') \mathscr{D}_{\nu\lambda}^k(s),$$

and s denotes the Euler angles of the body-fixed axes with respect to the space-fixed axes. For axial symmetry $\alpha'_{k0} = \beta_k$, $\alpha'_{k\,q\neq 0} = 0$, where β_k is the usual deformation parameter, and in this case

$$\delta R = R_0 \beta_k \sum_{k\nu} Y_k^{\nu *}(\theta, \phi) \mathscr{D}_{\nu 0}^k(\hat{s}). \qquad (7.26)$$

Now, comparing equations (7.13) and (7.24) we obtain for the first-order term in the interaction

$$V_{LM} = -i^L \beta_L R_0 \frac{d}{dr} V(r-R_0) \mathscr{D}_{M0}^L(s). \qquad (7.27)$$

This expression may now be placed between suitable wavefunctions for the collective states to give the nuclear matrix element of equation (7.14). For example, in the simple case of transitions between members of a $K=0$ rotational band the nuclear wavefunctions are given by

$$|JMK\rangle = \left[\frac{2J+1}{8\pi^2}\right]^{\frac{1}{2}} \mathscr{D}_{M0}^J(\hat{s})$$

and, starting from the ground state $J_L=0$, final states with $J_f=L$ can be excited with change of parity $(-1)^L$. The reduced matrix element is given by

$$\langle J_f \| V \| J_i \rangle = (J_f M_f K | V | J_i M_i K)/(J_i M_i LM | J_f M_f) \qquad (7.28)$$

and since

$$\int \mathscr{D}_{M_f 0}^{J_f^*} \mathscr{D}_{M0}^L \mathscr{D}_{M_i 0}^{J_i} d\Omega = \frac{8\pi^2}{2J_f+1} (J_i 0 L 0 | J_f 0)(J_i M_i LM | J_f M_f)$$

we have $$\langle J_f = L \| V \| J_i = 0 \rangle = -i^L \beta_L R_0 \frac{dV}{dr} (2L+1)^{-\frac{1}{2}} \qquad (7.29)$$

so that
$$A_L F_L(r) = -i^L \beta_L R_0 (2L+1)^{-\frac{1}{2}} \frac{dV}{dr}. \tag{7.30}$$

In the vibrational model the nuclear surface is represented by the dynamical deformation parameters α_{kq} (see Section 2.3),

$$\alpha_{kq} = (h\omega_k/2C_k)^{\frac{1}{2}} [b_{kq} + (-1)^2 b^*_{k-q}]$$

where b_{kq}, b^*_{kq} are the phonon annihilation and creation operators for a 2^k-pole oscillation with angular momentum k. The energy of the phonon is $h\omega_k$ and C_k is the restoring force parameter. Since the deformation is no longer static we may work directly in the space-fixed system so that

$$R(\theta, \phi) = R_0 [1 + \sum_{kq} \alpha^*_{kq} Y_k^{q*}(\theta, \phi)] \tag{7.31}$$

and the first-order term in the interaction becomes

$$V_{LM} = -i^L R_0 \alpha^*_{LM} \frac{dV}{dr}. \tag{7.32}$$

If the target nucleus has even A and ground state $J_i = 0$ this interaction will excite a single 2^L-pole phonon state with reduced matrix element

$$A_L F_L(r) = -i^L \left(\frac{h\omega_L}{2C_L}\right)^{\frac{1}{2}} R_0 \frac{dV}{dr}. \tag{7.33}$$

The collective model thus predicts a universal form factor F_L given by the derivative of the optical potential and fits to the data determine a single parameter β_L or $(h\omega_L/2C_L)^{\frac{1}{2}}$. There are no other free parameters because the optical potential is determined by fitting the elastic scattering. Because of these features, the extended optical model has become the standard method for the analysis of experimental data on inelastic scattering and has been used with great success for a wide range of projectiles, targets and incident energies [87]. An example of the results obtained for α-particle scattering is given in figure 7.2.

It may be noted that we have evaluated the interaction V_{LM} to first order. From equation (7.24), the second order interaction is of the form $\frac{1}{2}(\delta R)^2 (d^2/dr^2) V$, and an excitation arising from the single operation of the term is called a *direct second order excitation*. However, this is only a part of the second-order excitation because in DWBA we derive the transition matrix element in first order only, and therefore to calculate the full second-order excitation it is necessary to take account of second-order terms arising from the first-order interaction. This gives rise to a *multiple second order excitation* through the operator $\delta R(dV/dr)G_1^+\,\delta R(dV/dr)$ (see equation (3.144)) where G_1 is the Green's function which describes propagation through the nuclear medium represented by the optical potential. Several investigations have shown that interference between the multiple and direct excitations is important and must be taken into account in attempts to fit data on the excitation of two-phonon states. These calculations are usually carried out in the framework of the strong coupling approximation described in section 7.4.

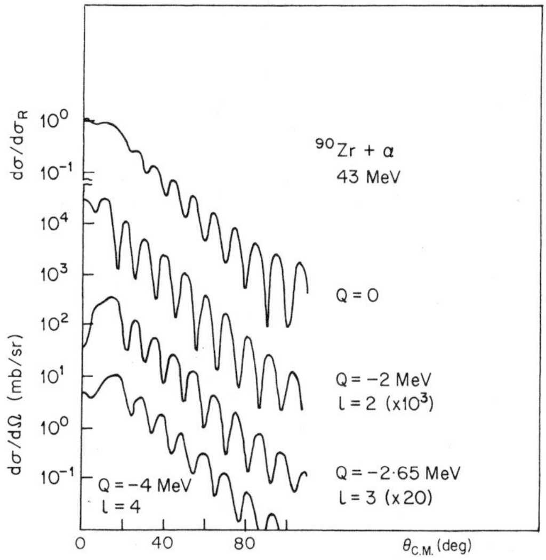

Figure 7.2 Calculated cross-sections for 43 MeV α-particle scattering from ^{90}Zr obtained using the optical model for elastic scattering and DWBA for the inelastic scattering. [From R. H. Bassel, G. R. Satchler, R. M. Drisko and E. Rost, *Phys. Rev.* **128,** 2693 (1962).]

The effective interaction defined in equation (7.12) or (7.14) may also be evaluated using a microscopic model in which the interaction potential $V(\boldsymbol{r}, \xi)$ is written as a sum of two-body interactions

$$V(\boldsymbol{r}, \xi) = \sum_{j=1}^{A} (-V_0 - V_1 \boldsymbol{\sigma}\cdot\boldsymbol{\sigma}_j) g(|\boldsymbol{r} - \xi_j|) \tag{7.34}$$

where

$$V_0 = V_{0\alpha} + V_{0\beta}\boldsymbol{t}\cdot\boldsymbol{t}_j \tag{7.35}$$

and similarly for V_1. The initial and final nuclear states must also be described in microscopic terms and this approach to inelastic scattering therefore provides a means of testing and extending our understanding of the structure of the nuclear states. The excitation of single-particle, weakly collective, and strongly collective states, are treated in the same manner and this provides an opportunity of investigating the underlying microscopic structure of a collective state. An example of the way in which many single-particle excitations contribute collectively to the form factor for a collective

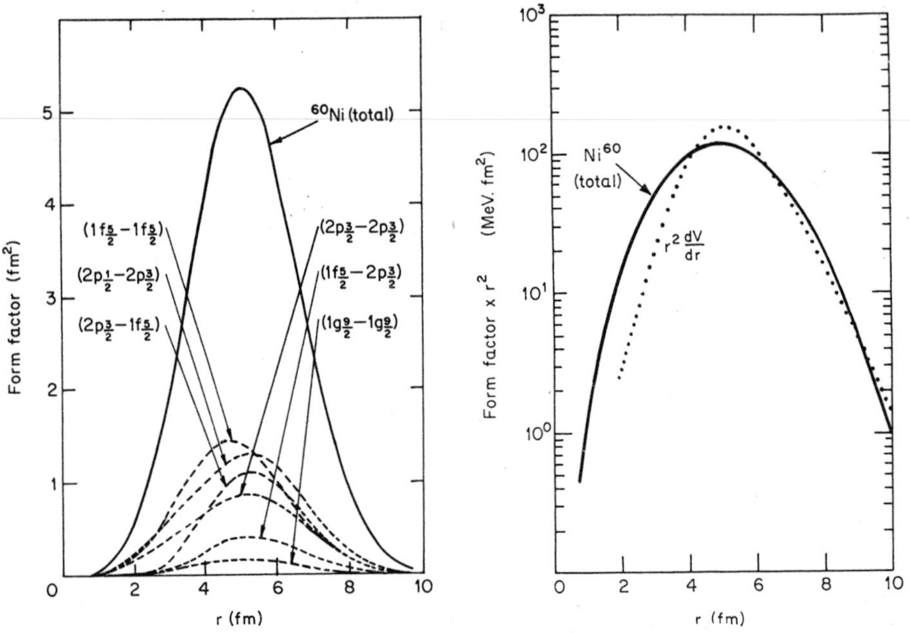

Figure 7.3 (**a**) Contribution of single-particle terms to the microscopic form factor for excitation of the 2^+ state in ^{60}Ni. (**b**) Comparison of the microscopic form factor with the derivative form factor given by the macroscopic model. [From N. K. Glendenning and M. Veneroni, *Phys. Rev.* **144,** 839 (1966).]

state excited by proton scattering is shown in figure 7.3. This figure also shows the similarity between the collective form factor calculated from a microscopic model for a strongly collective 2^+ state and the universal form factor of the macroscopic collective model. This similarity does not hold for the excitation of more complicated states such as 0^+ states.

The microscopic approach to inelastic scattering is in its early stages and there are still many uncertainties in the correct choice of the two-body interaction [88]. It can be seen from figure 7.3 that the effective interaction for proton inelastic scattering, both in the microscopic and the macroscopic model, extends well inside the nucleus. This means that proton scattering can be expected to be sensitive to the radial shape of the effective interaction as well as its strength and therefore to provide a test of the nuclear model used to construct the effective interaction. It also means that the accuracy and reliability of the distorted wave calculations depends on the extent to which the optical model wavefunctions give an accurate description of the penetration of the projectile into the interior of the nucleus. There are a number of effects which cast doubt on the reliability of the optical model wavefunctions; for example, complex projectiles such as the deuteron may be virtually excited during the scattering process, and it is known [89] that the scattering wavefunction generated by a non-local potential has a smaller amplitude in the interior region than the wavefunction generated by an equivalent local potential. This situation may be dealt with either by seeking a phenomenological optical potential which gives a reasonable description of both elastic and inelastic scattering or, more fundamentally, by including explicitly those additional effects which appear to be important. As discussed in Chapter 4, inelastic electron scattering is also sensitive to the details of the effective interaction. In contrast, the inelastic scattering of α-particles is sensitive to the strength but not to the shape of the effective interaction for reasons discussed in section 8.4.

7.4 Strong coupling approximation

In DWBA the distorted wavefunctions for the incoming and outgoing particles are generated using the optical potential which describes elastic scattering and the effect of the non-elastic channels is taken into account through the imaginary part of the optical potential. If, however, one or more inelastic channels are strongly coupled to the elastic channel it is not sufficiently accurate to take them into account through the optical potential when calculating the elastic scattering, nor is it sufficiently accurate to use DWBA to calculate the scattering into these inelastic channels. This strong coupling situation arises when strongly collective states are excited. It can easily be seen that this situation leads to coupled equations if we use the notation of section 6.2 and write the total scattering wavefunction as

$$\psi(\boldsymbol{r}, \xi) = \sum_{\beta} \Phi_\beta(\xi)\, \psi_\beta(\boldsymbol{r})$$

where $$[\varepsilon_\alpha - H(\xi)]\,\Phi_\alpha(\xi) = 0, \quad E - \varepsilon_\alpha = \frac{\hbar^2}{2\mu_\alpha} k_\alpha^2 .$$

The Schrödinger equation for ψ is

$$[E - T - H(\xi) - V(\boldsymbol{r}, \xi)]\,\psi(\boldsymbol{r}, \xi) = 0 \tag{7.36}$$

and multiplying on the left by Φ_α^* and integrating we find

$$\sum_\beta [(E - \varepsilon_\alpha - T)\delta_{\alpha\beta} - (\Phi_\alpha | V(\boldsymbol{r}, \xi) | \Phi_\beta)]\,\psi_\beta(\boldsymbol{r}) = 0 .$$

Thus we obtain a set (in principle an infinite set) of coupled equations for the ψ_β. The transition matrix element is given by

$$T_{fi} = \langle \phi_f \Phi_f | V | \psi^+ \rangle .$$

Thus if we neglect all the potential matrix elements except the optical potential $(\Phi_0 | V | \Phi_0)$ but take the other channels into account through an imaginary potential, the wavefunction ψ reduces to the distorted wavefunction χ and the transition matrix element reduces to DWBA.

In practice it is more convenient to expand the total wavefunction in terms of eigenfunctions of the total angular momentum of the entrance channel, as was done in section 3.7,

$$\psi(\boldsymbol{r}, \xi) = \sum_{jm} \psi_{jm}(\boldsymbol{r}, \xi) \tag{7.37}$$

and to express the wavefunction ψ_{jm} for each channel $(lZjm)$ as a superposition of elastic and inelastic scattering states,

$$\psi_{jm} = \frac{1}{r} u_{Zl}^j(r)\,\phi_{Zl}^{jm}(\hat{\boldsymbol{r}}, \xi) + \frac{1}{r} \sum_{\substack{Z' \neq Z \\ l'}} u_{Z'l'}^j\,\phi_{Z'l'}^{jm}(\hat{\boldsymbol{r}}, \xi) . \tag{7.38}$$

When the target is in its ground state the corresponding radial function u_{Zl}^j contains both incoming and outgoing waves but for the target in an excited state the corresponding function $u_{Z'l'}^j$ contains only outgoing waves. If we now substitute equations (7.37) and (7.38) into equation (7.36), multiply on the left by ϕ_{Zl}^{jm*} and integrate over the nuclear co-ordinates ξ and the angular co-ordinates of $\boldsymbol{r}$ we obtain

$$\left[\frac{d^2}{dr^2} + k_Z^2 - \frac{l(l+1)}{r^2}\right] u_{Zl}^j(r) = \frac{2\mu}{\hbar^2} \sum_{Z'l'} V_{Zl,Z'l'}^j(r)\,u_{Z'l'}^j(r) , \tag{7.39}$$

and similarly if we multiply on the left by $\phi_{J'l'}^{jm*}$ and integrate we obtain

$$\left[\frac{d^2}{dr^2} + k_{Z'}^2 - \frac{l'(l'+1)}{r^2}\right] u_{Z'l'}^j(r) = \frac{2\mu}{\hbar^2} \sum_{Z''l''} V_{Z''l'',Z'l'}^j(r)\, u_{Z''l''}^j(r) \quad (7.40)$$

where we have used the fact that the reduced mass is the same in all inelastic channels, and the coupling potentials are given by

$$V_{Zl,Z'l'}^j(r) = \langle \phi_{Zl}^{jm}(\hat{\mathbf{r}}, \xi) | V(\mathbf{r}, \xi) | \phi_{Z'l'}^{jm}(\hat{\mathbf{r}}, \xi) \rangle \,. \quad (7.41)$$

There is now no difficulty in including multiple excitation. As an example we consider the excitation of a nucleus with a spin-zero ground state by a spin zero projectile, so that $Z = J_i = 0$ and $Z' = J_f$, where J_f is the spin of the excited state. We assume that the states of the target nucleus follow a $0^+, 2^+, 4^+, 6^+ \ldots$ sequence. Then the excitation of the 4^+ state can proceed through the one-step process $0^+ \to 4^+$ and also through the two-step process $0^+ \to 2^+ \to 4^+$. Examination of equations (7.39)–(7.41) shows that the coupling potentials $V_{0l,4l''}$, $V_{4l'',0l}$ describe the direct one-step transition between the 0^+ and 4^+ states while the coupling potential $V_{4l',2l''}$ allows for multiple excitation from the 2^+ state. At the same time the inclusion of the coupling to the 4^+ state affects the wavefunction in the channel corresponding to excitation of the 2^+ state through $V_{2l',4l'''}$. Similarly, inclusion of the coupling to the 6^+ state would affect the wavefunction in the 4^+ channel. Strictly, of course, the set of coupled equations is infinite and it is necessary to assume that the set can be truncated after a limited number of levels. This procedure is usually referred to as a *Tamm-Dancoff approximation*, and the effect of the higher inelastic channels together with the other non-elastic channels are roughly taken into account through the imaginary part of the effective interaction. An example of results obtained in a coupled channels calculation and the competition between multiple and direct excitation of a 4^+ state is shown in figure 7.4.

The set of coupled radial integrations must be integrated numerically, and the mathematical and computational techniques involved have been described by Buck, Stamp and Hodgson [90]. The asymptotic behaviour of the radial functions is given by

$$u_{0l}^j \to \tfrac{1}{2}(k_0 r)[H_l^{(2)}(k_0 r) + S_{0l}^j H_l^{(1)}(k_0 r)] \quad (7.42)$$

$$u_{J_f l'}^j \to \tfrac{1}{2}(k_f r) S_{J_f l'}^j H_{l'}^{(1)}(k_f r) \quad (7.43)$$

where the H_l are spherical Hankel functions or Coulomb functions. The S-matrix elements are determined by matching the solutions of the coupled equations to the asymptotic forms, and the differential cross-sections for

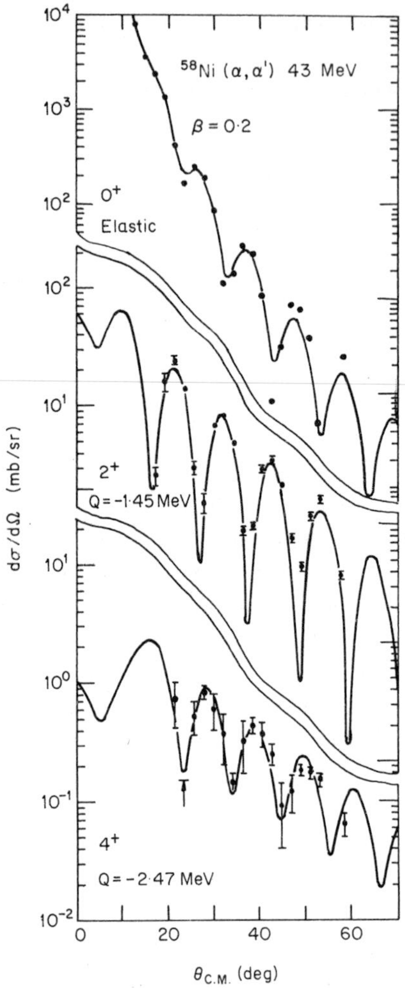

Figure 7.4 **(a)** Comparison of a coupled channels calculation for 43 MeV α-particle scattering from ^{58}Ni with the experimental data. The excitation is described in terms of the vibrational model.

elastic and inelastic scattering and the total absorption cross-section can then be calculated using equations (5.19) and (5.20). In the *strong coupling approximation* (SCA), as in DWBA, there are essentially no free parameters except those which occur in the model chosen for the nuclear wavefunctions, for example, the deformation parameters of the rotational model. The SCA

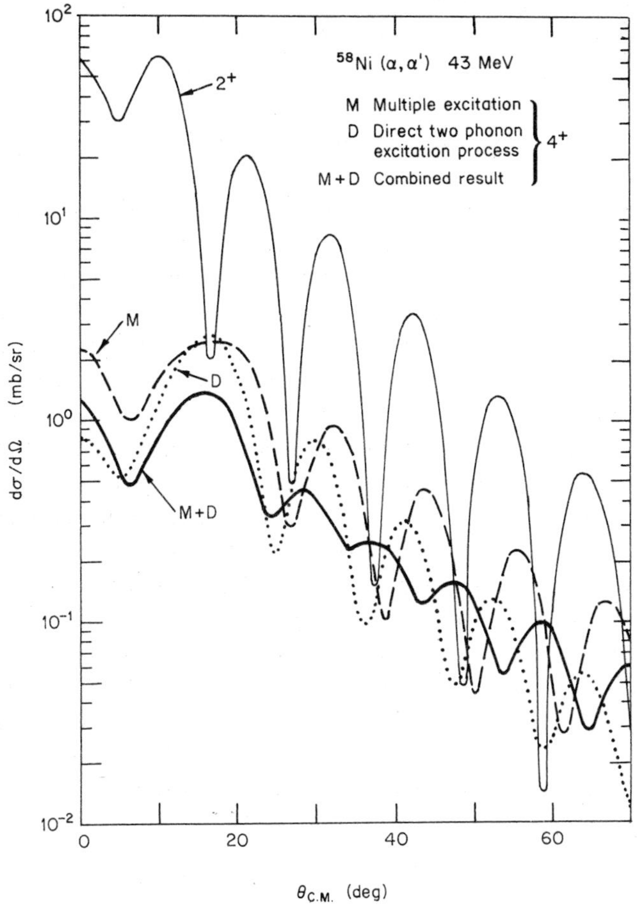

Figure 7.4 (b) The contribution of the multiple and direct two-phonon excitation processes to the calculated cross-section for the 4^+ state. [From B. Buck, *Phys. Rev.* **127,** 940 (1962).]

has the additional and very important feature that the wavefunctions for elastic and inelastic scattering are calculated simultaneously. The disadvantage of the SCA is a practical one, namely that the computation is lengthy especially if a search procedure is used. Fortunately, the failure of DWBA occurs in the lower energy region where the number of partial waves required is not so large as to make the SCA impracticable.

7.5 The charge-exchange reaction

All the methods described in the preceding sections can be used for calculations on charge exchange reactions. There are a number of charge-exchange reactions involving complex particles, such as the (^{3}He, ^{3}H) reaction, but for simplicity we consider here only the charge-exchange reactions involving nucleons, i.e. the (p, n) and (n, p) reactions.

The extension of the microscopic models is quite straightforward. The direct charge exchange process occurs through the $\boldsymbol{t}\cdot\boldsymbol{t}_j$ part of the two-nucleon interaction so that only the coefficients labelled β appear in the formulae for the cross-section. The (p, n) and (n, p) reactions on light nuclei have been studied in the energy region above 90 MeV and the data have been analysed in impulse approximation, but at the present time the resolution obtained in the experiments is not very satisfactory owing to the difficulties involved in producing and detecting neutron beams at these energies. The (p, n) reaction has been studied in much greater detail and for a wide range of nuclei at lower energies, and has been analysed in terms of a microscopic model using DWBA [88]. In the distorted wave theory there is no difficulty in taking account of the Coulomb distortion for the incoming proton as well as the nuclear distortion for both particles or of the difference in energies of the incoming and outgoing particles.

One of the most important features of the (p, n) reaction is the occurrence in all the energy spectra of a strongly excited state which stands out above the background [91], and which has the characteristic that the magnitude of the Q-value is given by the coulomb energy difference between two states of an isobaric spin multiplet, i.e.

$$-Q = \Delta E_c = c^2[-M(T, T_Z)+M(T, T_Z-1)-m(\mathrm{p})+m(\mathrm{n})]$$

where $m(\mathrm{p})$ and $m(\mathrm{n})$ are the mass of proton and neutron respectively. A typical neutron spectrum from the (p, n) reaction is shown in figure 7.5. The state of the residual nucleus so formed is identified as the *isobaric analogue* of the ground state of the target nucleus which has the same total angular momentum J, the same parity, and the same I-spin T as the target ground state but differs in T_Z. In light nuclei for which the neutron excess is small the strength of the excitation of the analogue state is not markedly different from the strength of excitation of other states in the residual nucleus which have different J, but in heavy nuclei where the neutron excess is large there are many particle-hole configurations which can contribute to the analogue state so that the excitation has a strong collective character.

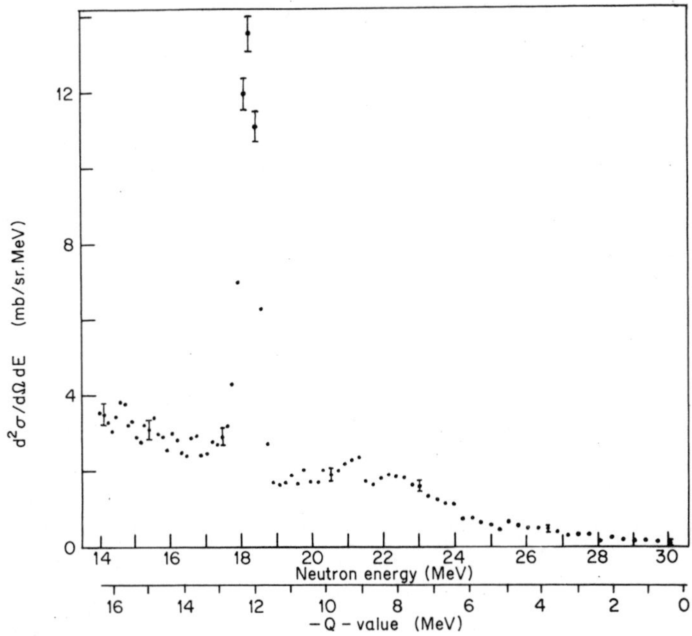

Figure 7.5 Energy spectrum for the ^{93}Nb(p,n) reaction at 30 MeV showing the strong excitation of the isobaric analogue state. [From C. J. Batty *et al.*, ref. 93.]

Because of the strong excitation of the analogue state it is appropriate to extend the collective models of inelastic scattering to describe also the excitation of the analogue state in charge-exchange reactions. If it is assumed that the coupling to other channels is negligible compared to the coupling between the proton and neutron channels, the total scattering wavefunction can be written as [92]

$$\psi = \psi_{\mathrm{p}}(\boldsymbol{r})\Phi_{\mathrm{pi}} + \psi_{\mathrm{n}}(\boldsymbol{r})\Phi_{\mathrm{nf}}$$

where Φ_{pi} describes the I-spin states of the proton and target and the internal structure of the target, and Φ_{nf} gives a similar description of the neutron and final state. We write the nucleon optical potential in the form

$$V_{\mathrm{opt}} = -U_0 - U_1(\boldsymbol{t}\cdot\boldsymbol{T}) - V_{\mathrm{so}} + V_{\mathrm{c}}$$

so that the Hamiltonian is

$$H = H(\xi) + T - U_0 - U_1(\boldsymbol{t}\cdot\boldsymbol{T}) - V_{\mathrm{so}} + (\tfrac{1}{2} - t_z)V_{\mathrm{c}} + (\tfrac{1}{2} + t_z)\Delta E_{\mathrm{c}}$$

and substituting for ψ in the Schrödinger equation

$$(H-E)\psi=0$$

we obtain the equations

$$(T-U_0-\tfrac{1}{2}TU_1-V_{so}+V_c-E)\psi_p+\tfrac{1}{2}(2T)^{\frac{1}{2}}U_1\psi_n=0$$
$$(T-U_0+\tfrac{1}{2}(T-1)U_1-V_{so}+\Delta E_c-E)\psi_n+\tfrac{1}{2}(2T)^{\frac{1}{2}}U_1\psi_p=0\,.$$

These equations can be solved by the same techniques (and often the same computer codes) that are used for the SCA description of inelastic scattering. Alternatively, if the Coulomb terms V_c and ΔE_c are neglected, the equations can be decoupled and solved by the techniques used for optical model analyses of elastic scattering. A further alternative follows from the assumption that the coupling term is weak so that the transition matrix element can be obtained in DWBA.

Detailed analyses of the (p, n) reaction leading to the analogue state by means of DWBA or the optical model yield considerable information about the symmetry term in the optical potential. In one of the more extensive of these analyses [93] the symmetry term is taken to be complex and the radial parts of the real and imaginary parts can have volume or surface-peaked shape, so that, using the notation of section 6.3, the coupling term has the form

$$\tfrac{1}{2}(2T)^{\frac{1}{2}}U_1=[V_1(r)+iW_1(r)]\,\frac{(N-Z)^{\frac{1}{2}}}{2A}\,.$$

It is also necessary to take account of the energy dependence of the potential so that the potentials differ in the two channels. Some typical results are shown in figure 7.6. These results favour a complex symmetry term whose real part is surface-peaked, but the shape of the imaginary part is not well determined. It is also found that different optical potentials which yield good fits to elastic scattering data sometimes yield different DWBA cross-sections for the (p, n) reaction.

The analyses of charge-exchange reactions in terms of a microscopic model using either DWBA or DWIA yield valuable information about the structure of the nuclear states involved. In DWIA the parameters of the two-body interaction are known since these are taken from the free interaction but as noted above there are as yet few data in the energy region where DWIA is applicable. Many microscopic analyses have been carried out at lower energies using DWBA, and by careful choice and comparison of the

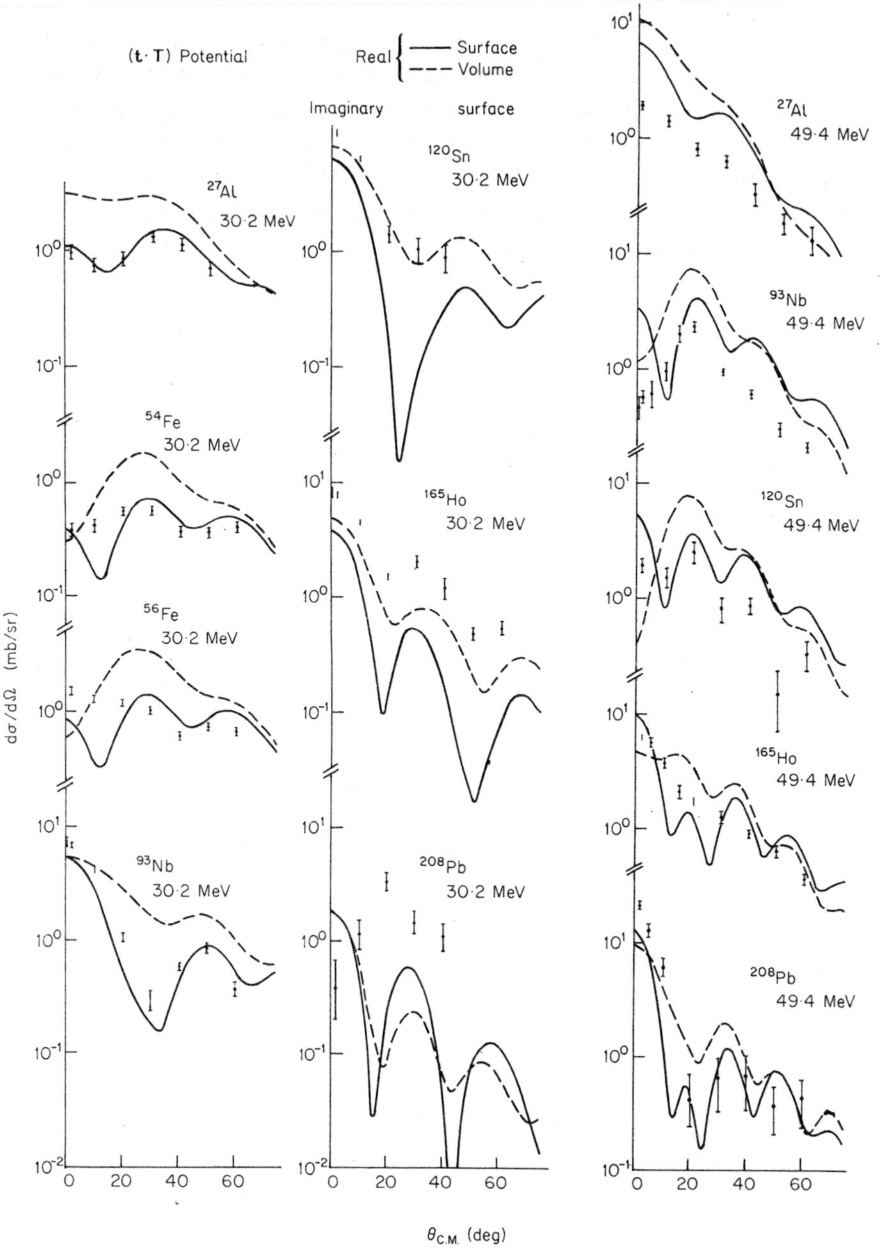

Figure 7.6 Cross-sections for the excitation of isobaric analogue states in the (p,n) reaction on a range of nuclei. [From C. J. Batty *et al.*, ref. 93.]

transitions it has been possible to study the parameters of the effective interaction (7.34). Information is also obtained on the single-particle wave-functions of the target nucleons involved in the charge-exchange process, and on the extent to which the *I*-spin *T* is a good quantum number. The charge-exchange reactions also provide direct experimental evidence on Coulomb energy differences and their dependence on *Z* and *A*. Reviews of these topics are given in reference [76].

8 Strong absorption

8.1 Strong absorption in elastic scattering

The optical model wavefunction for a spinless particle can be written in the form

$$\psi^+(\mathbf{k}, \mathbf{r}) = \sum_l i^l (2l+1) f_l(kr) P_l(\cos\theta) e^{i\sigma_l}$$

where f_l satisfies the boundary condition

$$f_l(kr) \xrightarrow[r\to\infty]{} \tfrac{1}{2}[H_l^{(2)}(kr) + \eta_l H_l^{(1)}(kr)] .$$

In section 3.6 we defined η_l as the reflection coefficient for the lth partial wave and showed that for a complex potential $|\eta_l| < 1$ for the low partial waves and tends to unity for the high partial waves. This means that at some distance outside the potential we have ingoing waves for small l and standing waves reflected from the angular momentum barrier for large l. If $|\eta_l| \approx 0$, or alternatively if the opacity $1 - |\eta_l|^2$ is close to its maximum value of unity, the nucleus is said to be *black* to this partial wave. If $|\eta_l|$ vanishes for all the low partial waves and then rises sharply to unity, we have a situation of *strong absorption* [94]. In this situation the scattering is similar to diffraction by a black disc, and may be described by approximate methods based on diffraction theory.

For strong absorption to occur it is not necessary for the imaginary part of the potential to be excessively large, nor for the optical model wavefunction to be negligible in the interior of the nucleus. An explanation of this effect has been given by Austern [95] in terms of the behaviour of the classical turning point. The turning point r_0 for the lth partial wave is defined through the relation

$$E = V_{\text{tot}}(r_0, l) = \text{Re } V_{\text{opt}}(r_0) + \frac{\hbar^2}{2\mu}\frac{l(l+1)}{r_0^2} \tag{8.1}$$

or

$$k^2(r_0) = l(l+1)/r_0^2 \tag{8.2}$$

where E is the projectile energy in the c.m. system and $k(r)$ is the effective wave number at the point r. Values of $V_{\text{tot}}(r, l)$, which is the total effective real potential including the Coulomb potential, are plotted in figure 8.1(a)

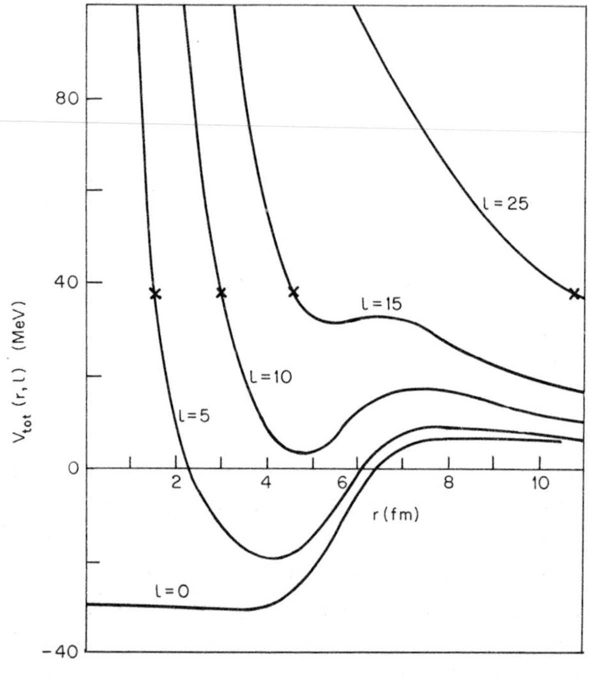

Figure 8.1 (**a**) The real part of the effective potential for various partial waves plotted for 42 MeV α-particles on ^{42}Ca using potential 1 of Table 8.1. The crosses mark the positions of the turning points.

for 42 MeV α-particles and the potential 1 of table 8.1 and the positions of the turning points are marked by a cross. For the low partial waves the turning point occurs well inside the nucleus, and this means that the absorptive part of the optical potential is able to act strongly on the low partial waves as they move in to r_0 and are reflected out again. Reflections from the surface are minimized by the gradual change in the potential in the surface region and the corresponding smooth change in $k(r)$. Strong reflection from the angular momentum barrier can occur if the absorptive part of the potential is very weak or if the barrier moves a long way outside the nucleus. The latter is the case for high partial waves as can be seen from figure 8.1 (a). For a few intermediate partial waves the turning point can be thought of as sampling the nuclear surface, so giving rise to the dependence of the scattering amplitude on the details of the potential. For these particular partial waves the turning points are much further inside the nucleus than they would be in the absence of a potential. Thus the real part of the potential

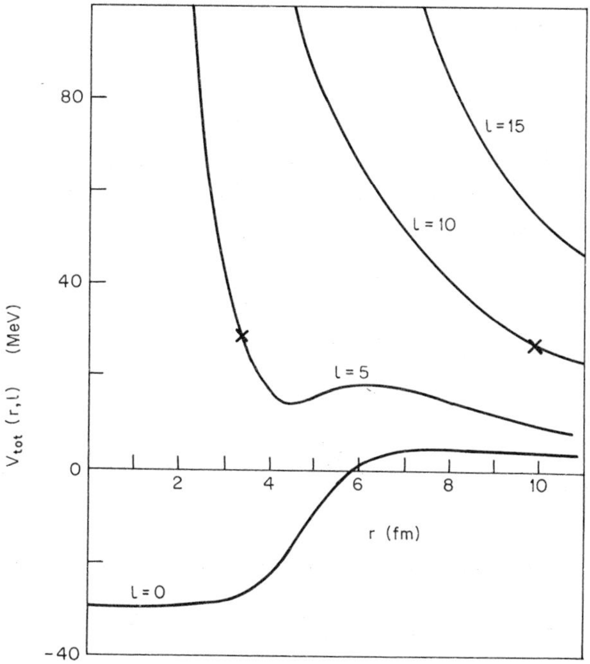

Figure 8.1 (b) The same quantity plotted for 29 MeV protons on ^{58}Ni using potential 4 of Table 6.1.

exerts a powerful refractive effect, causing these partial waves to be pulled in to a region where the absorptive part of the potential can be effective.

In figure 8.1(b) we have also plotted $V_{tot}(r, l)$ for 29 MeV protons and potential 4 of table 6.1. In this case the decrease in incident energy and in the projectile mass causes the turning point for a given partial wave to occur much further out, and this explains why $|\eta_l| \rightarrow 1$ at a much lower value of l for the 29 MeV protons than for 42 MeV α-particles.

As we have noted several times already, the elastic scattering data do not serve to determine the behaviour of the potential and of the optical model wavefunction in the interior region of the nucleus. Because of this, and because of the ambiguities in the optical potential for complex projectiles which are discussed in the next section, an alternative approach to elastic scattering has been developed in terms of a parametrization of the reflection coefficients. The simplest model is the *sharp cut-off model* in which it is assumed that all partial waves with l less than the cut-off angular momentum L are absorbed. Thus

$$\begin{aligned} \eta_l &= 0\,, \qquad l < L \\ \eta_l &= 1\,, \qquad l \geqslant L. \end{aligned} \tag{8.3}$$

Using the semi-classical relation $L+\frac{1}{2} \approx kR$, this model is equivalent to the assumption that all partial waves with impact parameter b less than R are completely absorbed. From equation (3.12) the scattering amplitude for an uncharged particle is given by

$$f(\theta) = \frac{i}{2k} \sum_{l=0}^{L-1} (2l+1) P_l(\cos\theta) \tag{8.4}$$

and to evaluate this we may use the small angle formula

$$P_l(\cos\theta) \approx J_0[(l+\tfrac{1}{2})\theta]\,, \qquad \theta \text{ small}\,, \tag{8.5}$$

and convert the sum to an integral*. The same result is obtained using equations (3.138) and (3.140), i.e.

$$f(\theta) = ik \int_0^R J_0(kb\theta)\, b\, db = ikR^2 \left[\frac{J_1(kR\theta)}{kR\theta}\right]. \tag{8.6}$$

For charged particles the scattering amplitude is given by

$$f(\theta) = f_c(\theta) + \frac{i}{2k} \sum_{l=0}^{L-1} (2l+1) e^{2i\sigma_l} P_l(\cos\theta)\,. \tag{8.7}$$

Exact evaluation of (8.7) can only be done numerically, but approximate analytic formulae have been given [37]. In this case, L and R are related through the formula for the turning point in the presence of the Coulomb potential

* A slightly better procedure [94] is to set $\eta_L = \frac{1}{2}$, and $\eta_l = 1$ for $l > L$, so that

$$f(\theta) = \frac{i}{2k} \sum_{l=0}^{L-1} (2l+1) P_l(\cos\theta) + \tfrac{1}{2}(2L+1) P_L(\cos\theta)\,.$$

$$k^2 = \frac{2\mu Z_1 Z_2 e^2}{\hbar^2 R} + \frac{L(L+1)}{R^2}$$

which gives
$$kR = \gamma \pm [\gamma^2 + L(L+1)]^{\frac{1}{2}} \tag{8.8}$$

or
$$L(L+1) = k^2 R^2 [1 - 2\gamma/kR] \tag{8.9}$$

where γ is the Coulomb parameter defined by equation (3.19).

The cross-sections given by equation (8.6) or (8.7) show pronounced diffraction patterns with periodicity determined by the parameter R. They give qualitative agreement with experiment for the elastic scattering of α-particles and heavy ions, but the oscillations are too pronounced and fall away too slowly at large angles. These shortcomings can be removed by relaxing the assumption of the sharp cut-off model to allow a smooth transition of $|\eta_l|$ from zero to unity. In one form of the *modified cut-off model* the reflection coefficient is written as $\eta_l = |\eta_l| e^{2i\delta_l}$ with

$$|\eta_l| = \left[1 + \exp\left(\frac{L-l}{\Delta}\right)\right]^{-1}, \quad \delta_l = \delta\left[1 + \exp\left(\frac{l-L}{\Delta_\delta}\right)\right]^{-1} \tag{8.10}$$

so that there are now four parameters L, Δ, Δ_δ, and δ, which may be varied to give agreement with the data. Some results for η_l obtained from optical model calculations on α-particle scattering are shown in figure 8.2. It can be seen that a smooth parametrization of $|\eta_l|$ is reasonably consistent with the picture presented by the optical model. (This is evidently not true for the weakly absorbed protons whose reflection coefficients are shown in figure 6.5.)

The most elaborate and successful calculations involve a *complex model* [96] for η_l. Frahn and Venter have used the parametrization

$$\text{Re}\,\eta_l = g_1(t) + \varepsilon_1[1 - g_1(t)] \tag{8.11}$$

$$\text{Im}\,\eta_l = \mu_1 \frac{dg_2}{dt} + \varepsilon_1[1 - g_1(t)] \tag{8.12}$$

where g_1 and g_2 are continuous functions of the variable $t = l + \frac{1}{2}$ which may typically be of Saxon-Woods form with half-way point T_1 and diffuseness Δ_1. By using different parameters for different values of j it is possible to extend this model to describe particles with spin, and since ε_1 and ε_2 can be non-zero the model can describe situations in which the opacity is substantially less than unity for the low partial waves. Springer and Harvey

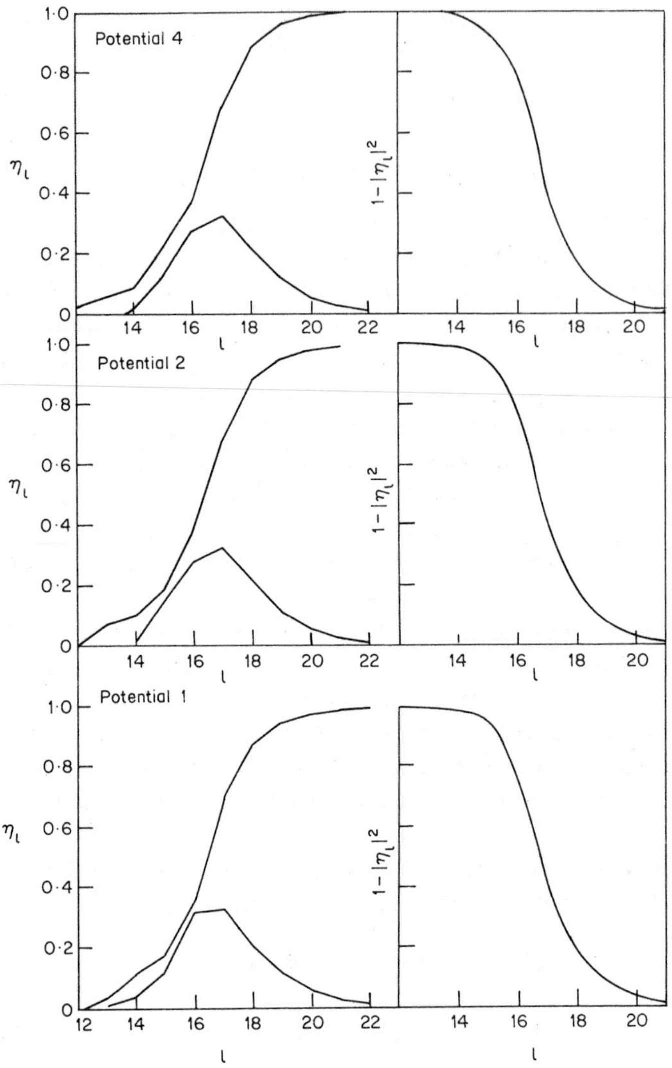

Figure 8.2 The reflection coefficients and transmission coefficients for 42 MeV α-particle scattering predicted by the potentials listed in Table 8.1. [From D. F. Jackson and C. G. Morgan, *Phys. Rev.* **175,** 1402 (1968).]

have added a first derivative to equation (8.11) and a second derivative to equation (8.12). Various forms of the complex model have been used with success to describe the scattering of complex particles at medium energies, and to describe the elastic scattering and polarization of nucleons at energies between 180 MeV and 20 GeV.

8.2 Ambiguities in the optical potential

The optical potentials for strongly absorbed particles show some interesting ambiguities. Sets of potentials can be found, whose depths increase by discrete amounts, which give essentially the same reflection coefficients and hence the same cross-sections for elastic scattering. The parameters of such a set of potentials are given in table 8.1 and the corresponding reflection coefficients are shown in figure 8.2. An interpretation of this phenomenon has been given [97] in terms of the WKB expression for the phase shift, which is given by (see equation (3.139))

$$\delta_l = \int_{r_0}^{\infty} \left[k^2 - \frac{2\mu}{\hbar^2} V_{\text{opt}} - \frac{l(l+1)}{r^2}\right]^{\frac{1}{2}} dr - \int_{r_0}^{\infty} \left[k^2 - \frac{l(l+1)}{r^2}\right]^{\frac{1}{2}} dr \, .$$

If the depths of the real and imaginary parts of the potential are changed in such a way that

$$\delta_l(r_0') = \delta_l(r_0) \pm n\pi \, , \qquad n = 1, 2, 3 \ldots \tag{8.13}$$

then the reflection coefficient η_l is unchanged for the lth partial wave. The δ_l for neighbouring partial waves are also unchanged to a good approximation, although for very shallow potentials it is not possible to satisfy equation (8.13) for all important partial waves and these consequently give a poorer fit to the data. Thus the scattering of strongly absorbed particles does depend on the interior region of the potential in such a way that there are discrete ambiguities in the parameters of the potential, and the elastic scattering data do not serve to resolve this ambiguity.

It was originally thought that fits to the elastic scattering data for α-particle scattering could be obtained for a wide range of parameters provided that the potential was unchanged in the surface region. For a Saxon-Woods potential this gives the conditions, due to Igo,

$$V_0 \, e^{(R_0 - r)/a} = \text{constant} \tag{8.14}$$

$$W_0 \, e^{(R_0' - r)/a'} = \text{constant} \, . \tag{8.15}$$

For a very small variation in the diffuseness parameter a these conditions reduce to the requirements that $V_0 \exp(R_0/a)$ and $W_0 \exp(R_0'/a')$ should be constant. It can be seen from table 8.1 that these quantities are roughly constant for potentials 1–4 and 9 which have similar values of a. Thus the con-

Table 8.1 Optical Potentials for Elastic α-Particle Scattering from ^{42}Ca at 42 MeV.* The equivalent radius R_{EQ} is given for the real part of the potential only.

	V_0 (MeV)	W_0 (MeV)	a (F)	a' (F)	r_0 (F)	r'_0 (F)	R_{EQ} (F)	$V_0 \exp(R_0/a)$	$W_0 \exp(R'_0/a')$	$R^{om}_{\frac{1}{2}}$ (F)	$V(R^{om}_{\frac{1}{2}})$	$W(R^{om}_{\frac{1}{2}})$
1	47·0	16·9	0·602	0·604	1·596	1·591	6·25	$4{\cdot}72 \times 10^5$	$1{\cdot}60 \times 10^5$	7·323	−2·34	−0·83
2	118·4	19·8	0·599	0·597	1·438	1·546	5·76	4·98	1·61	7·313	−2·44	−0·74
3	162·8	20·8	0·593	0·589	1·390	1·540	5·61	5·60	1·85	7·318	−2·44	−0·72
4	212·3	22·3	0·589	0·586	1·354	1·520	5·48	6·27	1·84	7·327	−2.45	−0·66
5	30·9	18·4	0·648	0·526	1·663	1·667	6·58	$2{\cdot}30 \times 10^5$	$11{\cdot}68 \times 10^5$	7·332	−2·58	−0·93
6	84·2	16·7	0·675	0·569	1·430	1·631	5·93	1·33	3·55	7·324	−2·50	−0·86
7	138·8	17·5	0·759	0·527	1·257	1·667	5·67	0·44	10·43	7·367	−2·63	−0·84
8	146·3	13·9	0·788	0·545	0·208	1·721	5·65	0·30	8·13	7·377	−2·55	−1·00
9	1336·0	160·4	0·603	0·605	1·038	1·191	4·65	$5{\cdot}30 \times 10^5$	$1{\cdot}50 \times 10^5$	7·367	−2·62	−0·77

* Taken from D. F. Jackson and C. G. Morgan, *Phys. Rev.* **175**, 1402 (1968).

ditions (8.14) and (8.15) remain valid although they must be modified to take account of the discrete ambiguities in V_0 and W_0. One of the consequences of condition (8.14) is that the halfway radius R_0 decreases as the depth V_0 increases, and hence the mean square radius and the equivalent radius of the potential are functions of V_0. This effect is clear from table 8.1, and implies that, because of the ambiguities in the phenomenological potentials for strongly absorbed particles, nuclear size information can not be extracted unambiguously from the size parameters of the potential. One way of resolving this ambiguity is the calculation of the relevant potentials, and there are two ways in which this has been done. Firstly, the potential may be obtained by averaging the appropriate nucleon optical potentials over the ground state of the projectile. For the deuteron optical potential this gives

$$V_{\text{opt}}(\boldsymbol{r}) = \int \phi_{\text{d}}^2(\boldsymbol{x})\left[V_{\text{n}}(\boldsymbol{r}+\tfrac{1}{2}\boldsymbol{x}) + V_{\text{p}}(\boldsymbol{r}-\tfrac{1}{2}\boldsymbol{x})\right] d\boldsymbol{x} \tag{8.16}$$

where ϕ_{d} is the ground state wavefunction for the deuteron, and V_{p} and V_{n} are the proton and neutron optical potentials taken at the appropriate energy. Such calculations indicate [98] that the depth of the real part of the potential for medium energy projectiles is roughly 50 A_p where A_p is the mass number of the projectile. Alternatively, the appropriate two-body interaction can be averaged over the ground state of the target nucleus (see section 6.3). For α-particles this procedure [80] yields a potential whose real depth is ~150–200 MeV which is consistent with the other method. Both these calculations are lowest-order approximations to the optical potential and can be improved by explicit inclusion of appropriate effects, such as deuteron break-up. As a calculation of the optical potential the first method is probably to be preferred as it deals with the real and imaginary parts of the potential on the same basis, but the second method offers a means of investigating the relation between the optical potential and the structure of the target nucleus.

In an alternative approach to the interpretation of the elastic scattering of strongly absorbed projectiles, the problem of the ambiguities in the optical potential is bypassed by the introduction of more fundamental size parameters. There is now a family of these *strong absorption radii* [94, 99]. But we shall limit the discussion here to two of these parameters, $R_{\frac{1}{2}}$ and $\bar{R}_{\frac{1}{2}}$. The radius $R_{\frac{1}{2}}$ is the radius calculated from equation (8.8) when the angular momentum is set equal to the value $L_{\frac{1}{2}}$ for which

$$\text{Re}\,\eta(L_{\frac{1}{2}}) = \tfrac{1}{2}\,. \tag{8.17}$$

Similarly, the radius $\bar{R}_{\frac{1}{2}}$ corresponds to the angular momentum $\bar{L}_{\frac{1}{2}}$ such that

$$1-|\eta(\bar{L}_{\frac{1}{2}})|^2=\tfrac{1}{2}. \tag{8.18}$$

The important feature of this approach is that the angular momenta $L_{\frac{1}{2}}$ and $\bar{L}_{\frac{1}{2}}$ can be obtained equally well from an optical model analysis or a parametrization of the reflection coefficients, and the values obtained should be independent of the ambiguities in the optical potential. In practice, the

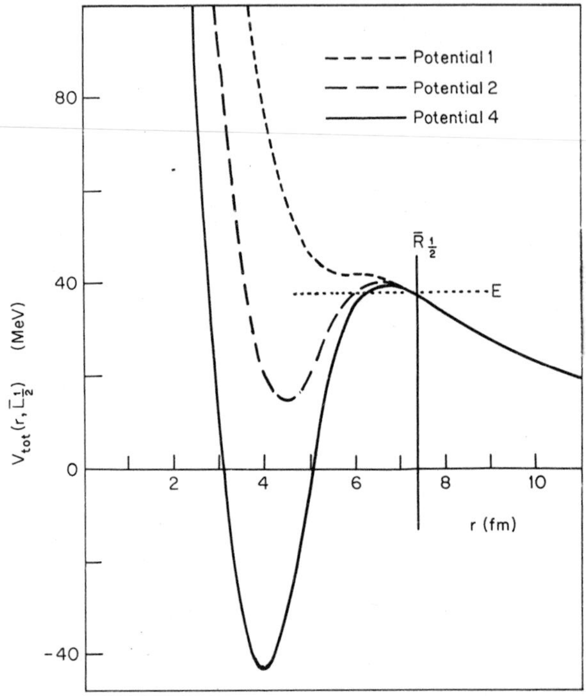

Figure 8.3 The real part of the effective potential evaluated at the critical angular momentum $\bar{L}_{\frac{1}{2}}$ for some of the potentials listed in Table 8.1.

variation in the values of $L_{\frac{1}{2}}$ obtained from optical model analyses are very much less than the variations in the corresponding optical parameters. Surveys of a wide range of data for ^{3}He, α-particle, and heavy ion scattering have indicated the strong absorption radius $R_{\frac{1}{2}}$ can be characterized by the formula [37, 94, 100]

$$R_{\frac{1}{2}}=r_0(A_1^{\frac{1}{3}}+A_2^{\frac{1}{3}})$$

where r_0 is a constant determined to have the value $1{\cdot}4 \sim 1{\cdot}5$ fm and A_1, A_2 are the mass numbers of the target and projectile. However, groups of neighbouring nuclei show substantial departures from this formula and there is good evidence that very precise determinations of $R_{\frac{1}{2}}$ can yield valuable nuclear size and nuclear structure information.

In figure 8.3 we have plotted the total potentials $V_{tot}(r, \bar{L}_{\frac{1}{2}})$ for several of the potentials given in table 8.1. This shows that the potentials which give

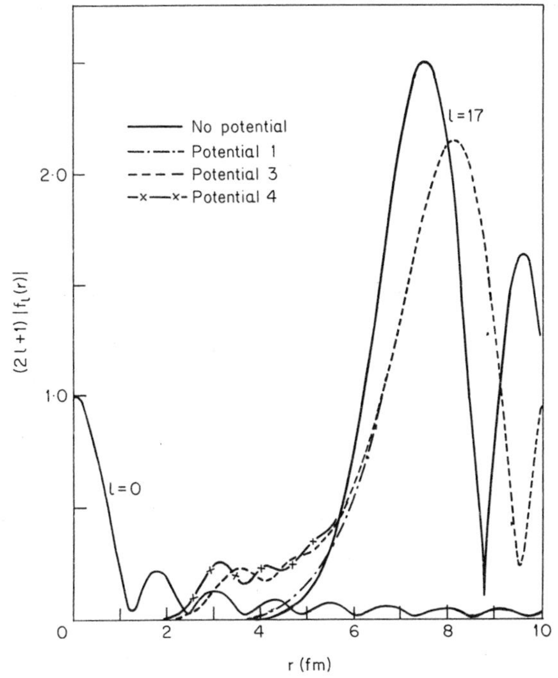

Figure 8.4 The radial parts of the wavefunctions with $l = 17$ for 42 MeV α-particles on ^{42}Ca using some of the potentials listed in Table 8.1 and for no potential. The zero potential result for $l = 0$ is also shown.

the same scattering are identical at and beyond $\bar{R}_{\frac{1}{2}}$. On the scale of the figure the difference between $R_{\frac{1}{2}}$ and $\bar{R}_{\frac{1}{2}}$ is negligible and it is not surprising therefore that the numerical values of the potentials at $R_{\frac{1}{2}}$, given in table 8.1, are almost constant. This ensures that the partial waves around $\bar{L}_{\frac{1}{2}}$ have the same reflection coefficients. For this calculation, the nearest integer to $L_{\frac{1}{2}}$ and $\bar{L}_{\frac{1}{2}}$ is 17, and in figure 8.4 we have therefore plotted the partial waves in the form $(2l+1)|f_l(kr)|$ for $l = 17$. The full line shows the result obtained when the potential is set equal to zero, so that $f_l(kr) = j_l(kr)$. The spherical Bessel function for $l = 0$ is also shown on the figure and by comparison it can be seen how the angular momentum barrier forces the higher partial waves away from the nucleus. As might be expected from the comparison of the potentials in figure 8.4, these potentials yield identical wavefunctions for $l = 17$ beyond $R_{\frac{1}{2}}$. In the intermediate region between 4 fm and 6 fm the shallowest potential yields results closest to the plane wave while the stronger attractive effect of the deeper potentials allows the wave to penetrate further into the interior of the nucleus, and this provides confirmation of the statement made at the beginning of this chapter, namely that for strong absorption to occur it is not necessary that the optical model wavefunction vanishes inside the nucleus. Further confirmation of this is provided by the total optical model wavefunctions shown in figure 8.5. From this figure it can be seen that on the bright side of the nucleus ($z < 0$) the wavefunctions show very similar behaviour and that the absorption effect begins in the vicinity of the strong absorption radius. On the dark side of the nucleus ($z > 0$) we see the focussing effect which was discussed in section 6.4 in connection with proton scattering. As the strength of the real part of the potential increases the focus increases in intensity and moves further inside the nucleus. At the same time secondary peaks are formed due to interference of the incident wave with a wave reflected from the far surface of the nucleus. For the very deep potential the focus moves further inside the nucleus, but the intensity is severely reduced owing to the large imaginary term in this potential, and only in this case can it be said that the probability for the α-particle to penetrate into the interior of the nucleus is negligible.

Comparison of figure 8.5 with figures 6.6 and 6.7 shows that for roughly the same depth of real potential, i.e. around 40–50 MeV, the focus observed in proton scattering is much stronger than that observed in α-particle scattering, but when the α-particle potential is increased to 150–200 MeV the intensity of the focus becomes comparable to that observed in proton scattering. This result again shows that it is important to take note of the ambiguities in the optical potential for strongly absorbed projectiles when making comparisons of the scattering of strongly absorbed and weakly absorbed projectiles.

8.3 Diffraction scattering

In the preceding section the strong absorption condition was formulated in angular momentum space through equation (8.3), and was connected with

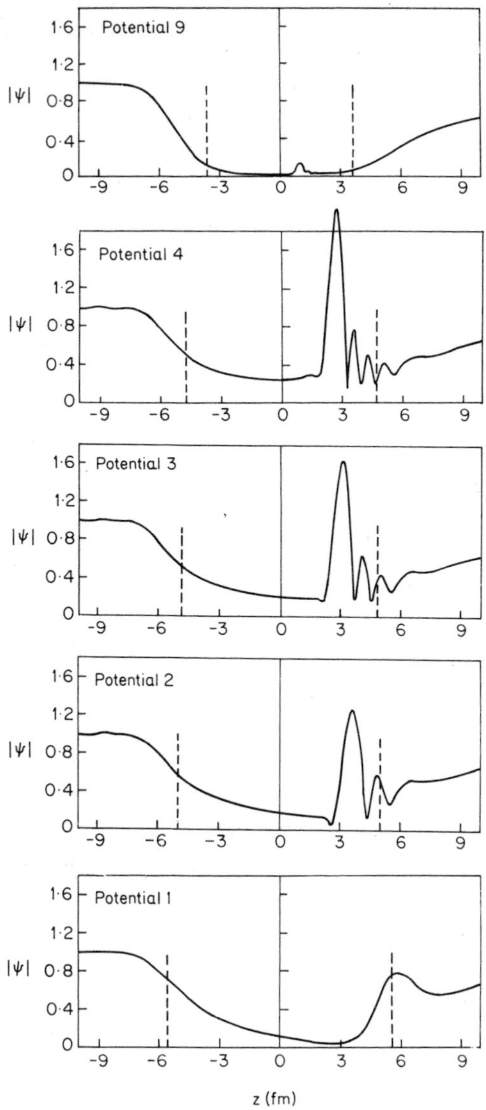

Figure 8.5 The moduli of the wavefunctions for 42 MeV α-particle scattering from ^{42}Ca calculated using some of the potentials listed in Table 8.1. The wavefunctions are calculated along the *z*-axis with the origin at the centre of the potential. The dotted lines indicate the position of the half-way radius of the potential. [From D. F. Jackson and C. G. Morgan, *Phys. Rev.* **175,** 1402 (1968).]

the interpretation in configuration space through the relation $L+\frac{1}{2}\approx kR$. A direct formulation in configuration space can be achieved by using classical diffraction theory.

The exact expression for the scattering amplitude is given by equation (3.100) as

$$f(\theta, \phi) = \frac{-\mu}{2\pi\hbar^2}\int e^{-i\mathbf{k}'\cdot\mathbf{r}}V(\mathbf{r})\psi^+(\mathbf{k}, \mathbf{r})d\mathbf{r}$$

and using the Schrödinger equation this can be transformed to

$$f(\theta, \phi) = -\frac{1}{4\pi}\int e^{-i\mathbf{k}'\cdot\mathbf{r}}(\nabla^2+k^2)\psi^+(\mathbf{k}, \mathbf{r})d\mathbf{r}$$

$$= -\frac{1}{4\pi}\int [e^{-i\mathbf{k}'\cdot\mathbf{r}}(\nabla^2\psi^+)-(\nabla^2 e^{-i\mathbf{k}'\cdot\mathbf{r}})\psi^+]d\mathbf{r}$$

where we have used the condition for elastic scattering, $k'^2=k^2$. The integration is over the volume within which $V(\mathbf{r})$ is non-zero, but by means of Green's theorem the integral can be converted to the integral over the surface S bounding this volume. Then

$$f(\theta, \phi) = -\frac{1}{4\pi}\int_S [e^{-i\mathbf{k}'\cdot\mathbf{r}}(\nabla\psi^+)-(\nabla e^{-i\mathbf{k}'\cdot\mathbf{r}})\psi^+]dS.$$

This is just Kirchhoff's integral which gives the mathematical formulation of Huygen's principle [94, 101]. Further, we can divide the surface S into an illuminated region I (aperture) and a shadow region J (screen) and apply Kirchhoff's assumptions for the diffraction of light by a small aperture, namely that the disturbance in the aperture is the same as the disturbance which would be observed if the screen were not present and that ψ and $\nabla\psi$ are zero on the screen itself. Thus we have the boundary conditions,

$$\text{on } I: \quad \psi^+ = e^{i\mathbf{k}\cdot\mathbf{r}}, \qquad \nabla\psi^+ = \nabla e^{i\mathbf{k}\cdot\mathbf{r}}$$

$$\text{on } J: \quad \psi^+ = 0, \qquad \nabla\psi^+ = 0,$$

and the scattering amplitude becomes

$$f(\theta, \phi) = \frac{ik}{4\pi}(1+\cos\theta)\int_I e^{i(\mathbf{k}-\mathbf{k}')\cdot\mathbf{r}}dS$$

$$\approx \frac{ik}{2\pi}\int_I e^{i(\mathbf{k}-\mathbf{k}')\cdot\mathbf{r}}dS, \text{ for small angles}. \tag{8.19}$$

It may be recalled that the term *Fraunhofer diffraction* is used when the source and the observer are a long way from the aperture [101], and that the corresponding diffraction patterns are calculated on the assumption that $e^{ik\phi(x, y)}$ may be expanded to first order in x and y.

The theory of diffraction scattering from nuclei was introduced by Blair, Drozdov, and Inopin to describe the elastic and inelastic scattering of strongly absorbed projectiles from collective nuclei [94]. In order to extend the theory given above to inelastic scattering it is necessary to make the *adiabatic approximation.* From equations (3.114) and (3.115) the exact expression for the transition matrix element is

$$T_{fi} = \langle \phi_f \Phi_f | V | \Psi^+ \rangle \tag{8.20}$$

$$= \left\langle \phi_f \Phi_f \right| V + V \frac{1}{E - H_0 - H(\xi) - V + i\varepsilon} V \left| \phi_i \Phi_i \right\rangle \tag{8.21}$$

where ϕ_i, ϕ_f are plane wave states for the initial and final momenta $\boldsymbol{k}$, $\boldsymbol{k}'$ and Ψ^+ is the exact scattering solution. The adiabatic approximation consists essentially in neglecting the presence of the nuclear Hamiltonian $H(\xi)$ in the Green's function and at the same time neglecting the energy loss of the projectile. The wavefunction Ψ^+ may then be approximated by

$$\Psi^+ \approx \Phi_f(\xi) \psi_\xi^+ (\boldsymbol{k}, \boldsymbol{r}) \tag{8.22}$$

where ψ^+ is a solution of the equation

$$[T + U_\xi(\boldsymbol{r}) - E] \psi_\xi^+ (\boldsymbol{k}, \boldsymbol{r}) = 0 \tag{8.23}$$

in which $U_\xi(\boldsymbol{r})$ is a generalized optical potential which contains the nuclear co-ordinates ξ as parameters. This gives

$$T_{fi} = \langle \Phi_f | t(\xi) | \Phi_i \rangle \tag{8.24}$$

$$t(\xi) = \langle \phi | U_\xi | \psi_\xi^+ \rangle \tag{8.25}$$

where $t(\xi)$ is the matrix element for elastic scattering of the projectile evaluated for fixed values of the co-ordinates ξ. In terms of the scattering amplitude we have

$$f(\theta, \phi) = \langle \Phi_f | f(\xi, \theta, \phi) | \Phi_i \rangle \, . \tag{8.26}$$

Thus in the adiabatic approximation, the nuclear motion is 'frozen' during the time of the collision and the nuclear states are assumed to be degenerate. This treatment is most likely to be valid for low-lying collective states and for projectiles whose energy is considerably greater than the spacings of these levels.

In the diffraction theory, the scattering amplitude $f(\xi, \theta, \phi)$ is given by equation (8.19) where the integration extends over the projection of the nuclear surface on a plane perpendicular to the incident direction $\boldsymbol{k}$. The z-axis is taken along the incident direction $\boldsymbol{k}$ and the scattered beam is taken to be in the x–z plane at an angle θ to the direction $\boldsymbol{k}$. To first order in the collective co-ordinates α_{lm} the required projection is bounded by the nuclear radius evaluated in the equatorial plane, i.e.

$$R\left(\frac{\pi}{2}, \phi\right) = R_0\left[1+\sum_{lm}\alpha_{lm}Y_l^m\left(\frac{\pi}{2}, \phi\right)\right] = R_0+\delta(\phi) \tag{8.27}$$

and the scattering amplitude is given by

$$f(\xi, \theta) = \frac{ik}{2\pi}\int_0^{2\pi} d\phi \int_0^{R_0+\delta(\phi)} e^{-ik\theta r\cos\phi}\, r\, dr$$

$$\approx \frac{ik}{2\pi}\int_0^{2\pi} d\phi \int_0^{R_0} e^{-ik\theta r\cos\phi}\, r\, dr + \frac{ik}{2\pi}\int_0^{2\pi} d\phi\, e^{-ik\theta R_0\cos\phi} R_0\delta(\phi) \tag{8.28}$$

where we have used the relations

$$|\boldsymbol{k}-\boldsymbol{k}'| \approx 2k\sin\theta/2 \approx k\theta$$

$$I[R_0+\delta(\phi)] = \int_0^{R_0+\delta(\phi)} e^{-ik\theta r\cos\phi}\, r\, dr \approx I(R_0)+\delta(\phi)\frac{dI}{dR_0}.$$

The first term in equation (8.28) can be evaluated using equation (3.137) and gives $ikR_0^2[J_1(kR_0\theta)/kR_0\theta]$ which is identical to the result previously obtained (equation (8.6)) for elastic scattering from a completely absorbing spherical nucleus. The cross-section arising from this term has the form

$$\frac{d\sigma}{d\Omega} = (kR_0^2)^2\left[\frac{J_1(kR\theta)}{kR\theta}\right]^2 \tag{8.29}$$

which is identical to the Airy pattern for the distribution of intensity due to Fraunhofer diffraction by a circular aperture.

In order to evaluate the second term in equation (8.28) the following formulae are needed [43],

$$\begin{aligned} Y_l^m\left(\frac{\pi}{2}, \phi\right) &= i^m\left(\frac{2l+1}{4\pi}\right)^{\frac{1}{2}}[l:m]\,e^{im\phi}, \qquad && l+m \text{ even}, \\ &= 0, && l+m \text{ odd}, \end{aligned} \tag{8.30}$$

where $$[l:m] = i^l \frac{[(l-m)!(l+m)!]^{\frac{1}{2}}}{(l-m)!!(l+m)!!} \qquad l+m \text{ even}, \tag{8.31}$$

and [17] $$\begin{aligned}\int_0^{2\pi} e^{iz\cos\phi} e^{im\phi}\, d\phi &= 2\pi i^m J_m(-z) \\ &= 2\pi(-i)^m J_m(z). \end{aligned} \tag{8.32}$$

The matrix elements of the α_{lm} can be evaluated as described in section 7.3. For example, with a spin-zero target and the vibrational model we have

$$\langle \Phi_f | \alpha_{lm} | \Phi_i \rangle = (h\omega_l/2C_l)^{\frac{1}{2}}.$$

Collecting these formulae together we obtain the cross-section for inelastic scattering. Thus the cross-section for excitation of a one-phonon state is given by

$$\frac{d\sigma}{d\Omega}(0^+ \to l) = (kR_0^2)^2 \sum_{\substack{m \\ l+m \text{ even}}} \left(\frac{2l+1}{4\pi}\right)\left(\frac{h\omega_l}{2C_l}\right) \frac{[(l-m)!(l+m)!]}{[(l-m)!!(l+m)!!]^2}$$

$$\times [J_{|m|}(kR_0\theta)]^2. \tag{8.33}$$

The cross-section for one-phonon excitation of state J_f from a ground state $J_i \neq 0$ is related to this basic cross-section by a statistical factor (see equation (7.20))

$$\frac{d\sigma}{d\Omega}(J_i \to J_f; l) = \frac{(2J_f+1)}{(2J_i+1)(2l+1)} \frac{d\sigma}{d\Omega}(0^+ \to l), \tag{8.34}$$

and a similar relation is obtained from the rotational model

$$\frac{d\sigma}{d\Omega}(J_i \to J_f; l) = (J_i K_i l 0 | J_f K_f)^2 \frac{d\sigma}{d\Omega}(0^+ \to l). \tag{8.35}$$

Some important predictions can be deduced from these expressions for the cross-sections for elastic and inelastic scattering [94]. The most important of these is the *Blair phase rule* which states that for sufficiently large θ the oscillations in the cross-section for an even parity excitation are out of phase with the oscillations in the cross-section for elastic scattering from the same nucleus, while the cross-section for an odd parity excitation is in phase with the elastic scattering. Since the parity of the excitation is $(-1)^l$ the rule follows from the condition that $l+m$ must be even and from the asymptotic behaviour of the Bessel functions J_m. Although this prediction is based on diffraction theory it is confirmed by experiment, and is reproduced in distorted wave calculations as can be seen from figure 7.2. The phase rule provides a valuable means of identifying the parity of excited states and distinguishing between states such as 2^+ and 3^-. For the excitation of two-phonon states the phase rule sometimes breaks down owing to the interference between multiple and direct second-order processes, and this effect is illustrated in figure 7.4. It also follows from equation (8.33) that the cross-section is zero at $\theta = 0$ if l is odd and is finite if l is even. This rule has been stated by Glendenning and by Kromminga and McCarthy, and again allows a determination of the parity of the excitation [94, 102]. In addition, the cross-section at small angles predicted from equation (8.33) is characteristic of the l-value so that measurements at small angles allow a determination of the multipolarity of the excitation.

The close connection between the strong absorption condition and classical diffraction theory explains why the experimental cross-sections for scattering of strongly absorbed projectiles show pronounced diffraction patterns. For weakly absorbed projectiles the assumption of a sharp surface is inappropriate and the experimental cross-sections do not show pronounced oscillations. The importance of the sharp surface is illustrated by the fact that the cross-section calculated in plane wave Born approximation from a sharp-surfaced potential sometimes gives reasonable agreement

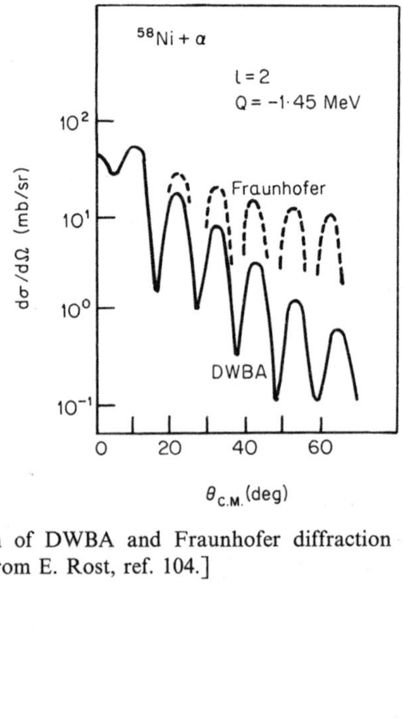

Figure 8.6 Comparison of DWBA and Fraunhofer diffraction calculations for 43 MeV α-particles on ^{58}Ni. [From E. Rost, ref. 104.]

with the data for scattering of strongly absorbed projectiles even though the use of plane waves is completely invalid in this situation.

The diffraction theory has been used to fit a great deal of data on the elastic and inelastic scattering of medium energy α-particles. A good fit to the data is obtained at small angles only, while at large angles the predicted diffraction pattern falls off too slowly. This effect is illustrated in figure 8.6. Nevertheless by fitting the spacing of the maxima and minima and the magnitude at small angles, values for the radius R_0 and the collective strength parameter β_l or $(\hbar\omega_l/2C_l)$ are determined. In this description of elastic and inelastic scattering what is being observed is essentially the shadow of the nuclear shape in the sense of physical optics [94, 103].

8.4 Distorted wave theory for strongly absorbed projectiles

The DWBA method described in section 7.3 is superior to the diffraction theory for inelastic scattering in that it permits an exact description of Coulomb scattering, does not require the adiabatic approximation, and is not limited to small angles. The effective interaction can be derived for any chosen nuclear model, but it is found that for strongly absorbed projectiles the angular variation of the cross-section is almost completely insensitive to the shape of the effective interaction and that the magnitude of the cross-section is determined by the magnitude of the effective interaction in the surface region. It appears that the angular distributions obtained using effective interactions calculated from a macroscopic model and from a microscopic model are almost identical, provided that these effective interactions are normalized so that their magnitudes are in agreement in the region of the strong absorption radius. This result can be understood in terms of the crucial importance of the partial waves with angular momenta near to $L_{\frac{1}{2}}$. We consider the radial integrals defined by equation (7.22)

$$U_{ll'L} = \int f_l(kr)\, F_L(r)\, f_{l'}(k'r)\, r^2\, dr$$

where $F_L(r)$ is the form factor of the effective interaction which may be thought of as spanning a fairly broad region of the nuclear surface. For the high partial waves $l, l' > L_{\frac{1}{2}}$ the strength of the angular momentum barrier causes the partial waves to be negligibly small in the region where $F_L(r)$ is significant so that the corresponding radial integrals are negligibly small. For the low partial waves which are strongly absorbed and have $\eta_l \approx 0$ there is essentially no outgoing wave, and in the lowest order WKB approximation the radial integrals are given by

$$U_{ll'L} \approx \int F_L(r) \exp[-ik_l(r) - ik_{l'}(r)]\, dr$$

where $k_l(r)$ is the local wave number. The phases of the exponential terms vary rapidly over the region spanned by F_L and the integrals average to zero. For the partial waves with $l, l' \approx L_{\frac{1}{2}}$ partial reflection does occur and the 'phase averaging' process is upset [104]. As can be seen from figure 8.5, these important partial waves overlap with the form factor most strongly in the region of the strong absorption radius and not at all in the interior of the nucleus. The form factor F_L may, or may not, be sharply localized in configuration space depending on the nuclear model chosen, but as a result of the strong absorption of the low partial waves the radial integrals are sharply localized in angular momentum space.

In the region of the nucleus to which the inelastic scattering of strongly absorbed projectiles is sensitive the wavefunctions of single-particle states have essentially similar radial behaviour, and for this reason transitions between them may be described in terms of a deformation of the nuclear surface through the extended optical model. Several comparisons of effective interactions derived from different nuclear models have confirmed this observation which explains the success of universal form factors of derivative form in DWBA calculations. (This effect can be seen in the results for the effective interactions for proton scattering shown in figure 7.3, but the weakly absorbed protons are sensitive to much more than the extreme tail of the effective interaction.)

The relation between the DWBA and the adiabatic method has been elucidated by Austern and Blair [105]. We give here a simplified discussion [94, 103] of their analysis, starting from the DWBA matrix element for a single excitation in a spin-zero target nucleus. For this discussion it is convenient to consider a more general form of the extended optical potential

$$V(R+\alpha(\hat{r}), r) = U(R, r) + W(R, \alpha(\hat{r}), r) \tag{8.36}$$

where U is the spherically symmetric optical potential which depends on the parameter R, and W is the perturbing potential which gives rise to excitations through the operator $\alpha(\hat{r})$. In the collective model, $\alpha(\hat{r})$ describes the displacement of the nuclear surface from the equilibrium position, but other interpretations are possible [105]. We expand W to first order in α and write the nuclear matrix element as

$$\langle LM \big| \alpha(\hat{r}) \frac{\partial U}{\partial R}(r, R) \big| 00 \rangle = C_1(L) \frac{\partial U}{\partial R}(r, R)\, Y_L^{M*}(\hat{r}), \tag{8.37}$$

and we also make the adiabatic approximation, so that $k' = k$. Then using equations (7.14), (7.16), (7.21), (7.23), and (8.37) the DWBA matrix element for inelastic scattering is given by

$$T_{fi}^{\text{in}} = (4\pi)(2L+1)^{\frac{1}{2}} C_1(L) \sum_{ll'} i^{l-l'} (2l'+1)^{\frac{1}{2}} e^{i(\sigma_l + \sigma_{l'})} Y_{l'}^{-M}(\theta, 0)$$

$$(l'0L0|l0)(l'-MLM|l0)\, U_{ll'} \tag{8.38}$$

where the radial integral is given by

$$U_{ll'} = k^{-2} \int_0^\infty u_{l'}(kr) \frac{\partial U}{\partial R}(r, R)\, u_l(kr)\, dr\,. \tag{8.39}$$

The exact transition matrix element for elastic scattering is obtained using equations (3.25), (3.32), and (3.116) as

$$T^{\text{el}} = \frac{2\pi\hbar^2}{\mu} \frac{i}{2k} (4\pi)^{\frac{1}{2}} \sum_l (2l+1)^{\frac{1}{2}} e^{2i\sigma_l} (\eta_l - 1)\, Y_l^0(\theta, 0)$$

where $\eta_l = \eta_l(R, \alpha)$ is the reflection coefficient for scattering from the complete potential V, and the contribution to elastic scattering due to the perturbing potential W is given by

$$T^{\text{el}} = \frac{2\pi\hbar^2}{\mu} \frac{i}{2k} (4\pi)^{\frac{1}{2}} \sum_l (2l+1)^{\frac{1}{2}} e^{2i\sigma_l} [\eta_l(R, \alpha) - \eta_l(R, 0)]\, Y_l^0(\theta, 0)\,.$$

(From the Gell-Mann-Goldberger relation, equation (3.125), the contributions to the transition matrix element from U and W are additive.) The difference between the reflection coefficients may be expanded in powers of α, and the adiabatic approximation can then be used to determine the inelastic matrix element, so that, to first order in α, we have

$$T_{fi}^{\text{in}} = \frac{iE}{2k^3} 4\pi C_1(0) \sum_l (2l+1)^{\frac{1}{2}} e^{2i\sigma_l} Y_l^0(\theta, 0) \frac{\partial \eta_l}{\partial R} \tag{8.40}$$

where we have used equation (8.24), and have set $Y_0^0(\hat{r}) = (4\pi)^{-\frac{1}{2}}$. The energy E is the kinetic energy of the projectile in the c.m. system. For the case when $L = 0$, equation (8.38) can be compared with equation (8.40) to yield the identity

$$k^2 U_{ll} = \frac{iE}{2k} \frac{\partial \eta_l}{\partial R}. \tag{8.41}$$

Thus the radial integrals which occur in DWBA can be connected, through the adiabatic approximation, with the derivatives of the elastic reflection coefficients.

It is assumed by Austern and Blair that in the case when $L \neq 0$, i.e. $l \neq l'$, the values of l and l' may be replaced by $\bar{l} = \frac{1}{2}(l + l')$ and that

$$k^2 U_{ll'} \approx \frac{iE}{2k} \frac{\partial \eta_{\bar{l}}}{\partial R}. \tag{8.42}$$

It is further assumed that $\eta_{\bar{l}}$ is a function of $\bar{l} - L_{\frac{1}{2}}$, as in all strong absorption models for the reflection coefficients. It can then be shown that

$$\frac{\partial}{\partial R} \eta_{\bar{l}} \approx -k \frac{\partial}{\partial \bar{l}} \eta(\bar{l} - L_{\frac{1}{2}}) \approx -k \frac{\partial}{\partial l'} \eta(l' - L_{\frac{1}{2}}). \tag{8.43}$$

These assumptions are valid for strongly absorbed projectiles because of the sharp localization in the radial integrals in angular momentum space around $\bar{l} \simeq L_{\frac{1}{2}}$. It can also be shown [105] that $\sum_l i^{l-l'} (l' 0 L 0 | l 0)(l' - M L M | l 0)$ is independent of l' if $l' \gg L$. Finally, using these approximations, neglecting Coulomb effects, putting $kR_{\frac{1}{2}} \simeq L_{\frac{1}{2}} + \frac{1}{2}$, $k\rho \approx l' + \frac{1}{2}$, and using the small angle approximation

$$Y_{l'}^{-M}(\theta, 0) \rightarrow \left[\frac{2l'+1}{4\pi}\right] J_M([l' + \tfrac{1}{2}]\theta)$$

we can convert the sum over l' to an integral, so that the matrix element (8.38) becomes

$$\begin{aligned} T_{fi}^{\text{in}} &\approx AC_1(L) \int k\rho \left[\frac{\partial}{\partial (k\rho)} \eta(k\rho - kR_{\frac{1}{2}})\right] J_M(k\rho\theta)\, d(k\rho) \\ &\approx AC_1(L) kR_{\frac{1}{2}} J_M(kR_{\frac{1}{2}}\theta) \end{aligned}$$

where A represents all the inessential constants. Thus, by a sequence of approximations based on a knowledge of the behaviour of the radial integrals and reflection coefficients for strongly absorbed particles, it is possible to derive the Fraunhofer diffraction formula from the DWBA matrix element.

9 Transfer and knock-out reactions

9.1 Basic properties of transfer and knock-out reactions

In the two preceding chapters we have considered only those direct interactions which lead to excitation of the target without change of mass number, i.e. inelastic scattering and charge-exchange. There is, however, a large group of direct reactions which involve the addition or removal of a single nucleon or a small group of nucleons from the target nucleus. It is convenient to classify these reactions into *transfer reactions* in which a nucleon or group of nucleons is transferred between the projectile and the target, and *knock-out reactions* in which the projectile knocks a nucleon or group of nucleons out of the nucleus giving rise, in general, to a multi-particle final state. The transfer reactions are normally subdivided into *stripping reactions* in which the projectile is stripped of x nucleons so that the target nucleus A has x nucleons added to it to make the residual nucleus $B(A+x)$, and *pick-up reactions* in which these x nucleons are picked up by the projectile so that the target nucleus makes the transition $B(A \neq x) \rightarrow A$. Thus knock-out and pick-up reactions give rather similar information about the effect of making holes in the target ground state while a stripping reaction gives information about the effect of adding particles.

The basic information obtainable from these reactions may be considered in terms of the conservation of energy and the conservation of total angular momentum. For the pick-up reaction

$$A + a \rightarrow b(a+x) + B(A-x) \tag{9.1}$$

we denote the kinetic energy of the projectile a and of the lightest reaction product b by E_a and E_b respectively. Using the standard definitions [1], the binding energy of particle b is given by

$$B_b = (m_a + m_x - m_b)c^2 , \tag{9.2}$$

the separation energy for the removal of particle x from nucleus A is given by

$$S_{xA} = (m_x + M_B - M_A)c^2 \tag{9.3}$$

and the Q-value of the reaction is given by

$$Q = E_b - E_a = (m_a + M_A - m_b - M_B)c^2 \tag{9.4}$$

where we have assumed that the recoil energy of the residual nucleus is negligible. Thus the Q-value for the pick-up reaction (9.1) is given by

$$Q = -S_{xA} + B_b . \tag{9.5}$$

Similarly, the Q-value for the stripping reaction

$$A + a(c+x) \rightarrow c + C(A+x) \tag{9.6}$$

is given by

$$Q = E_c - E_a = S_{xc} - B_a , \tag{9.7}$$

and the Q-value for the knock-out reaction

$$A + a \rightarrow a + x + B(A-x) \tag{9.8}$$

is given by

$$Q = E_a' + E_x - E_a = -S_{xA} , \tag{9.9}$$

where E_a' is the final kinetic energy of particle a. Thus a measurement of the energy spectrum of the light reaction product in a pick-up reaction, or the summed energy spectrum in a knock-out reaction, yields information about the separation energies of individual nucleons or groups of nucleons in the target nucleus, while a similar measurement for a stripping reaction yields information on separation energies in the residual nucleus. The occurrence of peaks in the energy spectra at Q-values whose magnitudes are greater than that for the ground state transition indicates the presence and energy of excited states of the residual nucleus. The total angular momentum of these states must be given by

$$\boldsymbol{J}_B = \boldsymbol{J}_A + \boldsymbol{j}_x \tag{9.10}$$

or

$$J_A + l_x + s_x \geqslant J_B \geqslant |J_A - |l_s - s_x|| \tag{9.11}$$

and hence if it is possible to identify j_x and l_x from the shape of the angular distribution, the reaction can be used to identify the spin and parity of the excited states of the residual nucleus.

As an example we consider some reactions on ^{12}C. The separation energies of a proton in ^{12}C, a neutron in ^{12}C, and a neutron in ^{13}C are 15·956 MeV, 18·721 MeV, and 4·947 MeV respectively, and the binding energy of the deuteron is 2·225 MeV. Thus the reactions $^{12}C(p, 2p)^{11}B$, $^{12}C(p, pn)^{11}C$, $^{12}C(p, d)^{11}C$, and $^{12}C(d, p)^{13}C$ have Q-values for the ground state transitions of −15·956 MeV, −18·721 MeV, −16·496 MeV, and +2·722 MeV, respectively. It follows that the deuteron stripping reaction can be

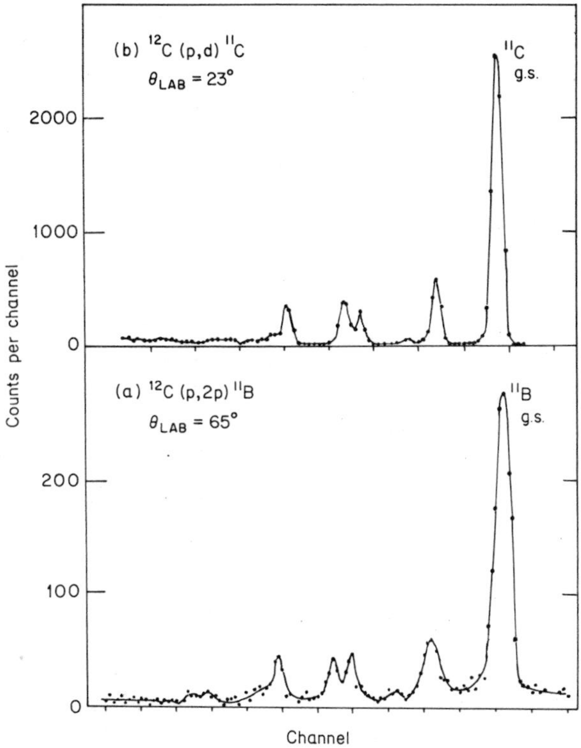

Figure 9.1 (**a**) A summed energy spectrum for the $^{12}C(p,2p)$ reaction at 50 MeV. (**b**) A deuteron energy spectrum for the reaction $^{12}C(p,d)$ at the same energy. [From H. G. Pugh, *et al.*, ref. 125.]

usefully investigated with a fairly low energy deuteron beam, whereas a proton beam of substantially higher energy is required for the other reactions. Some energy spectra for the $^{12}C(p, d)$ and $^{12}C(p, 2p)$ reactions are shown in figure 9.1. The spin assignments for the excited states are also shown, and it is interesting to note that since ^{12}C has a 0^+ ground state the occurrence of low-lying states with spin greater than $\frac{3}{2}^-$ indicates a failure either of the simple shell model or the one-step direct reaction mechanism. This particular problem is discussed further in section 9.5.

The detailed investigation of nuclear structure by means of these reactions is discussed in section 9.4. In sections 9.2 and 9.3 we derive the formulae used for the analysis of the data and discuss the validity of the approximations used.

9.2 Formalism for transfer reactions

For the two-body reaction $A(a, b)B$ the differential cross-section is given by

$$\frac{d\sigma}{d\Omega} = \frac{\mu_a \mu_b}{4\pi^2\hbar^4} \frac{k_b}{k_a} \frac{\sum |T_{ba}|^2}{(2J_A+1)(2S_a+1)} \tag{9.12}$$

where μ_a, μ_b are the reduced mass in the entrance and exit channels and the sum is over the magnetic quantum numbers M_A, M_B, m_a, m_b. The Hamiltonian for the system is

$$H = H_a + H_A + T_{aA} + V_{aA} \tag{9.13}$$

$$= H_b + H_B + T_{bB} + V_{bB} \tag{9.14}$$

where T_{aA}, T_{bB} are the kinetic energy operators for the relative motion of the particles in the entrance and exit channels, H_i is the internal Hamiltonian for particle i and V_{ij} is the total interaction between particles i and j. The transition matrix element in DWBA is then obtained from equation (3.142) as

$$T = \langle \chi_b^- \phi_b \Phi_B | V_{aA} - U_{aA} | \chi_a^+ \phi_a \Phi_A \rangle \tag{9.15}$$

where U_{aA} is the optical potential for elastic scattering in the entrance channel, ϕ_a, ϕ_b, Φ_A, Φ_A are the eigenstates of H_a etc. and $\chi^\pm$ are the distorted wavefunctions for particles a and b. Thus

$$H_A \Phi_A = \varepsilon_A \Phi_A\,, \quad H_a \phi_a = \varepsilon_a \phi_a \tag{9.16}$$

$$(T_{aA} + U_{aA})\chi_a^+ = (E - \varepsilon_a - \varepsilon_A)\chi_a^+\,, \tag{9.17}$$

and similarly for the final state.

In the case when the reaction $A(a, b)B$ involves pick-up of x nucleons the total interaction can be written as the sum,

$$V_{aA} = V_{aB} + V_{ax}$$

so that the transition matrix element becomes

$$T = \langle \chi_b^- \phi_b \Phi_B | V_{ax} + (V_{aB} - U_{aA}) | \chi_a^+ \phi_a \Phi_A \rangle .$$

It is usually assumed that the matrix element of $(V_{aB} - U_{aA})$ is essentially zero [106]. The argument for this is that the elastic effects of V_{aB} are largely contained in U_{aA} while the inelastic effects are small, and the assumption is probably most valid for heavy nuclei.

The increase in complexity in the matrix element for a rearrangement collision compared with inelastic scattering becomes apparent when we introduce the co-ordinates explicitly. We denote the internal co-ordinates, including spin co-ordinates, of the residual nucleus B by ξ, the internal co-ordinates of the projectile a by $\boldsymbol{\sigma}_a$, and the internal co-ordinates of the group of x nucleons by $\boldsymbol{\sigma}_x$. The target nucleus A is then described by the co-ordinates ξ, $\boldsymbol{\sigma}_x$, and the relative radial co-ordinate $\boldsymbol{r}_{xB}$, and the reaction product b is described by the co-ordinates $\boldsymbol{\sigma}_a$, $\boldsymbol{\sigma}_x$ and the relative co-ordinate $\boldsymbol{r}_{ax}$. If we neglect the spin-orbit terms in the optical potentials, the distorted wavefunctions are functions only of the separation distances of the colliding pair $\boldsymbol{r}_{aA}$ or $\boldsymbol{r}_{bB}$. Thus the transition matrix element can be written as

$$T = \langle \chi_b^- (\boldsymbol{k}_b, \boldsymbol{r}_{bB}) \phi_b(\boldsymbol{r}_{ax}, \boldsymbol{\sigma}_a, \boldsymbol{\sigma}_x) \Phi_B(\xi) | V_{ax}(\boldsymbol{r}_{ax}) | \times \chi_a^+ (\boldsymbol{k}_a, \boldsymbol{r}_{aA}) \phi_a(\boldsymbol{\sigma}_a) \Phi_A(\xi, \boldsymbol{\sigma}_x, \boldsymbol{r}_{xB}) \rangle . \tag{9.18}$$

But this can be simplified using the relations

$$\boldsymbol{r}_{aA} = \boldsymbol{r}_{ax} + \frac{B}{A} \boldsymbol{r}_{xB} \tag{9.19}$$

$$\boldsymbol{r}_{bB} = \boldsymbol{r}_{xB} + \frac{a}{b} \boldsymbol{r}_{ax} \tag{9.20}$$

where B/A represents M_B/M_A and a/b represents m_a/m_b, and by carrying out the integrations over ξ, $\boldsymbol{\sigma}_a$ and $\boldsymbol{\sigma}_x$, which gives

$$\int d\xi \, \Phi_B^*(\xi) \, \Phi_A(\xi, \boldsymbol{\sigma}_x, \boldsymbol{r}_{xB}) = \psi_x(\boldsymbol{r}_{xB}, \boldsymbol{\sigma}_x) \tag{9.21}$$

$$\int d\boldsymbol{\sigma}_a d\boldsymbol{\sigma}_x \phi_b^* (\boldsymbol{r}_{ax}, \boldsymbol{\sigma}_a, \boldsymbol{\sigma}_x) \psi_x(\boldsymbol{r}_{xB}, \boldsymbol{\sigma}_x) \phi_a(\boldsymbol{\sigma}_a) = C_{ax} \psi_x(\boldsymbol{r}_{xB}) \phi_b^* (\boldsymbol{r}_{ax}) \tag{9.22}$$

where C_{ax} is a constant introduced so that ϕ_b can be normalized to unity, and we have assumed that the particles a and x are in a relative S-state in b. The structure and properties of ϕ_b and ψ_x are discussed below and in section 9.4.

Using equations (9.19)–(9.22), the matrix element (9.18) becomes

$$T = C_{ax} \int d\boldsymbol{r}_{ax}\, d\boldsymbol{r}_{xB}\, \chi_b^{-*}\left(\boldsymbol{k}_b, \boldsymbol{r}_{xB} + \frac{a}{b}\boldsymbol{r}_{ax}\right) \phi_b^*(\boldsymbol{r}_{ax}) V_{ax}(\boldsymbol{r}_{ax}) \psi_x(\boldsymbol{r}_{xB})$$

$$\times\, \chi_a^+\left(\boldsymbol{k}_a, \boldsymbol{r}_{ax} + \frac{B}{A}\boldsymbol{r}_{xB}\right). \qquad (9.23)$$

We see from equation (9.23) that the general form for the matrix element for a pick-up reaction in DWBA contains a six-dimensional integral, and the same must be true for the matrix element for a stripping reaction. This integral can be reduced to a three-dimensional integral or a product of three-dimensional integrals in certain special cases. For example, in plane wave Born approximation, the distorted wavefunctions are replaced by

$$\chi_b^{-*} \simeq \exp\left(-i\boldsymbol{k}_b \cdot \boldsymbol{r}_{xB} - i\boldsymbol{k}_b \cdot \frac{a}{b}\boldsymbol{r}_{ax}\right), \quad \chi_a^+ \simeq \exp\left(i\boldsymbol{k}_a \cdot \boldsymbol{r}_{ax} + i\boldsymbol{k}_a \cdot \frac{B}{A}\boldsymbol{r}_{xB}\right),$$

so that the matrix element separates into

$$T = C_{ax} \int d\boldsymbol{r}_{xB}\, e^{-i\boldsymbol{P}\cdot\boldsymbol{r}_{xB}} \psi_x(\boldsymbol{r}_{xB}) \int d\boldsymbol{r}_{ax}\, \phi_b^*(\boldsymbol{r}_{ax}) V(\boldsymbol{r}_{ax}) e^{i\boldsymbol{K}\cdot\boldsymbol{r}_{ax}}, \qquad (9.24)$$

where

$$\boldsymbol{K} = \boldsymbol{k}_a - \frac{a}{b}\boldsymbol{k}_b, \quad \boldsymbol{P} = \boldsymbol{k}_b - \frac{B}{A}\boldsymbol{k}_a. \qquad (9.25)$$

Thus in PWBA the matrix element factorizes into two three-dimensional integrals. The first of these can be interpreted as the probability of finding particle x in nucleus A with momentum $\boldsymbol{P}$. The second integral may be written as

$$G(K) = \langle \phi_b(r) | V(r) | e^{i\boldsymbol{K}\cdot\boldsymbol{r}} \rangle = (2\pi)^{-3} \int d\boldsymbol{q}\, \langle \phi_b(\boldsymbol{r}') | e^{i\boldsymbol{q}\cdot\boldsymbol{r}'} \rangle$$

$$\times \langle e^{i\boldsymbol{q}\cdot\boldsymbol{r}} | V(\boldsymbol{r}) | e^{i\boldsymbol{K}\cdot\boldsymbol{r}} \rangle \qquad (9.26)$$

which has been interpreted by Chew and Goldberger as follows. The first factor in the integrand is the probability for the particles a and x to be found with relative momentum $\boldsymbol{q}$ in the bound state b and the second factor is the probability for the free particles a and x with initial relative momentum $\boldsymbol{K}$ to scatter into the relative momentum $\boldsymbol{q}$, i.e. $G(K)$ represents the probability that the momentum $\boldsymbol{P}$ of particle x can be combined with the momentum of the incident particle a through the interaction V_{ax} to form a relative momentum found in particle b. In order to evaluate $G(K)$ it is assumed that V_{ax} is the potential that binds a and x to form b, so that ϕ_b obeys the equation

$$\left[-\frac{\hbar^2}{2\mu_{ax}}\nabla^2+V_{ax}\right]\phi_b=\varepsilon_b\phi_b$$

where $-\varepsilon_b=\hbar^2\alpha^2/2\mu_{ax}$ is the binding energy of B_b of particle b. Hence

$$G(K) = -\frac{\hbar^2}{2\mu_{ax}}(K^2+\alpha^2)\int\phi_b^*(\boldsymbol{r})e^{i\boldsymbol{K}\cdot\boldsymbol{r}}d\boldsymbol{r}\,. \tag{9.27}$$

In DWBA the only way of reducing the matrix element (9.24) to a three-dimensional integral is to make a zero-range approximation. The usual way of doing this is to write [86]

$$V_{ax}(\boldsymbol{r}_{ax})\,\phi_b(\boldsymbol{r}_{ax})=D_0\,\delta(\boldsymbol{r}_{ax})=D_0\delta\left(\boldsymbol{r}_{aA}-\frac{B}{A}\boldsymbol{r}_{xB}\right) \tag{9.28}$$

so that the matrix element reduces to

$$T=C_{ax}D_0\int d\boldsymbol{r}\;\chi_b^{-*}(\boldsymbol{k}_b,\boldsymbol{r})\,\psi_x(\boldsymbol{r})\,\chi_a^+\left(\boldsymbol{k}_a,\frac{B}{A}\boldsymbol{r}\right). \tag{9.29}$$

From (9.27) it follows that

$$D_0=G(0)$$

and that the use of the zero-range approximation is equivalent to taking only the zero momentum component of $V\phi$. We may still interpret the integral in equation (9.29) as the probability of finding the particle x in nucleus A with certain momentum, in a range now determined by the local momenta introduced by the distortion.

In the (p, d) reaction, which is the most important of the pick-up reactions, the particles a, x, and b, are to be identified as proton, neutron, and deuteron, respectively. Thus V_{ax} is the p–n interaction and ϕ_b is the internal wavefunction of the deuteron. Then if ϕ_b is taken to be the Hulthen wavefunction for the S-state of the deuteron

$$\phi(r) = \left[\frac{\alpha\beta(\alpha+\beta)}{2\pi(\alpha-\beta)^2}\right]^{\frac{1}{2}}\frac{e^{-\alpha r}-e^{-\beta r}}{r}$$

where $\beta\approx 6.2\alpha$, $G(K)$ is given by [86]

$$G(K) = \frac{\varepsilon_b(8\pi)^{\frac{1}{2}}}{\alpha^{\frac{3}{2}}}\left(\frac{\alpha+\beta}{\beta}\right)^{\frac{3}{2}}\frac{\beta^2}{K^2+\beta^2} = G(0)\frac{\beta^2}{K^2+\beta^2}, \tag{9.30}$$

with

$$K=|\boldsymbol{k}_\mathrm{p}-\tfrac{1}{2}\boldsymbol{k}_\mathrm{d}|=[k_\mathrm{p}^2+\tfrac{1}{4}k_\mathrm{d}^2-k_\mathrm{p}k_\mathrm{d}\cos\theta]^{\frac{1}{2}} \tag{9.31}$$

and $$D_0^2 = \frac{8\pi\varepsilon_b^2}{\alpha^3}\left(\frac{\alpha+\beta}{\beta}\right)^3 \approx 1{\cdot}5\times 10^4\ \mathrm{MeV}^2.\mathrm{fm}^3. \qquad (9.32)$$

The error in the zero-range DWBA has been tested by exact finite-range calculations [107]. In the PWBA the finite-range correction is the transform of the two-body force and falls off rapidly with large momentum differences, i.e. large scattering angles, as can be seen from equations (9.27) and (9.31). The effect of distortion is to spread the momentum components of the wave-functions of the scattered particles, so introducing additional high and low

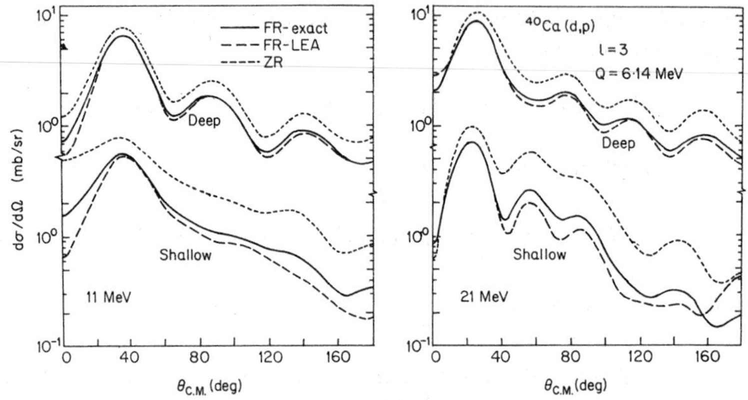

Figure 9.2 Comparison of finite range and zero range calculations for the ^{40}Ca(d,p) reaction at 11 MeV and 21 MeV. The shallow deuteron potential has $V_0 = 67$ MeV and the deep potential has $V_0 = 112$ MeV. [From J. K. Dickens *et al.*, ref. 107.]

momentum components. These additional low momentum components have an important effect at large angles where they prevent the cross-section in DWBA from falling as rapidly as in PWBA. Thus at medium energies and low Q-values the combined effect of the distortion and the finite-range correction is to reduce the cross-section fairly uniformly over the angular range. This effect is illustrated in figure 9.2. The high momentum components introduced by the distortion tend to be localized in the nuclear interior where the real optical potential is greatest, and consequently the effect of the finite-range correction is to reduce the contributions from the nuclear interior. Figure 9.3 shows the effect of multiplying the cross-section obtained in zero-range DWBA by the plane wave correction factor $|G(K)|^2$; this is clearly an unsatisfactory procedure because it is the combined effect of distortion and finite range which must be taken into account in the matrix

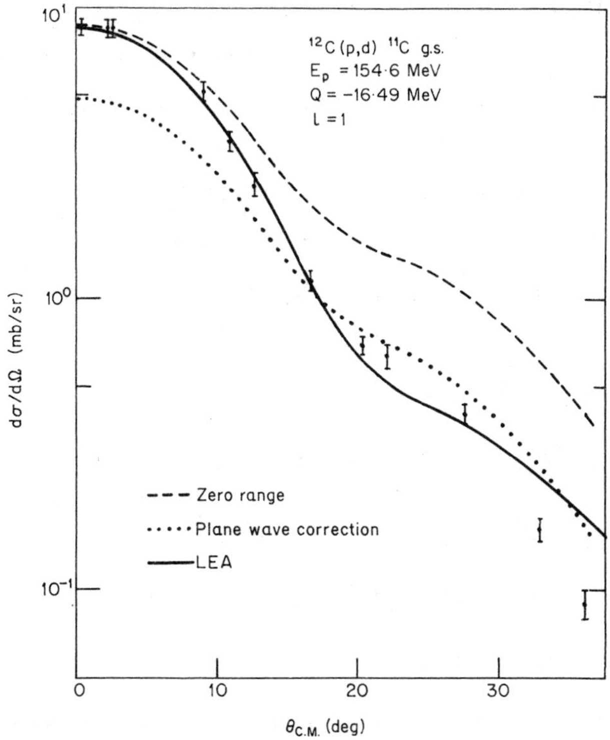

Figure 9.3 Comparison of zero range and approximate finite range calculations for the ^{12}C(p,d) reaction at 155 MeV. [From I. S. Towner, *Nucl. Phys.* **A93**, 145 (1967).]

element. On the other hand, the exact finite-range calculation is very lengthy, especially at energies in the region of 150 MeV where many partial waves are required, and for this reason some approximate methods for finite-range corrections have been developed.

The effective mass approximation and the local energy approximation (LEA) are equivalent in first order. We describe here the first-order LEA using the approach of Buttle and Goldfarb [108]. We consider the (p, d) reaction so that the factor a/b in equation (9.20) is $\frac{1}{2}$, and we set $B/A \approx 1$. The essential step in this approach is to make a Taylor expansion of the distorted waves,

$$\chi_p^+(\boldsymbol{k}_p, \boldsymbol{r}_{nB}+\boldsymbol{r}_{pn}) = \chi_p^+(\boldsymbol{k}_p, \boldsymbol{r}_{nB}) + (\boldsymbol{r}_{pn}\cdot\nabla)\chi_p^+(\boldsymbol{k}_p, \boldsymbol{r}_{nB}) + \tfrac{1}{2}(\boldsymbol{r}_{pn}\cdot\nabla)^2\chi_p^+(\boldsymbol{k}_p, \boldsymbol{r}_{nB}) + \dots$$

$$\chi_{\mathrm{d}}^{-*}(\boldsymbol{k}_{\mathrm{d}}, \boldsymbol{r}_{\mathrm{n}B}+\tfrac{1}{2}\boldsymbol{r}_{\mathrm{pn}}) = \chi_{\mathrm{d}}^{-*}(\boldsymbol{k}_{\mathrm{d}}, \boldsymbol{r}_{\mathrm{n}B}) + (\tfrac{1}{2}\boldsymbol{r}_{\mathrm{pn}}\cdot\boldsymbol{\nabla})\chi_{\mathrm{d}}^{-*}(\boldsymbol{k}_{\mathrm{d}}, \boldsymbol{r}_{\mathrm{n}B}) + \tfrac{1}{2}(\tfrac{1}{2}\boldsymbol{r}_{\mathrm{pn}}\cdot\boldsymbol{\nabla})^2\chi_{\mathrm{d}}^{-*}(\boldsymbol{k}_{\mathrm{d}}, \boldsymbol{r}_{\mathrm{n}B}) + \ldots$$

where the gradient operator operates with respect to $\boldsymbol{r}_{\mathrm{n}B}$. These expansions can be written symbolically as

$$\begin{aligned}\chi_{\mathrm{p}}^{+}(\boldsymbol{k}_{\mathrm{p}}, \boldsymbol{r}_{\mathrm{n}B}+\boldsymbol{r}_{\mathrm{pn}} &= \exp(\boldsymbol{r}_{\mathrm{pn}}\cdot\boldsymbol{\nabla}_{\mathrm{p}})\chi_{\mathrm{p}}^{+}(\boldsymbol{k}_{\mathrm{p}}, \boldsymbol{r}_{\mathrm{n}B})\\ &= \exp(i\boldsymbol{r}_{\mathrm{pn}}\cdot\boldsymbol{k}_{\mathrm{p}}^{\mathrm{op}})\chi_{\mathrm{p}}^{+}(\boldsymbol{k}_{\mathrm{p}}, \boldsymbol{r}_{\mathrm{n}B})\end{aligned} \tag{9.33}$$

$$\begin{aligned}\chi_{\mathrm{d}}^{-*}(\boldsymbol{k}_{\mathrm{d}}, \boldsymbol{r}_{\mathrm{n}B}+\tfrac{1}{2}\boldsymbol{r}_{\mathrm{pn}}) &= \exp(\tfrac{1}{2}\boldsymbol{r}_{\mathrm{pn}}\cdot\boldsymbol{\nabla}_{\mathrm{d}})\chi_{\mathrm{d}}^{-*}(\boldsymbol{k}_{\mathrm{d}}\cdot\boldsymbol{r}_{\mathrm{n}B})\\ &= \exp(i\tfrac{1}{2}\boldsymbol{r}_{\mathrm{pn}}\cdot\boldsymbol{k}_{\mathrm{d}}^{\mathrm{op}})\chi_{\mathrm{d}}^{-*}(\boldsymbol{k}_{\mathrm{d}}, \boldsymbol{r}_{\mathrm{n}B})\,.\end{aligned} \tag{9.34}$$

The matrix element (9.23) then becomes

$$T = (2\pi)^{\frac{3}{2}} C_{\mathrm{pn}} \int d\boldsymbol{r}_{\mathrm{n}B} D(\boldsymbol{K}_{\mathrm{op}})\chi_{\mathrm{d}}^{-*}(\boldsymbol{k}_{\mathrm{d}}, \boldsymbol{r}_{\mathrm{n}B})\psi_{\mathrm{n}}(\boldsymbol{r}_{\mathrm{n}B})\chi_{\mathrm{p}}^{+}(\boldsymbol{k}_{\mathrm{p}}, \boldsymbol{r}_{\mathrm{n}B}) \tag{9.35}$$

where

$$\boldsymbol{K}_{\mathrm{op}} = \boldsymbol{k}_{\mathrm{p}}^{\mathrm{op}} + \tfrac{1}{2}\boldsymbol{k}_{\mathrm{d}}^{\mathrm{op}} = -i(\boldsymbol{\nabla}_{\mathrm{p}} + \tfrac{1}{2}\boldsymbol{\nabla}_{\mathrm{d}}) \tag{9.36}$$

$$D(\boldsymbol{K}_{\mathrm{op}}) = (2\pi)^{-\frac{3}{2}} \int d\boldsymbol{r}_{\mathrm{pn}} D(\boldsymbol{r}_{\mathrm{n}})\exp(i\boldsymbol{r}_{\mathrm{pn}}\cdot\boldsymbol{K}_{\mathrm{op}}) \tag{9.37}$$

$$D(\boldsymbol{r}_{\mathrm{pn}}) = V_{\mathrm{pn}}(\boldsymbol{r}_{\mathrm{pn}})\phi_{\mathrm{d}}(\boldsymbol{r}_{\mathrm{pn}})\,. \tag{9.38}$$

Thus $D(\boldsymbol{K}_{\mathrm{op}})$ is the operator corresponding to the Fourier transform of $D(\boldsymbol{r})$, i.e. it replaces $G(K)$ in the plane wave theory. Using the Hulthen wave-function we have

$$D(\boldsymbol{K}_{\mathrm{op}}) = \frac{G(0)}{(2\pi)^{\frac{3}{2}}}\frac{\beta^2}{\beta^2+\boldsymbol{K}_{\mathrm{op}}^2} = \frac{G(0)}{(2\pi)^{\frac{3}{2}}}\left(1 - \frac{\boldsymbol{K}_{\mathrm{op}}^2}{\beta^2} + \frac{\boldsymbol{K}_{\mathrm{op}}^4}{\beta^4} - \ldots\right) \tag{9.39}$$

Buttle and Goldfarb restrict their analysis to terms of order $1/\beta^2$ and show by means of Green's theorem on the integral in equation (9.35) that $\boldsymbol{K}_{\mathrm{op}}^2$ may be replaced by $\frac{1}{4}\nabla_{\mathrm{d}}^2 - \frac{1}{2}\nabla_{\mathrm{p}}^2 - \frac{1}{2}\nabla_{\mathrm{n}}^2$ where $\boldsymbol{\nabla}_{\mathrm{n}}$ operates on $\psi_{\mathrm{n}}(\boldsymbol{r}_{\mathrm{n}B})$. The effect of this operator can be obtained from the equations

$$\nabla_{\mathrm{d}}^2\chi_{\mathrm{d}} = -\frac{4m}{\hbar^2}(T_{\mathrm{d}} - U_{\mathrm{d}B})\chi_{\mathrm{d}}\,, \quad \nabla_{\mathrm{p}}^2\chi_{\mathrm{p}} = -\frac{2m}{\hbar^2}(T_{\mathrm{p}} - U_{\mathrm{p}A})\chi_{\mathrm{p}}\,,$$

$$\nabla_{\mathrm{n}}^2\psi_{\mathrm{n}} = -\frac{2m}{\hbar^2}(-S_{\mathrm{n}B} - V_{\mathrm{n}B})\psi_{\mathrm{n}}\,,$$

so that the matrix element (9.35) becomes

$$T = C_{\text{pn}} G(0) \int d\boldsymbol{r}\, \chi_{\text{d}}^{-*}(\boldsymbol{k}_{\text{d}}, \boldsymbol{r}) \Lambda(r) \psi_{\text{n}}(\boldsymbol{r}) \chi_{\text{p}}^{+}(\boldsymbol{k}_{\text{p}}, \boldsymbol{r}) \tag{9.40}$$

where
$$\Lambda(r) = 1 - \frac{m}{\hbar^2 \beta^2} [U_{\text{d}B} - U_{\text{p}A} - V_{\text{n}B} - T_{\text{d}} - T_{\text{p}} - S_{\text{n}B}]$$

$$= 1 - \frac{m}{\hbar^2 \beta^2} [U_{\text{d}B} - U_{\text{p}A} - V_{\text{n}B} - B_{\text{d}}], \tag{9.41}$$

m is the nucleon mass and $V_{\text{n}B}$ is the real potential which binds the neutron to the core B in the target nucleus A. The same result is obtained for the stripping reaction.

The results obtained with the LEA shown in figures 9.2 and 9.3 indicate that this approximation reproduces most of the effects of the exact finite-range calculation and makes possible a satisfactory fit to the data.

It is customary to make a fractional parentage expansion for the wave-function Φ_A, i.e.

$$\Phi_{J_A}^{M_A}(\xi, \boldsymbol{\sigma}_x, \boldsymbol{r}_{xB}) = \sum_{jmJp} \mathscr{I}_{J_A J_P}(j) \Phi_{J_P}^{M_P}(\xi) \psi_j^m(\boldsymbol{r}_{xB}, \boldsymbol{\sigma}_x)(J_P M_p jm | J_A M_A) \tag{9.42}$$

so that the *overlap integral* of the initial and final state becomes

$$\psi_x(\boldsymbol{r}_{xB}, \boldsymbol{\sigma}_x) = \int d\xi\, \Phi_B^* \Phi_A$$

$$= \sum_{jm} \mathscr{I}_{J_A J_B}(j)(J_B M_B jm | J_A M_A) \psi_j^m(\boldsymbol{r}_{xB}, \boldsymbol{\sigma}_x). \tag{9.43}$$

We now assume that particle x moves within nucleus A with total angular momentum $\boldsymbol{j}$ composed of orbital angular momentum $\boldsymbol{l}$ and spin $\boldsymbol{s}$. If x is a single nucleon, then $\boldsymbol{s}$ is the intrinsic spin, but if x is a composite particle or cluster, $\boldsymbol{s}$ is the total internal angular momentum of the cluster and may take several values. The overlap integral becomes

$$\psi_x(\boldsymbol{r}_{xB}, \boldsymbol{\sigma}_x) = \sum_{\substack{jls \\ m\lambda\mu}} (J_B M_B jm | J_A M_A)(l\lambda s\mu | jm) \mathscr{I}_{J_A J_B}(lsj) R_{nlj}(r_{xB})$$

$$\times i^l Y_l^\lambda(\hat{\boldsymbol{r}}_{xB}) \psi_s^\mu(\boldsymbol{\sigma}_x). \tag{9.44}$$

The parentage coefficient $\mathscr{I}$ in the expansion (9.42) selects a particular cluster or nucleon. If there are N_{lj} identical particles x within A, the cross-section is proportional to the *spectroscopic factor* S [109], where*

$$S_{J_A J_B}(lsj) = N_{lj}[\mathscr{I}_{J_A J_B}(lsj)]^2 . \tag{9.45}$$

Similarly, if there are ν identical particles x within b, we have

$$\int \phi_b^*(\boldsymbol{r}_{ax}, \sigma_x, \sigma_a)\,\psi_x(\boldsymbol{r}_{xB}, \sigma_x)\,\phi_a(\sigma_a)\,d\sigma_a\,d\sigma_x = a(s)(s_a m_a s\mu|s_b m_b)\,\phi_b^*(\boldsymbol{r}_{ax})\,\psi_x(\boldsymbol{r}_{ax}) \tag{9.46}$$

and the cross-section is proportional to $\nu|a(s)|^2$. By comparing equation (9.46) with equation (9.22) we find

$$C_{ax}^2 = \nu[a(s)]^2 .$$

Thus $S_{ax} = C_{ax}^2$ is the spectroscopic factor for the overlap of the light nuclei a and b. Finally, the cross-section is obtained by using equations (9.12), (9.23), (9.43), (9.44), (9.45), and (9.46) and carrying out the sums over magnetic substates. For the (p, d) reaction this gives

$$\frac{d\sigma}{d\Omega} = \frac{3}{4}\,\frac{m^2 A(A-1)}{2\pi^2\hbar^4(A+1)^2}\,\frac{k_{\mathrm{d}}}{k_{\mathrm{p}}}\sum_{lj}\frac{S_{J_A J_B}(lj)}{2l+1}\sum_{\lambda}|B_{l\lambda}|^2 \tag{9.47}$$

where, for a finite range approximation,

$$B_{l\lambda} = \int d\boldsymbol{r}_{\mathrm{pn}}\,d\boldsymbol{r}_{\mathrm{n}B}\,\chi_{\mathrm{d}}^{-*}(\boldsymbol{k}_{\mathrm{d}}, \boldsymbol{r}_{\mathrm{n}B} + \tfrac{1}{2}\boldsymbol{r}_{\mathrm{pn}})\,\phi_{\mathrm{d}}^*(\boldsymbol{r}_{\mathrm{pn}})\,V_{\mathrm{pn}}(\boldsymbol{r}_{\mathrm{pn}})\,\psi_{\mathrm{n}}(\boldsymbol{r}_{\mathrm{n}B})$$
$$\times\,\chi_{\mathrm{p}}^{+}\left(\boldsymbol{k}_{\mathrm{p}}, \boldsymbol{r}_{\mathrm{pn}} + \frac{A-1}{A}\boldsymbol{r}_{\mathrm{n}B}\right), \tag{9.48}$$

or, in zero-range approximation,

$$B_{l\lambda} = D_0\int d\boldsymbol{r}\,\chi_{\mathrm{d}}^{-*}(\boldsymbol{k}_{\mathrm{d}}, \boldsymbol{r})\,R_{nlj}(r)\,Y_l^{\lambda}(\hat{\boldsymbol{r}})\,\chi_{\mathrm{p}}^{+}\left(\boldsymbol{k}_{\mathrm{p}}, \frac{A-1}{A}\boldsymbol{r}\right). \tag{9.49}$$

* The spectroscopic factor is sometimes expressed in terms of the *reduced width* θ^2 which is defined as the square of the overlap of the initial and final nuclear wavefunctions. Thus

$$\theta^2 = \left|\int \Phi_A \Phi_B^*\,d\xi\right|^2 = S(lj)\,\theta_0^2(lj)$$

where θ_0^2 is called the *single-particle reduced width* since by comparison with equation (9.43) it depends on the square of the single-particle wavefunction, i.e.

$$\theta_0(lj) = (J_B M_B jm|J_A M_A)\,\psi_j^m(\boldsymbol{r}_{xB}, \sigma_x) .$$

When evaluated at the surface θ^2 is proportional to γ^2, the reduced width defined for resonance reactions.

In this formalism the number N in equation (9.45) is the number of neutrons in the sub-shell labelled by nlj. Alternatively, the I-spin formalism can be used [109, 110], in which case the spectroscopic factor and fractional parentage coefficient are labelled by $T_A T_B$, the product C of I-spin C–G coefficients $(t_a m_{ta} t\ m_t | t_b m_{tb})$ and $(T_B M_{TB} t\ m_t | T_A M_{TA})$ appears in equation (9.43) and an additional factor C^2 appears in equation (9.47). The numerical values of S and $\mathscr{I}$ are different in the two cases, and the number N is now the number of active nucleons in the relevant subshell.

The cross-section for the (d, p) reaction can be obtained directly from equation (9.47) since it is the time-reversed reaction to (p, d). Using equation (3.84) we have

$$\frac{d\sigma}{d\Omega}\,\mathrm{dp} = \frac{(2s_\mathrm{p}+1)(2J_A+1)}{(2s_\mathrm{d}+1)(2J_B+1)}\,\frac{k_\mathrm{p}^2}{k_\mathrm{d}^2}\,\frac{d\sigma}{d\Omega}\,\mathrm{pd}$$

$$= \tfrac{2}{3}\,\frac{2J_A+1}{2J_B+1}\,\frac{k_\mathrm{p}^2}{k_\mathrm{d}^2}\,\frac{d\sigma}{d\Omega}\,\mathrm{pd}\,. \qquad (9.50)$$

The expressions (9.47) and (9.50) have been used with considerable success for the analysis of data, but for calculations at energies below ~60 MeV it is desirable to take account of the spin dependence of the distorted waves as the spin-orbit terms in the distorting potentials have an important effect on the j-dependence of the cross-sections (see section 9.4). The formalism for inclusion of spin-orbit distortion has been given by Goldfarb and Johnson [111]. Considerable simplification was achieved in the DWBA formalism through the assumption that particles a and x are in a relative S-state in b, but recent studies by Johnson and Santos [112] indicate that the D-state component of the deuteron makes a significant contribution to (p, d) and (d, p) reactions. More elaborate corrections arise from consideration of the exact matrix element (3.142a) for the transition in which the product $\chi_b^- \phi_b \Phi_B$ appearing in the DWBA matrix element (9.15) is replaced by the exact solution Ψ_{bB}^-. This leads to coupling between the elastic scattering and reaction channels.

It may be noted that when the DWBA formalism given above is applied to multi-nucleon transfer reactions, the interaction V_{ax} describes the interaction between the projectile and the 'lump' x. Alternatively, the interaction may be written as the sum of the interactions between the projectile and the individual nucleons in x. It is necessary to express the wavefunction of the nucleons forming x in terms of the internal co-ordinates and the co-ordinate

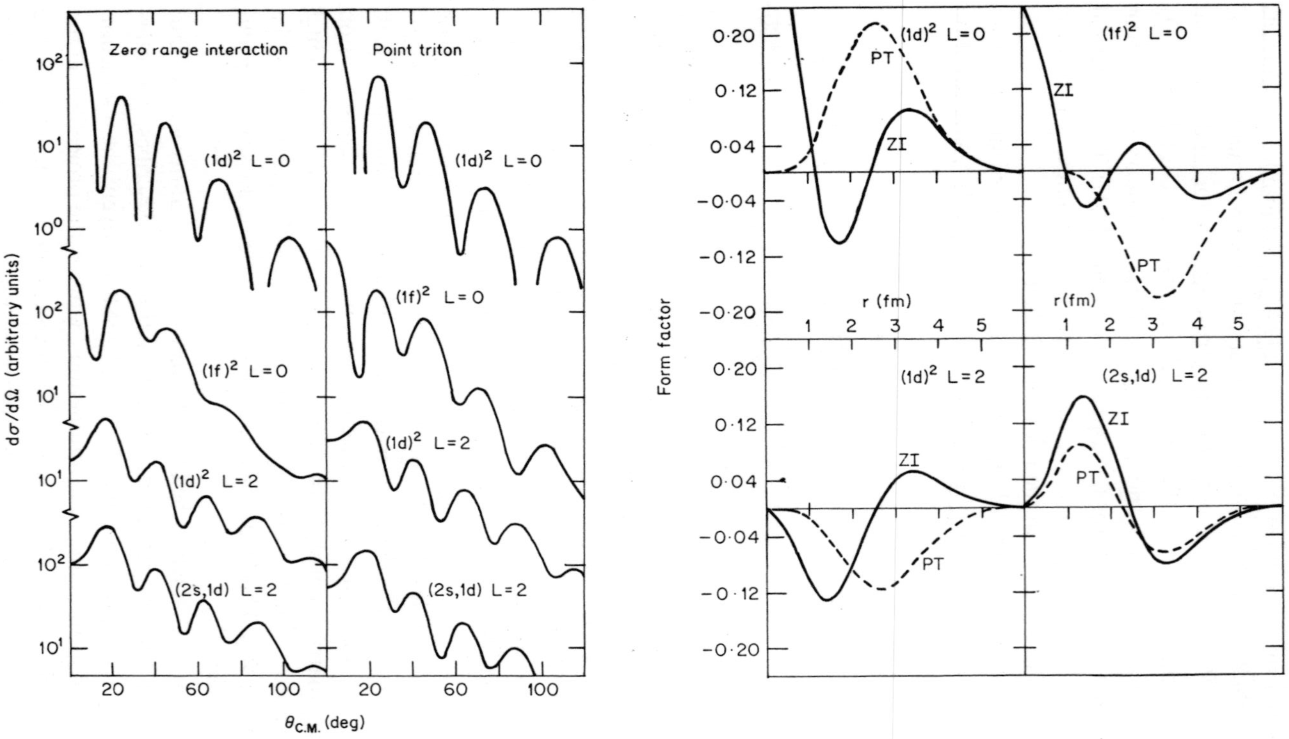

Figure 9.4 Comparison of cross-sections and form factors for the $^{40}Ca(p,t)$ reaction calculated using the zero range approximation and the point triton approximation. [From I. S. Towner and J. C. Hardy, ref. 114.]

of the centre of mass, and within the framework of the shell model this is normally done by using oscillator single-particle wavefunctions which are convenient for this purpose but incorrect (see section 2.2) or by expanding more realistic single-particle wavefunctions in terms of oscillator functions [113]. For a two-nucleon transfer reaction, the DWBA cross-section can be written in the form

$$\frac{d\sigma}{d\Omega} \propto \sum_{LST\lambda} f(S, T) |\sum_{l_1 l_2} [S(l_1 l_2; LSJT)]^{\frac{1}{2}} B_{L\lambda}(\theta)|^2$$

where $l_1 l_2$ are the angular momenta of the two nucleons, $LSJT$ are the quantum numbers of the pair, the function $f(S, T)$ depends on the particular reaction, and $S(l_1 l_2; LSJT)$ is the spectroscopic factor. Some calculations by Towner and Hardy [114] for the (p, t) reactions are shown in figure 9.4 which compare two zero-range approximations. The 'point triton' approximation involves a zero-range approximation for the triton wavefunction, and the 'zero-range interaction' involves a zero-range form for V_{ax}. The fact that these two approximations produce such different radial form factors but such similar cross-sections indicates that the contributions from the nuclear interior are not important.

9.3 Formalism for knock-out reactions

If the energy of the projectile is sufficiently high so that the reaction occurs by a direct one-step mechanism, knock-out reactions such as the (p, 2p) or (α, 2α) reactions may be described by a distorted wave formalism analogous to that used for transfer reactions. (If the one-step mechanism is not applicable the process is called sequential. Such processes are described in section 9.5.) In order to show the similarity between knock-out and transfer reactions we write the transition matrix element in the simplest possible way, using PWBA and neglecting spins. If $\boldsymbol{p}_0$, $\boldsymbol{q}_0$, $\boldsymbol{q}_1$ are the momenta of the incoming proton and the two outgoing protons in the lab system, this simple matrix element for the (p, 2p) reaction is given by [115]

$$T_{fi} = \int d\boldsymbol{r}_0 \, d\boldsymbol{r}_1 \, e^{-i\boldsymbol{q}_0 \cdot \boldsymbol{r}_0} e^{-i\boldsymbol{q}_1 \cdot \boldsymbol{r}_1} V(|\boldsymbol{r}_0 - \boldsymbol{r}_1|) e^{i\boldsymbol{p}_0 \cdot \boldsymbol{r}_0} \psi_p(\boldsymbol{r}_1)$$

where ψ_p is the overlap integral

$$\psi_p(\boldsymbol{r}_1) = \int d\xi \, \Phi^*_{A-1}(\xi) \Phi_A(\xi, \boldsymbol{r}_1), \tag{9.51}$$

and if we now substitute $\boldsymbol{r}=\boldsymbol{r}_0-\boldsymbol{r}_1$ the matrix element separates into

$$T_{fi}=\int d\boldsymbol{r}\, V(r)e^{i(\boldsymbol{p}_0-\boldsymbol{q}_0)\cdot \boldsymbol{r}} \int d\boldsymbol{r}_1\, e^{i(\boldsymbol{p}_0-\boldsymbol{q}_0-\boldsymbol{q}_1)\cdot \boldsymbol{r}_1}\psi_p(\boldsymbol{r}_1)\,. \tag{9.52}$$

The first of these integrals is the transform of the p–p interaction with respect to the momentum transfer $\boldsymbol{p}_0-\boldsymbol{q}_0$ which can be related to the matrix element for p–p scattering. From momentum conservation we have

$$\boldsymbol{p}_0=\boldsymbol{q}_0+\boldsymbol{q}_1+\boldsymbol{Q} \tag{9.53}$$

where $\boldsymbol{Q}$ is the recoil momentum of the residual nucleus, so that the second integral in equation (9.52) reduces to

$$g(Q)=\int d\boldsymbol{r}_1\, e^{i\boldsymbol{Q}\cdot \boldsymbol{r}_1}\psi_p(\boldsymbol{r}_1) \tag{9.54}$$

which is the probability of finding the proton in the nucleus with momentum Q. Comparison of equation (9.52) with equation (9.24) and its interpretation reveals the basic similarity between the direct interaction picture of the knock-out reaction and of the pick-up reaction, and the energy spectra shown in figure 9.1 indicate that this similarity is observed experimentally.

The plane wave formalism is often useful for a preliminary discussion of a reaction but it is never reliable for analysis of data. For a distorted wave description of knock-out reactions we would expect to use DWBA or DWIA according to the energy of the projectile, but an additional difficulty arises owing to the fact that the final state of a knock-out reaction is a three-body system consisting of two light outgoing particles and the recoiling residual nucleus. We now discuss the description of the final state, following the approach of Jackson and Berggren [116].

The Hamiltonian for the final state in a (p, 2p) reaction is given by

$$H_{A+1}=H_{A-1}(\xi)+T_0+T_1+T_C+V_{0C}+V_{1C}+V_{01} \tag{9.55}$$

where T_0, T_1, T_C are the kinetic energy operators for the outgoing protons and the residual nucleus C composed of $A-1$ nucleons, V_{0C}, V_{1C} are the interactions of the outgoing protons and the residual nucleus, and V_{01} is the effective interaction between the two protons. The total kinetic energy of the final state in the lab. system is

$$T=T_0+T_1+T_C=\frac{h^2}{2}[\boldsymbol{q}]^T[m^{-1}][\boldsymbol{q}] \tag{9.56}$$

where $[\boldsymbol{q}]$, $[\boldsymbol{q}]^T$ are column and row vectors composed of the elements $\boldsymbol{q}_0$, $\boldsymbol{q}_1$. $\boldsymbol{Q}$, and $[m^{-1}]$ is the matrix

$$[m^{-1}] = \begin{bmatrix} \frac{1}{m} & 0 & 0 \\ 0 & \frac{1}{m} & 0 \\ 0 & 0 & \frac{1}{(A-1)m} \end{bmatrix}$$

where m is the proton mass. In order to calculate the distorted wavefunctions we have to describe the motion of the outgoing particles in terms of suitable relative co-ordinates and the momenta conjugate to these co-ordinates. We let the basic set of co-ordinates be $\boldsymbol{r}_0$, $\boldsymbol{r}_1$, $\boldsymbol{R}_C$ referred to some arbitrary origin, and let the new set of vectors be $\boldsymbol{x}_1$, $\boldsymbol{x}_2$, $\boldsymbol{x}_3$ obtained by the transformation

$$[\boldsymbol{x}] = [a][\boldsymbol{r}]. \qquad (9.57)$$

Invariance of the action function for free particles requires that

$$[\boldsymbol{q}]^T.[\boldsymbol{r}] = [\boldsymbol{k}]^T.[\boldsymbol{x}]$$

so that

$$[\boldsymbol{q}]^T = [\boldsymbol{k}]^T[\boldsymbol{a}] \qquad (9.58)$$

and the kinetic energy can be written as

$$T = \frac{h^2}{2}[\boldsymbol{k}]^T[a][m^{-1}][a]^T[\boldsymbol{k}].$$

In general the matrix $[a][m^{-1}][a]^T$ is not diagonal, and there are terms of the form $\boldsymbol{k}_i\boldsymbol{k}_j\,(j \neq i)$ appearing in the kinetic energy which represent a coupling between the motions of the outgoing particles. The way in which this coupling appears depends on the choice of relative co-ordinates.

We first consider the *symmetric co-ordinates*

$$\boldsymbol{r}_{0C} = \boldsymbol{r}_0 - \boldsymbol{R}_C, \quad \boldsymbol{r}_{1C} = \boldsymbol{r}_1 - \boldsymbol{R}_C, \quad \boldsymbol{R}_{A+1} = \frac{1}{A+1}[\boldsymbol{r}_0 + \boldsymbol{r}_1 + (A-1)\boldsymbol{R}_C] \qquad (9.59)$$

and denote the conjugate momenta by $\boldsymbol{k}_{0C}$, $\boldsymbol{k}_{1C}$, $\boldsymbol{R}_{A+1}$. Then, using equations (9.53), (9.57) and (9.58) we have

$$\boldsymbol{k}_{0C} = \boldsymbol{q}_0 - \frac{1}{A+1}\boldsymbol{p}_0, \quad \boldsymbol{k}_{1C} = \boldsymbol{q}_1 - \frac{1}{A+1}\boldsymbol{p}_0, \quad \boldsymbol{K}_{A+1} = \boldsymbol{q}_0 + \boldsymbol{q}_1 + \boldsymbol{Q} \tag{9.60}$$

and

$$T = \frac{\hbar^2 A}{2m(A-1)}(k_{0C}^2 + k_{1C}^2) + \frac{\hbar^2}{m(A-1)}\boldsymbol{k}_{0C}\cdot\boldsymbol{k}_{1C} + \frac{\hbar^2}{2m(A+1)}K_{A+1}^2. \tag{9.61}$$

Thus the unperturbed Hamiltonian can be written in the form

$$H_{A+1} - V_{01} = H_{A-1} + T_{0C} + V_{0C} + T_{1C} + V_{1C} + T_{\text{coup}}, \tag{9.62}$$

and only if

$$T_{\text{coup}} = \frac{\hbar^2}{m(A-1)}\boldsymbol{k}_{0C}\cdot\boldsymbol{k}_{1C}$$

can be neglected can the final state wavefunction be written in the product form

$$\Psi^{\text{f}}_{A+1} = \Phi_{A-1}(\xi)\chi_f^-(\boldsymbol{k}_{0C}, \boldsymbol{r}_{0C})\chi_f^-(\boldsymbol{k}_{1C}, \boldsymbol{r}_{1C})e^{i\boldsymbol{K}_{A+1}\cdot\boldsymbol{R}_{A+1}}. \tag{9.63}$$

The neglect of T_{coup} can be justified in the limit of large A but is not valid in the plane wave limit.

An alternative choice of co-ordinates is given by the *di-proton co-ordinates*

$$\boldsymbol{r}_{01} = \boldsymbol{r}_0 - \boldsymbol{r}_1, \quad \boldsymbol{r}_{BC} = \tfrac{1}{2}\boldsymbol{r}_0 + \tfrac{1}{2}\boldsymbol{r}_1 - \boldsymbol{R}_C, \quad \boldsymbol{R}_{A+1}, \tag{9.64}$$

and proceeding as before we find

$$\boldsymbol{k}_{01} = \tfrac{1}{2}(\boldsymbol{q}_0 - \boldsymbol{q}_1), \quad \boldsymbol{k}_{BC} = \boldsymbol{q}_0 + \boldsymbol{q}_1 - \frac{2}{A+1}\boldsymbol{p}_0, \quad \boldsymbol{K}_{A+1} = \boldsymbol{q}_0 + \boldsymbol{q}_1 + \boldsymbol{Q}, \tag{9.65}$$

$$T = \frac{\hbar^2}{m}k_{01}^2 + \frac{\hbar^2(A+1)}{4(A-1)m}k_{BC}^2 + \frac{\hbar^2}{2(A+1)m}K_{A+1}^2. \tag{9.66}$$

In this case the unperturbed Hamiltonian becomes

$$\begin{aligned} H_{A+1} - V_{01} &= H_{A-1} + T_{01} + T_{BC} + V_{0C}(\boldsymbol{r}_{BC} + \tfrac{1}{2}\boldsymbol{r}_{01}) + V_{1C}(\boldsymbol{r}_{BC} - \tfrac{1}{2}\boldsymbol{r}_{01}) + T_{A+1} \\ &= H_{A-1} + T_0 + T_{BC} + [V_{0C}(r_{BC}) + V_{1C}(r_{BC})] \\ &\quad + V_{\text{coup}}(\boldsymbol{r}_{01}, \boldsymbol{r}_{BC}) + T_{A+1} \end{aligned} \tag{9.67}$$

where we have made a Taylor expansion of the potentials to give

$$V_{\text{coup}} = \sum_{n=1}^{\infty} \frac{(\tfrac{1}{2}\boldsymbol{r}_{01}\cdot\boldsymbol{\nabla}_{BC})^n}{n!}[V_{0C}(r_{BC}) + (-1)^n V_{1C}(r_{BC})], \tag{9.68}$$

and if V_{coup} can be neglected the final state wavefunction can be factorized into

$$\Psi^f_{A+1} = \Phi_{A-1}(\xi)\chi_f^-(\boldsymbol{k}_{BC}, \boldsymbol{r}_{BC})e^{i\boldsymbol{k}_{01}\cdot\boldsymbol{r}_{01}}e^{i\boldsymbol{K}_{A+1}\cdot\boldsymbol{R}_{A+1}}. \tag{9.69}$$

The factorization is exact in the plane wave limit, and is a valid approximation if the distorting potentials are small or if the interaction $V_{01}(\boldsymbol{r}_{01})$ is short-range.

Another possible set of co-ordinates are the *initial state co-ordinates*,

$$\boldsymbol{r}_{0A} = \boldsymbol{r}_0 - \frac{1}{A}[\boldsymbol{r}_1 - (A-1)\boldsymbol{R}_C], \quad \boldsymbol{r}_{1C}, \quad \boldsymbol{R}_{A+1}. \tag{9.70}$$

If we use these co-ordinates for the final state also, we find

$$\boldsymbol{k}_{0A} = \boldsymbol{q}_0 - \frac{1}{A}\boldsymbol{p}_0, \quad \boldsymbol{k}_{1C} = \boldsymbol{q}_1 + \frac{1}{A}\boldsymbol{q}_0 - \frac{1}{A}\boldsymbol{p}_0, \quad \boldsymbol{K}_{A+1} = \boldsymbol{q}_0 + \boldsymbol{q}_1 + \boldsymbol{Q}, \tag{9.71}$$

$$T = T_{0A} + T_{1C} + T_{A+1}. \tag{9.72}$$

In this case the coupling appears in the optical potential $V_{0C}(\boldsymbol{r}_{0A} + \boldsymbol{r}_{1C}/A)$ and again the unperturbed wavefunction can not be factorized unless terms of order $1/A$ are neglected. In the initial state the optical potential is defined to be $V_{0A}(\boldsymbol{r}_{0A})$ so that the unperturbed Hamiltonian is

$$H_{A+1} - V_{01} = H_A(\xi, \boldsymbol{r}_{1C}) + T_{0A} + V_{0A} + T_{A+1} \tag{9.73}$$

and the wavefunction factorizes in the usual way, i.e.

$$\Psi^i_{A+1} = \Phi_A(\xi, \boldsymbol{r}_{1C})\,\chi_i^+(\boldsymbol{k}_{0A}, \boldsymbol{r}_{0A})e^{i\boldsymbol{K}_{A+1}\cdot\boldsymbol{R}_{A+1}} \tag{9.74}$$

with $\boldsymbol{k}_{0A} = A\boldsymbol{p}_0/(A+1)$ and $\boldsymbol{K}_{A+1} = \boldsymbol{p}_0$. This explains why we have always been able to use product wavefunctions or linear combinations of product wavefunctions to describe a two-body system.

When we have chosen a factorized form for the final state wavefunction we can construct the transition matrix element in the usual way. Thus in zero-range DWBA, with $V(\boldsymbol{r}_{01}) = V_{01}\delta(\boldsymbol{r}_{01})$ we have

$$T_{fi} = V_{01}\int d\boldsymbol{r}\,\chi_f^*(\boldsymbol{k}_{0C}, \boldsymbol{r})\chi_f^{-*}(\boldsymbol{k}_{1C}, \boldsymbol{r})\psi_p(\boldsymbol{r})\chi_i^+\left(\boldsymbol{k}_{0A}, \frac{A-1}{A}\boldsymbol{r}\right) \tag{9.75}$$

if we use the symmetric co-ordinates, and

$$T_{fi} = V_{01}\int d\boldsymbol{r}\,\chi_f^{-*}(\boldsymbol{k}_{BC}, \boldsymbol{r})\psi_p(\boldsymbol{r})\,\chi_i^+\left(\boldsymbol{k}_{0A}, \frac{A-1}{A}\boldsymbol{r}\right) \tag{9.76}$$

if we use the di-proton co-ordinates. The overlap integral is defined as in equation (9.51), and the integral over the co-ordinate $\boldsymbol{R}_{A+1}$ leads to the δ-function $\delta(\boldsymbol{p}_0-\boldsymbol{q}_0-\boldsymbol{q}_1-\boldsymbol{Q})$ which expresses momentum conservation. It is clear that the first of these matrix elements, obtained using equation (9.63), is complicated by the presence of three distorted waves while the second, obtained using equation (9.69), contains only two distorted waves and resembles more closely the matrix element (9.29) for the pick-up reaction. In both cases the momenta $\boldsymbol{k}_{0C}$ and $\boldsymbol{k}_{1C}$, or $\boldsymbol{k}_{BC}$ depend on the scattering angles of the outgoing particles, through equations (9.60) and (9.65). This means that in a calculation of the cross-section the magnitudes of the momenta and the magnitudes of the energy-dependent optical potentials change at each angular point, and an accurate calculation is substantially more time-consuming than for a transfer reaction. A few calculations have been carried out using finite-range DWBA, but neglecting all corrections of order $1/A$. In this case the transition matrix element is given by

$$T_{fi}=\int d\boldsymbol{r}_{0C}d\boldsymbol{r}_{1C}\chi_f^{-}{}^*(\boldsymbol{k}_{0C},\boldsymbol{r}_{0C})\chi_f^{-}{}^*(\boldsymbol{k}_{1C},\boldsymbol{r}_{1C})V(|\boldsymbol{r}_{0C}-\boldsymbol{r}_{1C}|)\psi_p(\boldsymbol{r}_{1C})\chi_i^{+}(\boldsymbol{k}_{0A},\boldsymbol{r}_{0C}). \tag{9.77}$$

The transition matrix element in DWIA is given by equation (3.155). We assume that the transition operator can be approximated by a local but finite range operator $\tau(r_{01})$ and then, using the di-proton co-ordinates for the final state, we have

$$T_{fi}=\int d\boldsymbol{r}_{1C}d\boldsymbol{r}_{01}\chi_f^{-}{}^*(\boldsymbol{k}_{BC},\boldsymbol{r}_{1C}+\tfrac{1}{2}\boldsymbol{r}_{01})e^{-i\boldsymbol{k}_{01}\cdot\boldsymbol{r}_{01}}\tau(\boldsymbol{r}_{01})\psi_p(\boldsymbol{r}_{1C})$$
$$\times\chi_i^{+}\left(\boldsymbol{k}_{0A},\frac{A-1}{A}\boldsymbol{r}_{1C}+\boldsymbol{r}_{01}\right).$$

If it can be assumed that the p–p interaction is of sufficiently short range that the distortion functions do not change significantly over this range, the distorted wavefunctions can be written in the form

$$\begin{aligned}\chi_i^{+}(\boldsymbol{k}_{0A},a\boldsymbol{r}_{1C}+\boldsymbol{r}_{01})&=D^{+}(a\boldsymbol{r}_{1C}+\boldsymbol{r}_{01})e^{i\boldsymbol{k}_{0A}(a\boldsymbol{r}_{1C}+\boldsymbol{r}_{01})}\\&\simeq D^{+}(a\boldsymbol{r}_{1C})e^{i\boldsymbol{k}_{0A}\cdot(a\boldsymbol{r}_{1C}+\boldsymbol{r}_{01})}\\&\simeq\chi_i^{+}(\boldsymbol{k}_{0A},a\boldsymbol{r}_{1C})e^{i\boldsymbol{k}_{0A}\cdot\boldsymbol{r}_{01}},\end{aligned} \tag{9.78}$$

and similarly for $\chi_f^{-}{}^*$. The matrix element then separates into

$$T_{fi} = \int d\boldsymbol{r}_{01}\, e^{i(\boldsymbol{k}_{0A} - \boldsymbol{k}_{01} - \frac{1}{2}\boldsymbol{k}_{BC}).\boldsymbol{r}_{01}}\, \tau(\boldsymbol{r}_{01})$$

$$\times \int d\boldsymbol{r}_{1C}\, \chi_f^{-*}(\boldsymbol{k}_{BC}, \boldsymbol{r}_{1C})\, \psi_p(\boldsymbol{r}_{1C})\, \chi_i^{+}\left(\boldsymbol{k}_{0A}, \frac{A-1}{A}\boldsymbol{r}_{1C}\right). \qquad (9.79)$$

Thus the assumption used to derive equations (9.78) and (9.79) is equivalent to the assumption that the matrix elements of the transition operator are functions of momentum transfer and can be replaced by the value given by the asymptotic momenta [117]. The latter assumption was used and discussed in section 7.2. Because of this additional assumption, evidence that impulse approximation is valid for elastic scattering from a given target can not be taken as sufficient evidence for the validity of DWIA at the same energy.

The second integral in equation (9.79), or the equivalent integral if symmetric co-ordinates are used for the final state, will be referred to as the distorted momentum distribution and denoted by $g(\theta)$. The first integral can be related to the free p–p cross-section as follows. We write

$$T_{pp} = \int d\boldsymbol{r}_{01}\, e^{i(\boldsymbol{k}_{A} - \boldsymbol{k}_{01} - \frac{1}{2}\boldsymbol{k}_{BC})}\, \tau(\boldsymbol{r}_{01})$$

$$= \langle \bar{P}_1 | \tau_{01} | \bar{P}_0 \rangle$$

where

$$\bar{P}_0 = (\boldsymbol{k}_{0A} - \tfrac{1}{2}\boldsymbol{k}_{BC}) = \tfrac{1}{2}(\boldsymbol{p}_0 + \boldsymbol{Q}), \quad \bar{P}_1 = \boldsymbol{k}_{01} = \tfrac{1}{2}(\boldsymbol{q}_0 - \boldsymbol{q}_1), \qquad (9.80)$$

are the momenta of the scattered proton before and after collision in the proton–proton centre-of-mass system. The cross-section for free p–p scattering is given by [38]

$$\frac{d\sigma}{dE_1\, d\Omega_1} = \frac{2\pi m^2}{\hbar^4} \frac{|\bar{P}_1|}{|\bar{P}_0|} |T_{pp}|^2 \qquad (9.81)$$

where
$$\bar{E}_1 = \frac{\hbar^2}{2m} |\bar{P}_1|^2 = \tfrac{1}{4} \frac{\hbar^2}{2m} |\boldsymbol{q}_0 - \boldsymbol{q}_1|^2 = \tfrac{1}{4} E_0'$$

where E_0' is the relative energy in the lab. system after collision. Before collision the energy is

$$\bar{E}_0 = \frac{\hbar^2}{2m} |\bar{P}_0|^2 = \tfrac{1}{4} \frac{\hbar^2}{2m} |\boldsymbol{p}_0 + \boldsymbol{Q}|^2 = \tfrac{1}{4} E_0''.$$

Because energy is required to release the second proton from the target nucleus, it follows that $E_0' \neq E_0''$, i.e. the matrix element T_{pp} is off the energy-shell. The scattering angle in the p–p system is given by

$$\bar{P}_0 \bar{P}_1 \cos\bar{\theta} = \bar{\boldsymbol{P}}_0 . \bar{\boldsymbol{P}}_1 = \tfrac{1}{2}(\boldsymbol{p}_0 + \boldsymbol{Q}) . \tfrac{1}{2}(\boldsymbol{q}_0 - \boldsymbol{q}_1) .$$

Now, if the two outgoing protons are scattered at equal angles on opposite sides of the incident beam $(\boldsymbol{q}_0 - \boldsymbol{q}_1)$ is perpendicular to $(\boldsymbol{p}_0 + \boldsymbol{Q})$. Thus $\bar{\theta} = 90°$ and the p–p scattering occurs only in singlet even states. If the struck proton is moving towards the incident proton the relative momentum is greater than $\boldsymbol{p}_0$, and conversely if the two protons are moving apart. For lab. energies below 200 MeV the free p–p cross-section increases with decreasing relative energy so that the effect of the nucleon momentum in the nucleus is to depress the cross-section for $|\boldsymbol{p}_0 + \boldsymbol{Q}| > p_0$ relative to that for $|\boldsymbol{p}_0 + \boldsymbol{Q}| < p_0$.

The overlap integral is evaluated as in section 9.2. The cross-section for the (p, 2p) reaction is given by

$$\frac{d\sigma^3}{d\Omega_1 d\Omega_2 dE} = \frac{2\pi m^3}{\hbar^6} \frac{q_0 q_1}{p_0} (2\pi)^{-3} \sum |T_{fi}|^2 \qquad (9.82)$$

and using equations (9.44), (9.45), (9.46), (9.80) and (9.81) and carrying out the spin summations we have

$$\frac{d\sigma^3}{d\Omega_1 d\Omega_2 dE} = \frac{4m}{\hbar^2} \frac{q_0 q_1}{p_0} \frac{|\boldsymbol{p}_0 + \boldsymbol{Q}|}{|\boldsymbol{q}_0 - \boldsymbol{q}_1|} \frac{d\sigma}{d\Omega} \mathrm{pp}(E_0', \bar{\theta})(2\pi)^{-3} \sum_{nlj} B_{nlj}(\theta) \qquad (9.83)$$

$$\sum_{nlj} B_{nlj}(\theta) = \tfrac{1}{2} \sum_{nlj} \frac{S_{J_i J_f}(lsj)}{2l+1} (T_f M_{T_f} \tfrac{1}{2} - \tfrac{1}{2} | T_i M_{T_i})^2 \sum_m |g_{nlj}^m(\theta)|^2 . \qquad (9.84)$$

The validity and range of applicability of these formulae have been studied by a number of authors [118]. Figure 9.5 shows a comparison made by Lim and McCarthy of the cross-section for the ^{12}C(p, 2p) reaction calculated in DWIA and the distorted wave *t*-matrix approximation which is obtained when a suitable form for $\tau(r_{01})$ is chosen and a full finite range calculation is carried out. These calculations were made using symmetric co-ordinates for the final state and for the case when $\bar{\theta} = 90°$. At the lower energies there is an increasing discrepancy due to the contribution from p–p scattering in triplet odd states which is allowed even for $\bar{\theta} = 90°$ when the range of interaction is correctly treated, but at energies of 150 MeV and above the DWIA gives a satisfactory description of the finite-range effects through the momentum dependence of T_{pp}. For the same reaction, Jain has compared the effect of

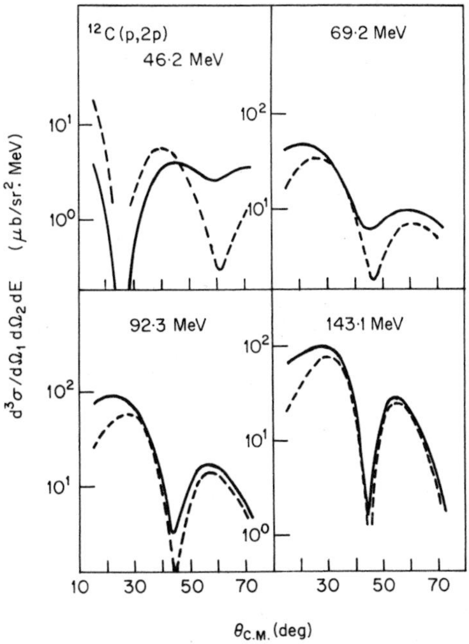

Figure 9.5 Comparison of distorted wave calculations for the $^{12}C(p,2p)$ reaction using the impulse approximation (dashed line) and a finite range *t*-matrix (full line). [From K. L. Lim and I. E. McCarthy, *Nucl. Phys.* **88,** 433 (1966).]

using the symmetric co-ordinates and neglecting T_{coup} and of using the di-proton co-ordinates and neglecting V_{coup}. The comparison shown in figure 9.6(a) indicates that these two methods yield results in substantial agreement. The most serious deficiency in the DWIA arises from the uncertainty about the correct energy at which the free p–p cross-section should be evaluated. In figure 9.6(b) the free cross-section is evaluated at the relative energy before and after scattering in the ^{12}C(p, 2p) reaction for incident energies of 160 MeV and 460 MeV and it is clear that at 160 MeV a major uncertainty is introduced.

The formalism outlined above can of course be applied to other knock-out reactions, such as (e, e′p) (see section 4.2), (p, pd), (p, pα), (α, 2α) and so on. Thus for the (p, pd) reaction the cross-section is given by

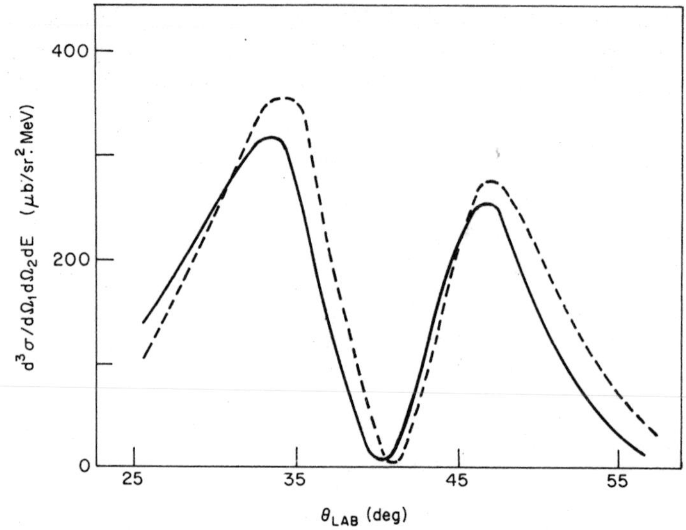

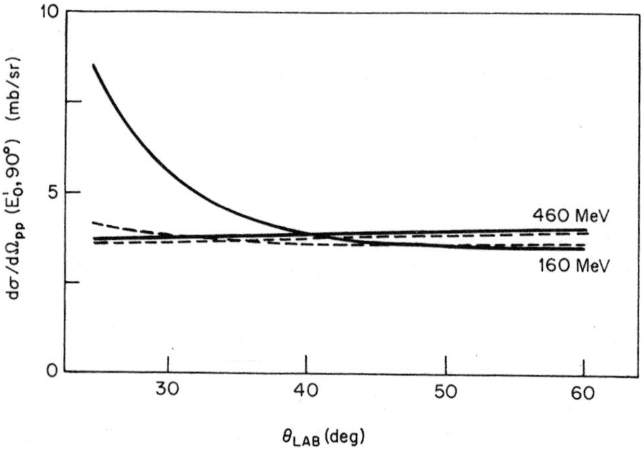

Figure 9.6 (**a**) Comparison of calculations for the $^{12}C(p,2p)$ reaction at 460 MeV using the symmetric co-ordinates for the final state (dashed line) and the di-proton co-ordinates (full line). (**b**) Comparison of the free proton–proton cross-section at 90° for relative energies E_0' before scattering (dashed line) and after scattering (full line). [From B. K. Jain, ref. 118.]

$$\frac{d^3\sigma}{d\Omega_1 d\Omega_2 dE} = \frac{9m}{2\hbar^2}\frac{q_0 q_d}{p_0}\frac{|\boldsymbol{p}_0+\frac{1}{2}\boldsymbol{Q}|}{|\boldsymbol{q}_0-\frac{1}{2}\boldsymbol{q}_d|}\frac{d\sigma}{d\Omega_{pd}}(E_0', \bar{\theta})|G(Q, q)|^2 \qquad (9.85)$$

where $\boldsymbol{q}_d$ is the momentum of the outgoing deuteron and $\boldsymbol{q}$ is the momentum transfer $\boldsymbol{p}_0-\boldsymbol{q}_0$. The structure of the function $G(Q, q)$ depends on the assumptions made about the basic interaction [119]. If, for example, it is assumed that the interaction between the incident proton and the nucleons which subsequently form the deuteron is identical to the interaction with a free deuteron so that the interaction can be represented in impulse approximation as $t_{pd}(q)\delta(\boldsymbol{r}_0-\boldsymbol{r}_d)$, and if the wavefunction of the target nucleus is expanded in the form

$$\Phi_A^i=\sum_{\alpha\beta\gamma}C^{\alpha\beta\gamma}\phi_d^\alpha(\boldsymbol{r}_{12})\psi_d^\beta(\boldsymbol{r}_{dC})\Phi_C^\gamma(\xi) \tag{9.86}$$

where the constant C represents the relevant fractional parentage coefficients and C–G coefficients, we find in PWIA

$$\begin{aligned} G(Q,q) &= \int \phi_d^{k*}(\boldsymbol{r}_{12})e^{i\boldsymbol{Q}\cdot\boldsymbol{r}_{dC}}\Phi_C^{f*}(\xi)\,\Phi_A^i(\xi,\boldsymbol{r}_{12},\boldsymbol{r}_{dC})d\boldsymbol{r}_{12}d\boldsymbol{r}_{dC}d\xi \\ &= \sum_{\alpha\beta}C^{\alpha\beta f}\int d\boldsymbol{r}_{12}\phi_d^{k*}(\boldsymbol{r}_{12})\,\phi_d^\alpha(\boldsymbol{r}_{12})\int d\boldsymbol{r}_{dC}e^{i\boldsymbol{Q}\cdot\boldsymbol{r}_{dC}}\psi^\beta(\boldsymbol{r}_{dC}) \\ &= \sum_\beta C^{k\beta f}\,G^\beta(Q) \end{aligned} \tag{9.87}$$

where k refers to the bound state of the deuteron. Alternatively, if it is assumed that the interaction is sensitive to the internal structure of the two-nucleon system so that it must be represented by a three-body transition operator, i.e. something of the form $t_{pp}\delta(\boldsymbol{r}_0-\boldsymbol{r}_1)+t_{pn}\delta(\boldsymbol{r}_0-\boldsymbol{r}_1)$, we have

$$\begin{aligned} G(Q,q) &= [F_d(q)]^{-1}\int \phi_d^{k*}(\boldsymbol{r}_{12})\Phi_C^{f*}(\xi)e^{i\boldsymbol{Q}\cdot\boldsymbol{r}_{dC}}e^{i\frac{1}{2}\boldsymbol{q}\cdot\boldsymbol{r}_{12}}\Phi_A^i(\xi,\boldsymbol{r}_{12},\boldsymbol{r}_{dC})d\boldsymbol{r}_{12}d\boldsymbol{r}_{dC}d\xi \\ &= [F_d(q)]^{-1}\sum_{\alpha\beta}C^{\alpha\beta f}\int d\boldsymbol{r}_{12}e^{i\frac{1}{2}\boldsymbol{q}\cdot\boldsymbol{r}_{12}}\phi_d^{k*}(\boldsymbol{r}_{12})\phi_d^\alpha(\boldsymbol{r}_{12})\int d\boldsymbol{r}_{dC}e^{i\boldsymbol{Q}\cdot\boldsymbol{r}_{dC}}\psi_d^\beta(\boldsymbol{r}_{dC}) \\ &= [F_d(q)]^{-1}\sum_{\alpha\beta}C^{\alpha\beta f}\,F^{\alpha k}(q)\,G^\beta(Q)\,, \end{aligned} \tag{9.88}$$

where F_d is the deuteron form factor

$$F_d(q)=F^{kk}(q)=\int d\boldsymbol{r}_{12}e^{i\frac{1}{2}\boldsymbol{q}\cdot\boldsymbol{r}_{12}}|\phi_d^k(\boldsymbol{r}_{12})|^2\,, \tag{9.89}$$

so that

$$G(Q,q)=\sum_\beta C^{k\beta f}G^\beta(Q)+\sum_{\alpha\neq k}C^{\alpha\beta f}\sum_\beta F^{\alpha k}(q)\,G^\beta(Q)\,. \tag{9.90}$$

Obviously, the inclusion of distortion in the formulae given above is a far from trivial matter and hence, although the (p, pd) reaction makes possible a study of two-hole states in nuclei and of clustering effects, the detailed analysis of the data is not simple.

9.4 Nuclear structure studies with transfer and knock-out reactions

As noted in section 9.1, measurements of the energy spectra for transfer and knock-out reactions yield the locations of the states in the residual nucleus formed by these reactions and values for the separation energies of the transferred and knocked-out particles, and measurements of the angular distribution for excitation of each of these states yield information about the transferred angular momentum from the shape of the distribution and about the strengths of the transitions from the magnitude.

The single-particle state formed by adding a single nucleon to a nucleus A is not necessarily an eigenstate of the residual nucleus but the single-particle strength may be distributed over several states of the residual nucleus. The distribution of the single-particle strength is measured by the spectroscopic factors for the transitions to these final states, since the definition in equation (9.45) yields the physical interpretation of the spectroscopic factor as the squared amplitude of a fragment of a single-particle state. The experimental spectroscopic factor is determined by fitting the experimental differential cross-section with the appropriate distorted wave formula and can then be compared with the theoretical spectroscopic factor derived from a suitable nuclear model [109]. Thus, a stripping experiment $A \rightarrow B(A+n)$ yields information about the strength and distribution of n-particle states in the final nucleus B and how their properties compare with those in other nuclei. In a pick-up or knock-out reaction $B(A+n) \rightarrow A$ a similar study is made of hole states. The latter reactions also give information about the ground state of the target nucleus and may be sensitive to small admixtures which would not show up in inelastic scattering. With $n > 2$ it is possible to reach members of isobaric spin multiplets which are not stable.

In certain special cases the spectroscopic factor for single-nucleon transfer takes a particularly simple form. In a pick-up reaction, N_{lj} is the number of nucleons in the sub-shell labelled by lj from which the transferred nucleon is taken. If the single nucleon is removed from the unfilled shell of a closed-shell-plus-one target, the coefficient of fractional parentage is unity and so is the spectroscopic factor. Similarly, if in a stripping reaction one nucleon is added to a closed-shell nucleus the c.f.p. and the spectroscopic factor are again unity. In other cases, the spectroscopic factor must be calculated using a suitable nuclear model and coupling scheme. The theoretical spectroscopic factors obey *sum rules*, which are obtained by taking the sum over some or all of the quantum numbers of the final states [109]. For example, using JJT coupling the sum rule for pick up is

$$\sum_{J_P} S_{J_A J_P}(lj) = N_{lj} \tag{9.91}$$

and for stripping is

$$\sum_{J_P T_P} (2J_p+1)(2T_p+1) S_{J_A J_p}(lj) = (N - N_{lj}+1)(2J_A+1)(2J_A+1) \tag{9.92}$$

where $N = 2(2j+1)$ is the total number of nucleons the sub-shell lj can hold. A true single-particle or single-hole state is identified as one which contributes a large proportion of the total strength as measured by the sum rule.

It is evident that the value of transfer reactions for detailed nuclear structure studies depends very strongly on the accuracy with which the experimental spectroscopic factor can be determined. The effect of some of the approximations in the distorted wave theory has been discussed in the two preceding sections, but at energies above 50 MeV the most serious source of uncertainty at present is lack of knowledge of the optical potentials. A more fundamental problem, not so far mentioned, is the construction of the overlap integral [120]. We note that the first of equations (9.16) can be written in the form

$$H_A \Phi_{J_A}^{M_A} = (H_B + T_{xB} + V_{xB} + H_x)\Phi_{J_A}^{M_A} = \varepsilon_A^{J_A} \Phi_{J_A}^{M_A}$$

where $\Phi_{J_A}^{M_A}$ is given by equation (9.42). If we now multiply on the left by $\Phi_{J_B}^{M_B*}$ and integrate over the co-ordinates of the residual nucleus B we obtain a Schrödinger equation for the overlap integral

$$(T_{xB} + V_{xB})\psi(\mathbf{r}_{xB}, \boldsymbol{\sigma}_x) = (\varepsilon_A^{J_A} - \varepsilon_B^{J_B} - \varepsilon_x^j)\psi(\mathbf{r}_{xB}, \boldsymbol{\sigma}_x). \tag{9.93}$$

From this equation we may deduce the following properties of the overlap integral. (i) The asymptotic behaviour of the radial part of ψ_x is governed by the separation energy $S_{xA} = -(\varepsilon_A^{J_A} - \varepsilon_B^{J_B} - \varepsilon_x^j)$ for the break-up of nucleus A into particle x and the residual nucleus B in the definite states specified by J_B and j, so that the asymptotic behaviour can be written as

$$R_{nlj}(r) \to \frac{e^{-\alpha r}}{r}, \qquad \alpha^2 = \frac{2\mu}{\hbar^2}|S_{xA}|. \tag{9.94}$$

(ii) The radial co-ordinate r_{xB} is referred to the centre-of-mass of nucleus B, not of A. (iii) The description of the target nucleus A is translation invariant since H_A is expressed entirely in terms of relative co-ordinates so that the overlap integral does not describe any centre-of-mass motion.

Because of the properties listed above, the single-particle wavefunctions ψ_j appearing in equation (9.43) are not the same as the wavefunctions generated in a shell model potential for the nucleus A. If, however, it can be assumed that the interaction V_{xB} can be represented by an effective one-body potential, the ψ_j can be generated in some finite potential whose depth is adjusted to give the correct separation energy for each transition. In some DWBA codes for single-nucleon transfer the radial parameters of this potential are taken to be the same as those of the proton optical potential while in other codes they are adjustable parameters. For (p, d) and (p, 2p) reactions on light nuclei in the 150–200 MeV region some success has been achieved by taking those potential parameters which yield proton wavefunctions in agreement with elastic electron scattering (see section 4.2). This procedure is valid if $1/A$ corrections and residual interactions are negligible and is therefore basically a lowest-order approximation, although it may be argued that by adjusting the parameters of the effective potential to fit experimental data, some higher-order effects are taken into account in an approximate way. As might be expected, this approach is least successful for deformed nuclei such as ^{9}Be and for nuclei in the middle of shells such as ^{28}Si. Since the states of the residual nucleus are not degenerate, there must be residual interactions not included in a spherically symmetric one-body potential for V_{xB} and these give rise to configuration mixing and coupled equations for the overlap integrals, although equation (9.94) remains valid for the asymptotic behaviour. A number of prescriptions which aim to give a more accurate description of the overlap integral in the interior region are currently being studied [120]. It should be noted that the theory of the overlap integral described above is strictly valid when the final state of the residual nucleus is bound, and special techniques be adopted when the final state is unbound [121].

A more detailed study of the shape of an angular distribution should give information on the angular momentum dependence of the overlap integral. For convenience we discuss this by referring to the plane wave formulae (9.24) and (9.52) although it must be stressed that plane wave calculations are never sufficiently accurate or reliable for precise comparison with the data. Examination of these formulae shows that the cross-section has a maximum when the momentum P or Q is zero if ψ_x has predominantly s-state components but it has a minimum (zero) if ψ_x contains no s-state components. Further, if we assume that the reaction is sharply localized at some radius R and replace $\psi_x(\mathbf{r})$ by $\delta(\mathbf{r}-\mathbf{R})$ we find that the cross-section is proportional to $|j_l(QR)|^2$ and the angular distribution is characterized by the location of the maxima and minima of the spherical Bessel function. Although this behaviour is very much modified in more accurate calculations the sensitivity to the angular momentum persists.

An example of the sensitivity of the shapes of angular distributions for transfer reactions to the angular momentum transfer l is shown in Figure 9.7. In addition, there are particular effects at different angles which allow a distinction to be made between transitions with total angular momentum transfer $j=l+\frac{1}{2}$ and $j=l-\frac{1}{2}$. (For a given reaction ψ_j has a definite parity so that there is no mixing of l values for a given j.) Some of these effects are shown in Figure 9.8; for $l=1$ there is a sharp minimum in the $p_{\frac{1}{2}}$ angular distribution at large angles which does not occur in the $p_{\frac{3}{2}}$ distribution, and for higher l values there are effects at forward and intermediate angles. Experimentally, these effects provide valuable additional information for the identification of the spin and parity of the final state [122]. Theoretically, they present a challenge to the DWBA theory. A j-dependence in the theoretical cross-section can arise from the spin-orbit terms in the optical potential and in the single particle potential for ψ_j, from the difference between the separation energies, and from configuration mixing. In addition, it must be noted that the angular momentum transfer is the same as the angular momentum of the transferred particle x in the nucleus because we have assumed that the particles a and x are in a relative S-state in b. A particular situation where this is not true arises when the D-state of the deuteron is taken into account so that additional values of angular momentum transfer l are possible, and this also leads to j-dependent effects [112]. At the present time, the only well-established angular momentum dependence in knock-out reactions occurs at zero recoil momentum where, as can be seen from equation (9.54), the cross-section must have a minimum if $l\neq 0$. This effect is illustrated in Figures 9.5 and 9.6(a).

A particular experiment will be sensitive to a restricted range of the momentum $\mathbf{P}$, defined in equation (9.25) for transfer reactions, or $\mathbf{Q}$, defined in equation (9.53) for knock-out reactions. In the case of a transfer reaction the range of P is determined by the Q-value of the reaction. For example, if we consider the (p, d) reaction and neglect terms of order 1/A we have

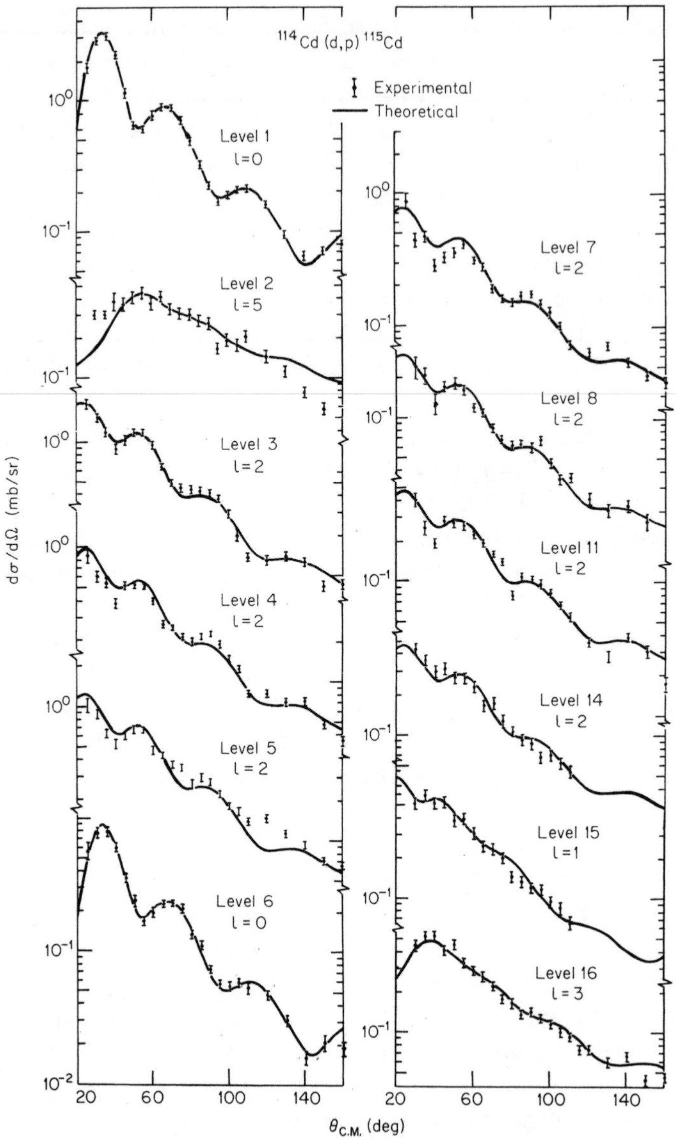

Figure 9.7 Comparison of experimental data and theoretical predictions for the ^{114}Cd(d,p) reaction at 12 MeV. [From G. R. Satchler, ref. 106.]

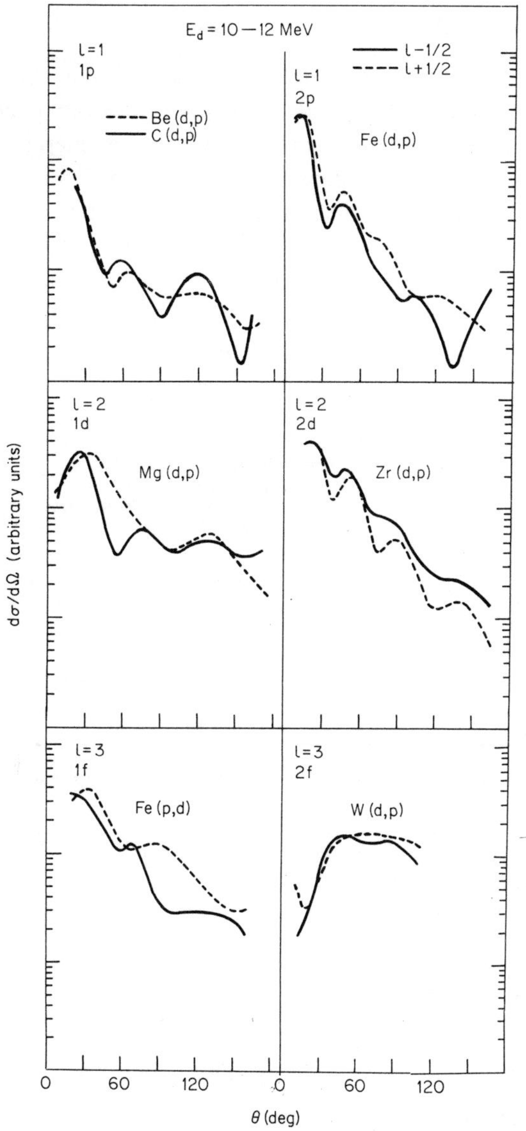

Figure 9.8 A summary of *j*-dependent effects in (d,p) reactions. The curves represent the trend of the experimental data for different *j*-values. [From J. P. Schiffer, ref. 122.]

$$P = [k_d^2 + k_p^2 - 2k_d k_p \cos\theta]^{\frac{1}{2}}, \qquad k_d = \sqrt{2}\, k_p[1 + Q_d/E_p]^{\frac{1}{2}},$$

where Q_d is the Q-value of the reaction. Thus assuming that $Q_d \lesssim 0$, the region of small P can be investigated using low or medium energy protons, while the region of larger P can be studied with proton energies $E_p \gg Q_d$, i.e. the same reaction at different energies is sensitive to different momentum components of the single-particle wavefunction, and the higher energy reaction is more sensitive to the interior region in configuration space. In the knock-out reactions the presence of two fast particles in the final state allows greater freedom in the range of recoil momentum $\boldsymbol{Q}$. If we take the scattering angles of the two outgoing protons to be θ_1, and θ_2 and the angle

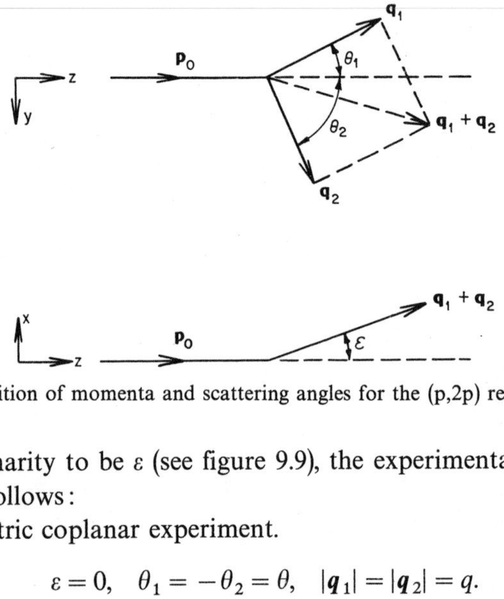

Figure 9.9 Definition of momenta and scattering angles for the (p,2p) reaction.

of non-coplanarity to be ε (see figure 9.9), the experimental events can be classified as follows:

(a) Symmetric coplanar experiment.

$$\varepsilon = 0, \quad \theta_1 = -\theta_2 = \theta, \quad |\boldsymbol{q}_1| = |\boldsymbol{q}_2| = q.$$

(b) Symmetric non-coplanar experiment.

$$\varepsilon \neq 0, \quad \theta_1 = -\theta_2 = \theta, \quad |\boldsymbol{q}_1| = |\boldsymbol{q}_2| = q.$$

(c) Non-symmetric coplanar experiment.

$$\varepsilon = 0, \quad |\boldsymbol{q}_1| \text{ and } \theta_1 \text{ fixed}, \quad \boldsymbol{q}_0 \text{ variable}.$$

(In experiments with fixed counters these events are often studied in separate experiments, but when spark chambers or a bubble chamber are used a mixture of these events and also non-symmetric, non-coplanar events is observed.) In the symmetric coplanar experiment the angle θ_m at which the recoil momentum Q is zero is obtained from equations (9.10) and (9.53) as

$$\cos\theta_m = p_0/2q = [\tfrac{1}{2}E_p/(E_p - S_{pA})]^{\frac{1}{2}}\,. \qquad (9.95)$$

Thus for $S_{pA} = 0$, $\theta_m = 45°$, while for $S_{pA} > 0$, $\theta_m < 45°$, and hence as θ is increased from small angles up to 90° the recoil momentum $Q = p_0 - 2q\cos\theta$ changes from negative to positive values. With this choice of geometry the knock-out reactions are particularly sensitive to the low momentum components of the single-particle wavefunction. For example, in the symmetric coplanar ^{12}C(p, 2p)^{11}B reaction with 155 MeV protons the range of Q covered is roughly -0.8 fm^{-1} to $+0.8$ fm^{-1}, whereas in the ^{12}C(p, d)^{11}C reaction at the same energy the range of P covered is 0.7 fm^{-1} to 2.7 fm^{-1}. It follows that at this energy the states rich in low momentum components, e.g. 1s states, will show up much more strongly in the (p, 2p) reaction than in the (p, d) reaction.

The measurement of proton separation energies in the (p, 2p) and (e, e′p) reactions has an important influence on the calculation of the single-particle states in a shell model potential. As noted in section 4.2, it is possible to find a single local potential which will yield the correct separation energies for the least bound protons and also yield a proton distribution which will fit the electron scattering data. However, as more data on separation energies became available it was found [46] that it is not possible to fit the separation energies for the lower levels, e.g. the 1p states in ^{40}Ca, unless different potential parameters are used for different proton shells in the same nucleus. This energy dependence in the bound state potential is similar to the energy dependence observed for the optical potential and arises for the same reason, that we are attempting to represent a fundamentally non-local interaction by a local potential. This can be seen from the discussion of the Hartree-Fock potential in section 2.4. (It may be noted that the values obtained for the separation energy for the 1p state in ^{40}Ca from the (e, e′p) experiment and several (p, 2p) experiments are in good agreement. The values for 1s and 1p states in heavier nuclei obtained from (e, e′p) and (p, 2p) experiments are not in agreement at the present time.)

Single-particle potentials for neutrons have been derived by fitting the separation energies for the least bound states. The separation energies for deeply bound neutron states are in general not known. Unfortunately, this procedure has not so far produced single-particle wavefunctions which fit the data on neutron pick-up reactions such as (p, d) and (d, t) [123].

9.5 Sequential processes and core excitation

In the preceding sections we have assumed that transfer and knock-out reactions proceed through a direct one-step mechanism, and this assumption is evidently adequate in a very large number of reactions. There is, however some experimental evidence that a two-step mechanism can make a significant contribution in certain cases. The following are examples of two-step processes:

(i) ^{40}Ca(α, 2α). $\quad {}^{40}\mathrm{Ca}+\alpha \rightarrow {}^{40}\mathrm{Ca}^{*}+\alpha'$
$\quad\quad\quad\quad\quad\quad\quad\quad \hookrightarrow {}^{36}\mathrm{Ar}+\alpha$

(ii) ^{11}B(p, 3α). $\quad {}^{11}\mathrm{B}+\mathrm{p} \rightarrow {}^{12}\mathrm{C}^{*}$
$\quad\quad\quad\quad\quad\quad\quad\quad \hookrightarrow {}^{8}\mathrm{Be}+\alpha$
$\quad\quad\quad\quad\quad\quad\quad\quad\quad \hookrightarrow \alpha+\alpha$

(iii) ^{12}C(p, d). $\quad {}^{12}\mathrm{C}+\mathrm{p} \rightarrow {}^{12}\mathrm{C}^{*}+\mathrm{p}' \rightarrow {}^{11}\mathrm{C}^{*}+\mathrm{d}$

(iv) ^{52}Cr(d, p). $\quad {}^{52}\mathrm{Cr}+\mathrm{d} \rightarrow {}^{52}\mathrm{Cr}^{*}+\mathrm{d}' \rightarrow {}^{53}\mathrm{Cr}^{*}+\mathrm{p}$
$\quad\quad\quad\quad\quad\quad\quad\quad \rightarrow {}^{53}\mathrm{Cr}+\mathrm{p} \rightarrow {}^{53}\mathrm{Cr}^{*}+\mathrm{p}'.$

Processes (i) and (ii) are usually referred to as *sequential processes* [124] and can be distinguished from one-step processes by a careful study of the kinematics of the emitted particles. Reaction (i) appears to proceed through an excited state of ^{40}Ca at about 18.8 MeV while reaction (ii) proceeds through the ground state and first excited state of ^{8}Be. Processes (iii) and (iv) are examples of what is called *double excitation* or *core excitation*, the extra step in the mechanism being inelastic scattering in the entrance or exit channel. Reaction (iii) has been invoked [125] to explain the strong excitation of the $\frac{5}{2}^-$ and $\frac{7}{2}^-$ states of ^{11}C in the ^{12}C(p, d) reaction which can be seen in figure 9.1. In a one-step mechanism, the pick-up of a $1p$ neutron from ^{12}C leads to $\frac{3}{2}^-$ and $\frac{1}{2}^-$ states and the strong excitation of $\frac{5}{2}^-$ and $\frac{7}{2}^-$ states would indicate an admixture of $1f$ neutrons substantially in excess of that predicted from nuclear structure calculations. However, if the first 2^+ state in ^{12}C is excited by inelastic scattering and a $1p$ neutron is then picked up from this level, states in ^{11}C with spins from $\frac{1}{2}^-$ to $\frac{7}{2}^-$ can be excited. Strong excitation of $\frac{5}{2}^-$ and $\frac{7}{2}^-$ states in ^{11}B is observed in the ^{12}C(p, 2p) and ^{12}C(d, ^{3}He) reactions and can be interpreted in the same way [125]. In reaction (iv) one-step stripping to the $\frac{7}{2}^-$ state of ^{53}Cr at 1.28 MeV is inhibited but a two-step process can occur through inelastic excitation of the 2^+ state of ^{52}Cr at 1.43 MeV followed by stripping to the $\frac{7}{2}^-$ state of ^{53}Cr or through stripping to the $\frac{3}{2}^-$ ground state of ^{53}Cr followed by inelastic excitation [126].

The theoretical description of core excitation [127] requires a generalization of the standard DWBA theory so that the distorted waves describe both elastic and inelastic scattering from the target and residual nucleus. The distorted waves are then solutions of a set of coupled equations which may be set up as in the strong coupling approximation for inelastic scattering described in section 7.4. This represents another attempt to include some of the effects which are omitted in DWBA when the exact wavefunction Ψ is replaced by the product $\chi\phi\Phi$.

9.6 Exchange processes

Apart from statements that the nuclear wavefunctions should be properly antisymmetrized we have so far ignored the implications of antisymmetrization for the description of nuclear reactions. Recognition that the nucleons of the projectile are indistinguishable from the nucleons of the target leads to additional terms in the DWBA amplitude, and the way that such terms arise in the pick-up reaction A(p, d)B is indicated diagrammatically in figure 9.10. The first process shown in the figure is the *direct pick-up* process which we have already described, while the second process is the corresponding exchange process. The latter is frequently referred to as *knock-out*, although it is evidently very different from the direct knock-out process described in section 9.3. The third process is usually called *heavy-particle knock-out*; it is not an exchange process but arises from the interaction of the incident proton p_0 and the nuclear core and will contribute if the core is not inert. Finally, the fourth process is the exchange process corresponding to heavy-particle knock-out and is usually called *heavy-particle pick-up*. The DWBA matrix elements for these four terms are given by [128]

$$
\begin{aligned}
T_{fi}^{(1)} &= \langle \Phi_B \chi_{\mathrm{d}}^- | V_{\mathrm{p_0 n}} | \Phi_A \chi_{\mathrm{p}}^+ \rangle \\
T_{fi}^{(2)} &= \langle \Phi_B \chi_{\mathrm{d}}^- | P_{01} V_{\mathrm{p_0 d}} | \Phi_A \chi_{\mathrm{p}}^+ \rangle \\
T_{fi}^{(3)} &= \langle \Phi_B \chi_{\mathrm{d}}^- | V_{\mathrm{p_0} B} - U_{\mathrm{p_0} A} | \Phi_A \chi_{\mathrm{p}}^+ \rangle \\
T_{fi}^{(4)} &= \langle \Phi_B \chi_{\mathrm{d}}^- | P_{01} (V_{P_1 B} - U_{P_1 A}) | \Phi_A \chi_{\mathrm{p}}^+ \rangle \ ;
\end{aligned}
$$

where P_{01} exchanges protons p_0 and p_1, and the protons and neutrons have been treated separately. If there are N_p protons in the target which can parti-

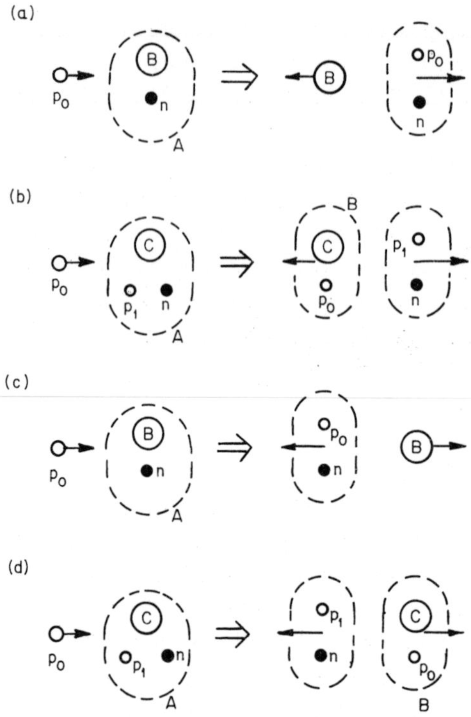

Figure 9.10 Diagrammatic representation of the direct and exchange terms which contribute to a (p,d) reaction.

cipate, then the total matrix element is obtained by multiplying the exchange matrix elements by $-N_p$ and adding them to the direct matrix elements. (The minus sign arises because the protons are fermions. If the exchange process involved a boson as the projectile and N identical bosons in the target the multiplication factor would be $+N$.)

In the case of heavy-particle pick-up the incident proton has least momentum change if the final nucleus continues in the forward direction, so that the distribution of the emitted deuterons is peaked at large scattering angles. For this reason this process is frequently invoked to explain backward peaks observed in the differential cross-sections for (p, d) and similar reactions. Unfortunately, the behaviour of the exchange and knock-out processes is still a subject of some uncertainty and controversy because, although it is possible to write formal expressions for the exact matrix elements, it is difficult to introduce the approximations necessary for calculations in a consistent way.

10 High energy nuclear reactions

10.1 Proton scattering and reactions above 400 MeV

In recent years, there has been an increasing interest in the scattering of high energy projectiles from nuclei. In this section we consider scattering and reactions due to protons in the energy region 400 MeV to 20 GeV. In particular, we consider the extent to which the theories and approximations introduced in earlier chapters remain applicable as the projectile energy is increased, whether the increased energy or different experimental techniques suggest or demand a different theoretical approach to the data, and what additional information about nuclear structure can be obtained.

For incident energies above about 150 MeV, it is necessary to take account of departures from non-relativistic kinematics. This can be done within the framework of the non-relativistic Schrödinger equation by modifying the optical potential and the wave number or, if the spin-orbit potential can be neglected, the Schrödinger equation can be replaced by the Klein-Gordon equation. Alternatively, the semi-classical or high energy approximation can be used to obtain approximate expressions for the elastic scattering wavefunctions and the phase shifts. These expressions are, in fact, identical with those given in equations (3.134) and (3.141) except that $\hbar v$ must be calculated using relativistic kinematics. It is then possible to use equation (3.141) to calculate the phase shifts and reflection coefficients and hence to calculate the cross-sections and polarization using the standard expressions derived in Chapter 3, or to use the full semi-classical approximation (3.135) for the cross-section. In either method there is no difficulty about including the Coulomb and spin-orbit parts of the optical potential, but it must be remembered that the semi-classical approximation is not valid for large scattering angles. Some results obtained from the first of these two semi-

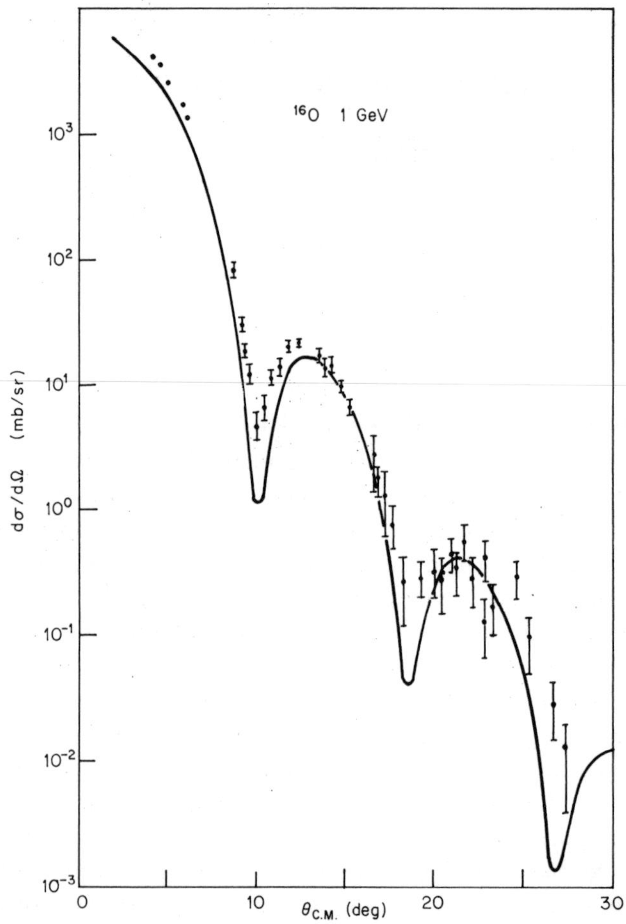

Figure 10.1 (**a**) The cross-section calculated for 1 GeV proton scattering from ^{16}O using a semi-classical approximation for the phase shifts. Coulomb scattering is not included.

classical methods for the elastic scattering of 1 GeV protons from ^{16}O are shown in figure 10.1. At this energy there is a large number of open channels which give rise to a large magnitude for the imaginary part of the optical potential. This effect, together with the diffraction-like appearance of the elastic scattering data and the smooth behaviour of the reflection coefficients, has prompted the use of strong absorption techniques for the analysis of elastic and inelastic scattering in this energy region.

Several analyses of elastic scattering in terms of the phenomenological

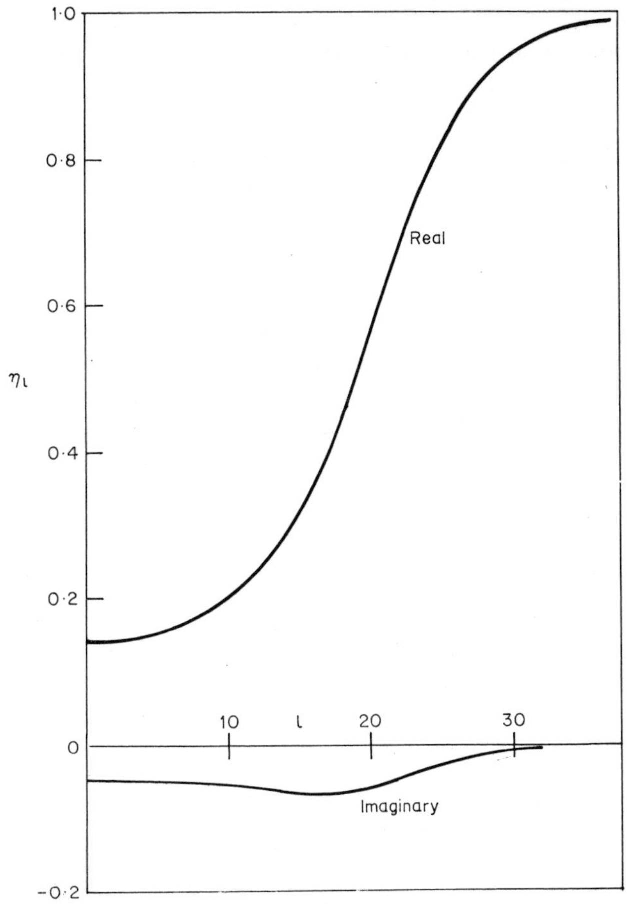

Figure 10.1 (b) The reflection coefficients calculated for 1 GeV proton scattering from ^{16}O using a semi-classical approximation for the phase shifts.

optical potential have been carried out. The analyses at incident energies between 100 MeV and 1 GeV [129] indicate that the real part of the optical potential changes sign at or before an energy of 400 MeV and is repulsive at higher energies. This is supported by an analysis at 20 GeV using the strong absorption model [130].

For the energy range under consideration the validity of the impulse approximation should be improved so that the phenomenological optical potential for nucleon–nucleus scattering can be related to nucleon–nucleus scattering through the relativistic forms of equations (6.55) to (6.63). The relativistic modification of these equations is discussed by Kerman, McManus and Thaler [36] and yields, to a good approximation,

$$V(r) = -Vf(r) - iWg(r) = -\frac{A}{(2\pi)^2}\frac{\hbar^2 c^2}{E_{\text{LAB}}}\frac{k_{\text{LAB}}}{k_0}\int e^{-i\boldsymbol{q}\cdot\boldsymbol{r}} A_\alpha(q) F(q)\, d\boldsymbol{q}\,, \quad (10.1)$$

$$\approx -2\pi A \frac{\hbar^2 c^2}{E_{\text{LAB}}}\frac{k_{\text{LAB}}}{k_0} A_\alpha(0)\,\rho(r)\,, \quad (10.2)$$

where the second form applies, as before, to the case of large A, and k_{LAB}, k_0 are the momenta corresponding to the total energies E_{LAB}, E_0 of the incident nucleon in the lab. system and the two-nucleon centre-of-mass system respectively. A similar modification can be made to the spin-orbit terms. In

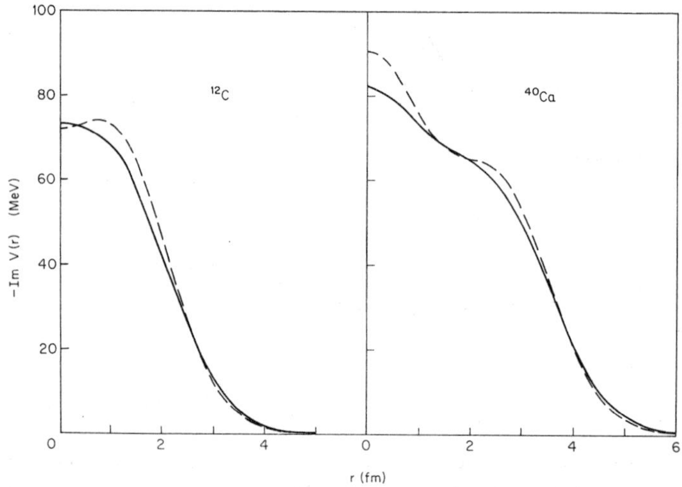

Figure 10.2 Comparison of the optical potentials for 1 GeV proton scattering from ^{4}He and ^{40}Ca calculated using the exact impulse approximation (full line) and the large A approximation (dashed line).

figure 10.2 the results obtained from equations (10.1) and (10.2) for 1 GeV protons on ^{40}Ca and ^{4}He are compared. It can be seen that even for ^{40}Ca the discrepancy is not negligible and that care must be exercised in the interpretation of results obtained from equation (10.2) for medium mass nuclei.

Using the approximate form (10.2) the phase shift function $\chi(\boldsymbol{b})$ defined by equation (3.136) becomes

$$\chi(\boldsymbol{b}) = \frac{2\pi A}{k_0} A_\alpha(0) \int_{-\infty}^{\infty} \rho(\boldsymbol{b}, z)\, dz \tag{10.3}$$

$$= \frac{4\pi A}{k} A_\alpha(0) \int_{-\infty}^{\infty} \rho(\boldsymbol{b}, z)\, dz \tag{10.4}$$

where k is the momentum of the incident nucleon in the nucleon–nucleus centre-of-mass system. The function

$$T(b) = A \int_{-\infty}^{\infty} \rho(\boldsymbol{b}, z)\, dz \tag{10.5}$$

has been called by Glauber the *thickness function* of the nucleus [131]. The complete transition matrix element for elastic scattering can also be expressed in terms of the thickness function. The exact expression for the transition matrix element is

$$T = \int e^{i\boldsymbol{k}'.\boldsymbol{r}}\, V(\boldsymbol{r})\, \psi^+(\boldsymbol{k}, \boldsymbol{r})\, d\boldsymbol{r}$$

which in the semi-classical approximation becomes (see section 3.9)

$$T = -i\hbar v \int_0^{2\pi} d\phi \int_0^{\infty} b\, db\, e^{i(\boldsymbol{k}-\boldsymbol{k}').\boldsymbol{b}}\, G(\boldsymbol{b}) \tag{10.6}$$

where the function $G(\boldsymbol{b})$ is

$$G(\boldsymbol{b}) = 1 - e^{i\chi(\boldsymbol{b})} = 1 - e^{-xT(\boldsymbol{b})}, \tag{10.7}$$

x is a constant given by

$$x = -i \frac{2\pi}{k_0} A_\alpha(0) \tag{10.8}$$

$$\approx \tfrac{1}{2}\sigma_{\text{tot}}, \quad \text{if} \quad |\text{Im}\, A_\alpha(0)| \gg |\text{Re}\, A_\alpha(0)|^2, \tag{10.9}$$

and in the last line we have used the optical theorem (3.69) to introduce the total cross-section for nucleon–nucleon scattering.

If the potential $V(\boldsymbol{b}, z)$ can be replaced by a sum of interactions between the incident nucleon and the target nucleons, as is done for example in the microscopic description of elastic and inelastic scattering at lower energies, the phase shift function is given by

$$\chi(\boldsymbol{b}-\boldsymbol{s}) = -\frac{1}{\hbar v} \int_{-\infty}^{\infty} \sum_j V(\boldsymbol{b}-\boldsymbol{s}_j, z)\, dz = \sum_j \chi_j(\boldsymbol{b}-\boldsymbol{s}_j) \tag{10.10}$$

where s_j is the component of the nucleon co-ordinate r_j in the b-plane. The transition matrix element for elastic or inelastic scattering is then given by [132]

$$T_{n0} = -i\hbar v \int d^2 b\, e^{i\boldsymbol{q}\cdot\boldsymbol{b}} \langle n | 1 - \exp\left(i \sum_j \chi_j(\boldsymbol{b}-\boldsymbol{s}_j)\right) | 0 \rangle \qquad (10.11)$$

where we have used equations (10.6) and (10.10) and have replaced the function $G(\boldsymbol{b})$ defined in equation (10.7) by

$$G(\boldsymbol{b}) = \int \Phi_n^*(\boldsymbol{r}_j)\, G(\boldsymbol{b}, \boldsymbol{s}_j)\, \Phi_0(\boldsymbol{r}_j) d\boldsymbol{r}_j$$

which is equivalent to a calculation of the scattering from fixed potentials at the points $\boldsymbol{r}_j$ each weighted by $\Phi_n^*(\boldsymbol{r}_j)\,\Phi_0(\boldsymbol{r}_j)$. Now

$$1 - \exp\left[i \sum_j \chi_j(\boldsymbol{b}-\boldsymbol{s}_j)\right] = 1 - \prod_{j=1}^{A} \exp\left[i\chi_j(\boldsymbol{b}-\boldsymbol{s}_j)\right] = 1 - \prod_{j=1}^{A} \left[1 - \Gamma_j(\boldsymbol{b}-\boldsymbol{s}_j)\right]$$
$$= \sum_j \Gamma_j(\boldsymbol{b}-\boldsymbol{s}_j) - \sum_{j\neq k} \Gamma_j(\boldsymbol{b}-\boldsymbol{s}_j)\,\Gamma(\boldsymbol{b}-\boldsymbol{s}_k) + \sum_{j\neq k\neq l} \Gamma_j \Gamma_k \Gamma_l - \ldots \qquad (10.12)$$

Equation (10.12) represents a multiple-scattering expansion for the operator $\Gamma(\boldsymbol{b}-\boldsymbol{s}) = 1 - e^{i\Sigma\chi_j}$, and by evaluating the contributions to the matrix element (10.11) term by term it is possible to estimate the contribution from single-scattering, double-scattering, etc. It may be noted that the semi-classical approximation allows only successive small angle scattering from different nucleons, as illustrated in figure 10.3(a). The type of scattering

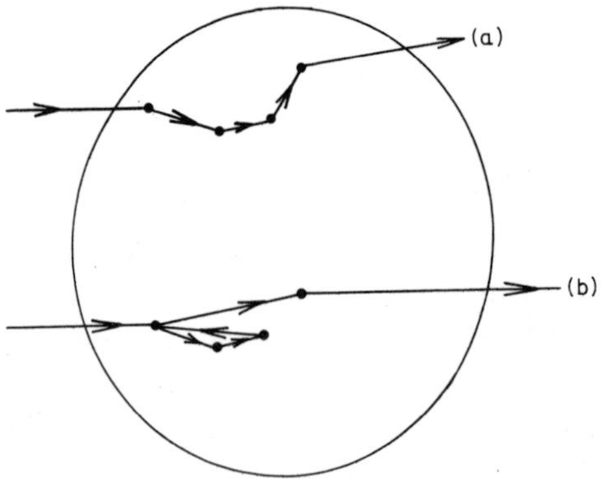

Figure 10.3 Diagrammatic representation of multiple scattering terms included (case a) and excluded (case b) in Glauber's multiple scattering formalism.

process illustrated in figure 10.3(b) involves at least one large angle scattering and is therefore not compatible with the semi-classical approximation [133]. If we use the density functions defined in equations (2.42) and (2.48) we obtain, for elastic scattering,

$$\begin{aligned} G(\boldsymbol{b}) &= \langle 0|1-e^{i\Sigma\chi_j}|0\rangle \\ &= A\int d^2s\,du\,\rho(\boldsymbol{r})\,\Gamma_j(\boldsymbol{b}-\boldsymbol{s}) \\ &\quad -A(A-1)\iint d^2s\,du\,d^2s'\,du'\,C(\boldsymbol{r},\boldsymbol{r}')\,\Gamma_j(\boldsymbol{b}-\boldsymbol{s})\,\Gamma_k(\boldsymbol{b}-\boldsymbol{s}')+\ldots \end{aligned} \tag{10.13}$$

where $\boldsymbol{r}$, $\boldsymbol{r}'$ have components $(\boldsymbol{s}, u)$ and $(\boldsymbol{s}', u')$ respectively, and $\Gamma_j(\boldsymbol{b}-\boldsymbol{s})$ is given by the Fourier transform of the semi-classical formula for the scattering amplitude, i.e.

$$\Gamma_j(\boldsymbol{b}-\boldsymbol{s}) = \frac{1}{2\pi i k_0}\int d^2q'\,e^{-i\boldsymbol{q}'\cdot(\boldsymbol{b}-\boldsymbol{s})}A_\alpha(q'). \tag{10.14}$$

For 1 GeV elastic scattering from ^{12}C and ^{16}O, Lee and McManus [134] have compared the cross-section calculated using the optical potential given by equation (10.2) with the cross-section obtained using the multiple scattering expansion. They find that there is good agreement with the data and good agreement between the two calculations, as indeed there should be provided that equation (10.2) is a good approximation for the optical potential. They have also compared the results obtained for a microscopic description of inelastic scattering to a definite final state using the multiple scattering expansion, with a standard distorted wave calculation using the distorted wavefunctions given by the WKB approximation, and again find satisfactory agreement. In all these calculations the real part of $A_\alpha(q)$ contributes only in the region of the diffraction minima where it causes a substantial filling-in. In the distorted wave calculation the transition density used also fitted the data at 150 MeV; thus for the region of momentum transfer covered by these data the results at 150 MeV and 1 GeV are consistent and can be described by the same reaction mechanism. The 1 GeV elastic scattering data for ^4He have been studied in some detail using the multiple scattering expansion [135], but no unambiguous information on short-range correlations has so far emerged.

The multiple scattering formalism has been applied to the process of pion production by high energy protons which is discussed in the next section. Extension of the formalism to other reactions does not appear to be a very easy task. Fortunately, the DWIA formalism for reactions can readily be extended to the high energy region using the distorted wavefunctions (3.132) and (3.158) given by the WKB approximation. A study of the symmetric coplanar (p, 2p) reaction between 400 MeV and 1 GeV has been carried out using this method. It is found [136] that the proton single-particle wavefunctions derived from analyses of electron scattering and spectroscopic factors derived from suitable nuclear models which lead to reasonable agreement with the data in the energy region 155–185 MeV also yield agreement at 460 MeV. For the ^{12}C (p, 2p) reaction the calculations also yield distorted momentum distributions in qualitative agreement with the data at 387 MeV and 1 GeV, which again confirms the similarity in the mechanism of reactions occurring between 150 MeV and 1 GeV. The effect of the distortion in these calculations is to introduce additional momentum components and to cause a reduction in the magnitude of the cross-section. At energies above 400 MeV the former effect is of little importance and the data therefore yield a much clearer picture of the momentum distribution of the struck proton than is obtained in this and other reactions at lower energies. Some measurements on the (p, pd) reaction at 1 Gev have also been made. The most significant feature of the data is that the cross-section is proportional to the cross-section for free p–d scattering which suggests that the quasi-elastic mechanism described in section 9.3 remains valid at this energy [137]. Some controversy surrounds other features of the data and is unlikely to be resolved until detailed calculations have been made.

In connection with the application of impulse approximation to elastic scattering and of DWIA to inelastic scattering and reactions it must be remembered that the assumption that the motion of the target nucleon can be neglected forms an essential step in the derivation of the formulae (see sections 6.5 and 7.2). Thus the meaningful investigation of high momentum components and short-range correlations calls for an improved approximation for the interaction. Further, at sufficiently high energies the whole concept of an optical potential and distorted wave theory becomes of doubtful validity.

The scattering of 20 GeV protons from a range of nuclei has been studied at CERN [138], and these results serve to illustrate some of the problems which arise at these high energies. For the medium and heavy nuclei studied, Cu, Pb, U, the small angle scattering shows a typical diffraction pattern with some filling-in of the minima. On the assumption that these data correspond to elastic scattering only, Frahn and Wiechers [130] have analysed the data in terms of a strong absorption model and have reached the conclusion that the filling-in is due to a repulsive real part in the optical potential. The values they find for the strong absorption radii are in agreement with those obtained at lower energies, and the corresponding critical angular momenta $L_{\frac{1}{2}}$ extend from 350(Cu) to 660(U). This indicates the advantage of approximate calculations through the strong absorption model or the semi-classical approximation compared with exact partial wave analyses. However, the momentum resolution of the scattered protons in the CERN experiment was such that the data could include some inelastic scattering

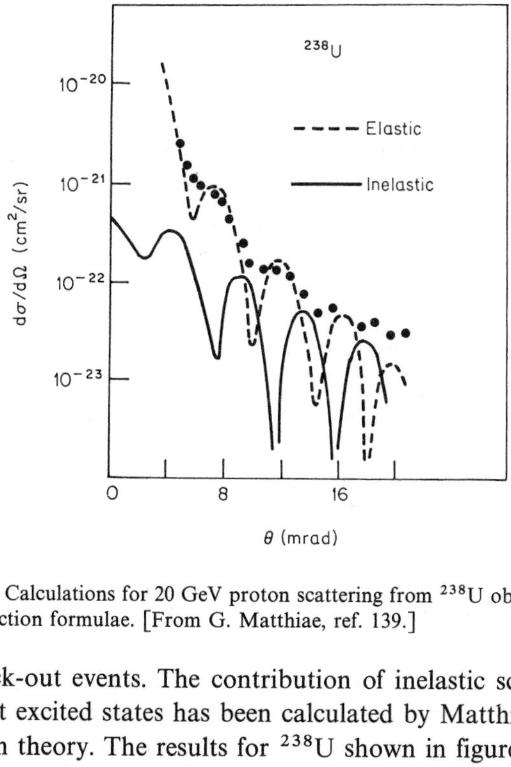

Figure 10.4 Calculations for 20 GeV proton scattering from ^{238}U obtained using the Fraunhofer diffraction formulae. [From G. Matthiae, ref. 139.]

and knock-out events. The contribution of inelastic scattering to certain important excited states has been calculated by Matthiae [139] using the diffraction theory. The results for ^{238}U shown in figure 10.4 indicate that inelastic scattering can make a substantial contribution in the regions of the minima. For the light nuclei studied, the small angle data are also consistent with diffraction scattering, but for these light nuclei the secondary maxima of the diffraction patterns are submerged in the background and at larger values of momentum transfer the slope of the cross-section is characteristic of free nucleon–nucleon scattering.

The difficulty of obtaining good resolution at these high energies and the possibility of high excitation of the nucleus suggests that the interpretation of the data might more profitably proceed through examination of summed cross-sections rather than cross-sections to individual final states. In DWIA the transition matrix element is given by

$$T_{n0} = \langle n|\sum_j t_j(q)\,\delta(\boldsymbol{r}-\boldsymbol{r}_j)\,\chi^{-*}(\boldsymbol{k}',\boldsymbol{r})\chi^{+}(\boldsymbol{k},\boldsymbol{r})|0\rangle \tag{10.15}$$

so that the sum over all excited states is

$$\sum_{n\neq 0}|T_{n0}|^2 = |\bar{t}(q)|^2\left[\sum_{n=0}^{\infty}|\langle n|\sum_j \chi^{-*}(\boldsymbol{k}',\boldsymbol{r}_j)\,\chi^{+}(\boldsymbol{k},\boldsymbol{r}_j)|0\rangle|^2 - |\langle 0|\sum_j \chi^{-*}(\boldsymbol{k}',\boldsymbol{r}_j)\chi^{+}(\boldsymbol{k},\boldsymbol{r}_j)|0\rangle|^2\right]$$

where $\bar{t}(q)$ is the matrix element for nucleon–nucleon scattering averaged over the initial and final nuclear spin and I-spin. Using plane waves and closure this sum reduces to equation (2.51). A simple form for the distorted waves can be obtained from equations (3.132), (3.158), (10.4), and (10.9), i.e.

$$\chi^{-*}(\boldsymbol{k}',\boldsymbol{r})\,\chi^{+}(\boldsymbol{k},\boldsymbol{r}) \approx e^{i\boldsymbol{q}\cdot\boldsymbol{r}}\,e^{-\frac{1}{2}\sigma_{\text{tot}}T(\boldsymbol{b})} \tag{10.16}$$

and again using closure we have

$$\sum_{n\neq 0}|T_{n0}|^2 = |\bar{t}(q)|^2\left[\int T(\boldsymbol{b})\,e^{-\sigma_{\text{tot}}T(\boldsymbol{b})}\,d^2 b + \text{functions of } q\right]. \tag{10.17}$$

Comparing equation (10.17) with equation (2.51) we see that the number

$$N(A) = \int T(\boldsymbol{b})\,e^{-\sigma_{\text{tot}}T(\boldsymbol{b})}\,b\,db\,d\phi \tag{10.18}$$

has replaced the total number of nucleons A and represents a crude estimate of the reduction in magnitude of the cross-section due to the absorptive part of the optical potential. Thus, to lowest order, equation (10.17) predicts that the variation of the summed cross-section with q should be determined by the free nucleon-nucleon cross-section, as is observed in the data at 20 GeV. The values of $N(A)$ deduced from these data are consistent with the values calculated from equation (10.18) using the density distributions deduced from analyses of elastic electron scattering [131]. However, the calculated values of $N(A)$ are sensitive to variations in the shape of the matter distribution [140]. For incident protons in the GeV region the integrand $T(b)e^{-\sigma T(b)}$ peaks approximately 0.5 fm beyond the halfway radius of the matter distribution, whereas for incident pions in the GeV region the integrand peaks at the halfway radius and spans the transition region. Thus measurements of $N(A)$ in proton-nucleus and pion–nucleus scattering provide information which is complementary to that obtained from electron scattering. In addition, a detailed study of the q-dependence of the summed cross-section in conjunction with more precise data could yield information on the higher density functions although corrections must be made to the impulse approximation and the semi-classical approximation in order to obtain reliable results at large momentum transfer.

10.2 Pion interactions with nuclei

The production of secondary pion beams of reasonable intensity from high energy proton accelerators has made feasible the study of pion interactions with nuclei. The interest in these interactions arises from the peculiar properties of the pion [3, 40]. The pion is a boson of spin zero and negative intrinsic parity, it is distinguishable from the nucleon, it has I-spin of unity, and exists in three charged states π^+, π^0, π^-. The π^+ and π^- form a particle-antiparticle pair with rest mass 273 times the electron mass m_e, while the π^0 has a mass of 264 m_e. The pion is therefore a rather light particle compared with the nucleon (in fact it is the lightest strongly interacting particle), and carries approximately 140 MeV of mass-energy. Because of the role of the pion as the exchanged quantum or field particle in the nucleon–nucleon interaction, the pion can be absorbed by one or more nucleons, and because of the existence of three charge states it can participate in charge-exchange reactions involving the exchange of one or two units of charge.

The elastic scattering of pions of energy between 20 MeV and 150 MeV has been studied, using the optical model. The simplest form for the optical potential is obtained using the impulse approximation as described in section 6.5. However, the pion–nucleon interaction differs from the nucleon–nucleon interaction in that it is rather weak and is dominated by s- and p-waves up to a few hundred MeV. In these circumstances the transition matrix element for the pion–nucleon interaction can be written as

$$\langle \boldsymbol{k}'|\tau|\boldsymbol{k}\rangle = a + c\boldsymbol{k}\,.\,\boldsymbol{k}' + [b + d\boldsymbol{k}\,.\,\boldsymbol{k}']\boldsymbol{t}_{\mathrm{N}}\,.\,\boldsymbol{t}_{\pi} \tag{10.19}$$

where we have neglected the motion of the target nucleon, $\boldsymbol{t}_\pi$ and $\boldsymbol{t}_{\mathrm{N}}$ are I-spin operators for the pion and nucleon respectively, and $\boldsymbol{k}\,.\,\boldsymbol{k}' = k^2 \cos\theta$. The coefficients are determined from free pion–nucleon scattering. The optical potential then becomes

$$\alpha U(\boldsymbol{r}) = a\rho(r) - c\boldsymbol{\nabla}.[\rho(r)\boldsymbol{\nabla}] + \frac{1}{A}\boldsymbol{t}_\pi.\boldsymbol{T}\{b\rho_t(t) - d\boldsymbol{\nabla}.[\rho_t(r)\boldsymbol{\nabla}]\} \qquad (10.20)$$

where $\boldsymbol{T}$ is the I-spin operator for the nucleus, ρ_t is the density of the nucleons which contribute to a non-zero I-spin i.e. the density of the excess neutrons, and α is a constant which non-relativistically is just $(-2\pi\hbar^2 A/\mu_0)^{-1}$. This non-local optical potential is the simplest potential having the required properties and there are a number of corrections and additional terms which are discussed by Ericson and Ericson [141]. For elastic scattering this potential is inserted into the Klein-Gordon equation as the fourth component of a four-vector. The details of the calculation are described by Auerbach *et al.* [142] who show that satisfactory agreement with the existing data can be obtained using parameters obtained from pion–nucleon scattering and nuclear density distributions consistent with electron scattering.

The presence of the $\boldsymbol{t}_\pi.\boldsymbol{T}$ term in the pion-nucleus optical potential gives rise to the possibility of charge-exchange reactions of the form $(\pi^\pm, \pi^0)$ in complete analogy with the (p, n) reaction. These reactions may be described using the optical model or using DWBA as discussed in section 7.5. There are as yet very few experimental data on these reactions but an examination [143] of the available data [144] for 180 MeV pions suggests that the total cross-section for (π^+, π^0) on light nuclei is approximately 2.5 times that for the (p, n) reaction at a comparable energy and that the reaction mechanism is similar in the two cases. The double charge-exchange reaction $(\pi^\pm, \pi^\mp)$ can be brought about by the $\boldsymbol{t}_\pi.\boldsymbol{T}$ term acting in second order or by an additional term in the optical potential of the form $(\boldsymbol{t}_\pi.\boldsymbol{T})^2/A^2$. There is a great deal of interest in the double exchange process since it involves a $\Delta T=2$ transition between nuclear states. The experiments at pion energies around 180 MeV have so far only been performed on light nuclei and the cross-sections are disappointingly low [143]. For pion energies in the range 2–6 BeV the results of Lundy *et al.* [145] suggest that double charge exchange is not an important mechanism and that the (π^-, π^+) reaction proceeds through single-nucleon resonance or boson production in the nucleus.

Additional information about the pion–nucleus interaction can be obtained from studies of pionic atoms and pionic x-rays. Since the pion has spin zero, the levels and wavefunctions of the pion in an extranuclear bound state are given by the Klein-Gordon equation and not the Dirac equation, but otherwise much of the theory of pionic atoms closly resembles that for muonic atoms. The important difference is that in addition to the electromagnetic interaction with the protons the pion interacts with both the protons and the neutrons in the nucleus through the strong interaction. An accurate description of the level shifts and widths determined from x-ray data requires the inclusion in the pion-nucleus potential of terms arising from the interaction of a pion with two nucleons [146].

The absorption or capture of pions by nuclei can be separated into two processes, radiative capture, and non-radiative capture which we will refer to simply as absorption. Radiative capture of negative pions [146],

$$\pi^- + A \rightarrow \gamma + B$$

is of considerable theoretical interest because it can be related to the corresponding μ^- capture,

$$\mu^- + A \rightarrow \nu_\mu + B\,,$$

and can give rise to the excitation of negative parity states with $T=1$ starting from target states with $J=0^+$, $T=0$, but except for very light nuclei this process represents only a few per cent of the total capture rate.

The simplest way to approach the non-radiative process is to assume that the primary process is pion absorption by a single nucleon,

$$\pi + \mathrm{N} \rightarrow \mathrm{N} \tag{10.21}$$

but an examination of the conditions for energy and momentum conservation shows that this process can only take place if the target nucleon has a rather high momentum. For a pion absorbed at rest the required momentum is approximately $(2m_\pi m_\mathrm{N} c^2)^{\frac{1}{2}}$ or 500 MeV/c (2.5 fm^{-1}), and studies of single-particle momentum components in nuclei (see sections 9.5 and 10.1) indicate that the abundance of such high momentum components is small. In an alternative approach, suggested by Brueckner, Serber and Watson [147], the absorption is assumed to occur on two nucleons through the basic reaction

$$\pi + \mathrm{N} + \mathrm{N} \rightarrow \mathrm{N} + \mathrm{N}\,. \tag{10.22}$$

For nucleons emitted in opposite directions, the conservation of energy and momentum requires a relative momentum in the nucleus of $\sim$750 MeV/c, and such a momentum is found at inter-nucleon distances of the order of 0·4 fm. This result suggests that pion absorption processes leading to the emission of two nucleons, such as (π^-, 2n) or (π^+, 2p), are sensitive to the behaviour of the wavefunction describing the relative motion of two nucleons at short distances inside the nucleus, and hence can yield information about short-range correlations. A number of calculations have indeed been carried out to seek this information. In the majority of the calculations the pion–nucleon interaction is described by the non-relativistic pseudoscalar-pseudovector form [40]

$$H = \frac{G}{2m_{\mathrm{N}}c}\sum_{i=1}^{A}\left[\boldsymbol{\sigma}.\nabla_{\pi}.\boldsymbol{t}.\boldsymbol{\phi} - \frac{m_{\pi}}{m_{\mathrm{N}}}\boldsymbol{t}.\boldsymbol{\phi}\,\boldsymbol{\sigma}.\nabla_{\mathrm{N}}\right]_i \tag{10.23}$$

where G is the coupling constant, $\boldsymbol{t}$, $\boldsymbol{\sigma}$ are the nucleon I-spin and spin operators, $\boldsymbol{\phi}$ is the pion field operator and the sum runs over all the nucleons. The operator $\boldsymbol{t}.\boldsymbol{\phi}$ can be expanded to give

$$\begin{aligned}\boldsymbol{t}.\boldsymbol{\phi} &= t_1\phi_1 + t_2\phi_2 + t_3\phi_3 \\ &= \sqrt{2}t_-\phi_+ + \sqrt{2}t_+\phi_- + t_3\phi_3\end{aligned} \tag{10.24}$$

where $t_{\pm}$ are the raising and lowering operators for the I-spin so that the first term creates a positive pion or destroys a negative pion and converts a proton into a neutron, the second term creates a negative pion or destroys a positive pion and converts a neutron into a proton, and the third term creates or destroys a neutral pion leaving the nucleon unchanged. Thus the first term in equation (10.24) contributes to the (π^-, 2n) reaction and the second term to the (π^+, 2p) reaction. By making the transformation $\boldsymbol{r}=\boldsymbol{r}_1-\boldsymbol{r}_2$, $\boldsymbol{R}=\frac{1}{2}(\boldsymbol{r}_1+\boldsymbol{r}_2)$, the interaction (10.23) can be rearranged in terms of sums and differences of the operators for the pair of nucleons 1 and 2, and this rearranged interaction is then used in conjunction with simple shell model wavefunctions of oscillator form which are amenable to the same transformation of co-ordinates. Additional high momentum components may then be introduced by modifying the part of the transformed wavefunction which describes the relative motion of the pair of nucleons. Modification of the momentum components arising in the simple model also occurs quite naturally as a result of the interaction of the two nucleons after emission (final state interaction) and scattering of the pion between the two nucleons before absorption (pion rescattering), and in the case of reactions initiated by fast pions also by distortion of the incoming pion wave by the nucleus. These effects [146] have been studied separately in a number of calculations.

An alternative approach to two-nucleon capture, so far applied only to very light nuclei, is based on a phenomenological description of the interaction (10.22). It is assumed that the two nucleons are in a relative S-state and that the interaction is short-range. The effects of short-range correlations are then absorbed into several phenomenological amplitudes which are determined from analysis of pion production in nucleon–nucleon collisions. This model does not provide any information about the short-range behaviour of the relative wavefunction of the pair but does allow investigation of the longer range behaviour of the wavefunction of the centre-of-mass of the pair of nucleons. In this respect, this model resembles the quasi-deuteron model of the (γ, np) reaction and the corresponding treatment of the (p, pd) reaction which leads to equation (9.87), while the description of the (π, 2N) reactions based on the interaction (10.23) is analogous to the description of the (p, pd) reaction leading to equation (9.88) [119]. Like the (p, pd) reaction, the (π, 2N) reactions yield information on the excitation of two hole states in nuclei, and qualitative interpretation of the energy spectra for light nuclei obtained by Charpak *et al.* and the data of Muirhead *et al.* can be made in terms of the shell model [148].

Total cross-sections for knock-out reactions of the form (π, πn) have been obtained by measuring the activity of the residual nucleus [144]. The simplest interpretation of the data is based on a quasi-elastic mechanism, illustrated in figure 10.5(a), similar to that used to interpret the (p, 2p) reaction. The interpretation of the nuclear transition is then identical to that for a single-nucleon transfer or knock-out reaction and Kolybasov [149] has used the spectroscopic factors appropriate to the $^{12}C(p, d)$ reaction in a simple plane wave calculation for the $^{12}C(\pi^-, \pi^- n)$ reaction which yields satisfactory agreement with the data of Reeder and Markowitz for pion energies above 200 MeV. However, Chivers *et al.* [144] have compared the total cross-sections for the (π, πn) reaction on light nuclei initiated by positive and negative pions, and their results cast doubt on the simple quasi-elastic mechanism.

In order to study the quasi-elastic knock-out process we have to examine the primary pion–nucleon interaction. From the values of the I-spin of the nucleon ($t=\frac{1}{2}$, $t_z=\pm\frac{1}{2}$) and of the pion ($t=1$, $t_z=\pm 1, 0$) it is evident that the pion–nucleon system can exist in states with total I-spin T of $\frac{3}{2}$ or $\frac{1}{2}$, and the wavefunction for the system can be written as a linear combination of the I-spin terms $\psi(T, T_z)$

$$\psi(\pi^- n) = \psi(\tfrac{3}{2}, -\tfrac{3}{2}) \tag{10.25}$$

$$\psi(\pi^+ n) = (\tfrac{2}{3})^{\frac{1}{2}}\,\psi(\tfrac{1}{2}, \tfrac{1}{2}) - (\tfrac{1}{3})^{\frac{1}{2}}\,\psi(\tfrac{3}{2}, \tfrac{1}{2}) \tag{10.26}$$

$$\psi(\pi^0 p) = (\tfrac{1}{3})^{\frac{1}{2}}\,\psi(\tfrac{1}{2}, \tfrac{1}{2}) + (\tfrac{2}{3})^{\frac{1}{2}}\,\psi(\tfrac{3}{2}, \tfrac{1}{2}) \tag{10.27}$$

where we have used convention A for the I-spin (see section 2.2) so that positive particles have positive t_z, and the coefficients are the Clebsch-Gordan

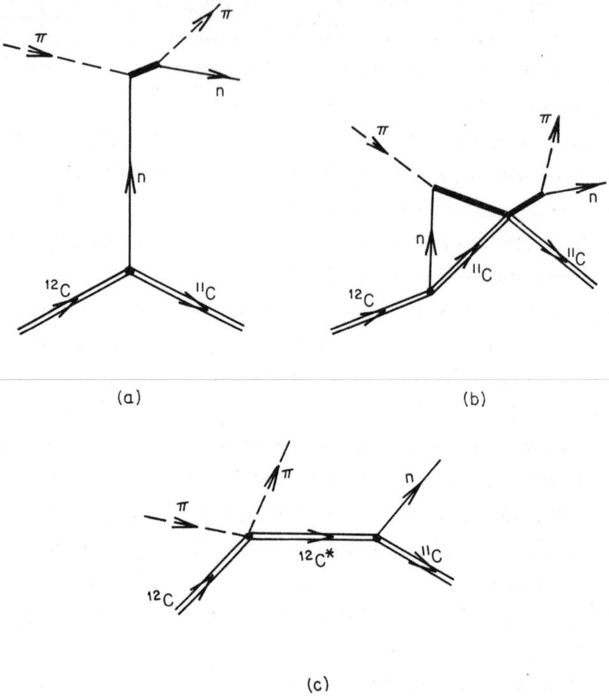

Figure 10.5 Diagrams for the processes which might contribute to the $^{12}C(\pi,\pi n)C^{11}$ reaction.

coefficients which combine the *I*-spin values to a definite total *I*-spin. The scattering of a pion-nuclear system can be described by a scattering amplitude *f* which may also be characterized by the total *I*-spin. Thus the outgoing scattered wave can be represented by

$$\begin{aligned} \psi(\pi^- n) &\to f(\tfrac{3}{2})\,\psi(\tfrac{3}{2}, -\tfrac{3}{2}) \\ \psi(\pi^+ p) &\to f(\tfrac{3}{2})\,\psi(\tfrac{3}{2}, +\tfrac{3}{2}) \end{aligned} \tag{10.28}$$

where the second line follows from charge independence, and

$$\begin{aligned} \psi(\pi^+ n) &\to (\tfrac{2}{3})^{\frac{1}{2}} f(\tfrac{1}{2})\,\psi(\tfrac{1}{2}, \tfrac{1}{2}) - (\tfrac{1}{3})^{\frac{1}{2}} f(\tfrac{3}{2})\,\psi(\tfrac{3}{2}, \tfrac{1}{2}) \\ &\to (\tfrac{2}{3})^{\frac{1}{2}} f(\tfrac{1}{2})[(\tfrac{2}{3})^{\frac{1}{2}}\,\psi(\pi^+ n) + (\tfrac{1}{3})^{\frac{1}{2}}\,\psi(\pi^0 p)] \\ &\quad - (\tfrac{1}{3})^{\frac{1}{2}} f(\tfrac{3}{2})[(\tfrac{2}{3})^{\frac{1}{2}}\,\psi(\pi^0 p) - (\tfrac{1}{3})^{\frac{1}{2}}\,\psi(\pi^+ n)] \\ &\to \tfrac{1}{3}[f(\tfrac{3}{2}) + 2f(\tfrac{1}{2})]\,\psi(\pi^+ n) - (\tfrac{2}{9})^{\frac{1}{2}}[f(\tfrac{3}{2}) - f(\tfrac{1}{2})]\,\psi(\pi^0 p) \end{aligned} \tag{10.29}$$

where the expressions for $\psi(\tfrac{1}{2}, \tfrac{1}{2})$ and $\psi(\tfrac{3}{2}, \tfrac{3}{2})$ have been obtained by elimination from equations (10.26) and (10.27). Thus the total cross-sections for pion–nucleon scattering are given by

$$\begin{aligned}
\sigma(\pi^- n \to \pi^- n) &\equiv \sigma(\pi^+ p \to \pi^+ p) = |f(\tfrac{3}{2})|^2 \\
\sigma(\pi^+ n \to \pi^+ n) &\equiv \sigma(\pi^- p \to \pi^- p) = \tfrac{1}{9}|f(\tfrac{3}{2}) + 2f(\tfrac{1}{2})|^2 \\
\sigma(\pi^+ n \to \pi^0 p) &\equiv \sigma(\pi^- p \to \pi^0 n) = \tfrac{2}{9}|f(\tfrac{3}{2}) - f(\tfrac{1}{2})|^2
\end{aligned}$$

and this yields the ratio

$$\frac{\sigma(\pi^+ n \to \pi^+ n) + \sigma(\pi^+ n \to \pi^0 p)}{\sigma(\pi^- n \to \pi^- n)} = \frac{|f(\frac{3}{2}) + 2f(\frac{1}{2})|^2 + |f(\frac{3}{2}) - f(\frac{1}{2})|^2}{3|f(\frac{3}{2})|^2} \tag{10.30}$$

$$= \tfrac{1}{3}, \quad \text{if} \quad f(\tfrac{1}{2}) = 0\,. \tag{10.31}$$

Now Chivers *et al.* find the ratio

$$\frac{^{12}C(\pi^+, \pi^+ n) + {}^{12}C(\pi^+, \pi^0 p)}{^{12}C(\pi^-, \pi^- n)} \simeq 1{\cdot}0$$

at a pion energy of 180 MeV which is the energy of the (3, 3) pion–nucleon resonance due to the formation of the nucleon isobar with $J = \frac{3}{2}$, $T = \frac{3}{2}$. Assuming a quasi-elastic mechanism and assuming that the $T = \frac{1}{2}$ amplitude is negligible at the resonance, the predicted value for the ratio is $\frac{1}{3}$, from equation (10.31), and this is in conflict with the experimental result not only for ^{12}C but also for ^{16}O and ^{14}N. A number of explanations for this discrepancy have been put forward. It is possible that the (π, 2N) primary interaction contributes but this would disturb the interpretation of the (π^+, π^0) reaction. It is also possible that there is a rescattering of the isobar as shown in figure 10.5(b), and this mechanism has been used by Dalkarov [150] to fit the data of Reeder and Markowitz below 200 MeV. A further possibility is that the data include some events corresponding to a core-excitation process such as that shown in figure 10.5(c) which would invalidate the quasi-elastic assumption. Perhaps the most simple explanation arises from the fact that, because of the momentum of the struck nucleon in the nucleus, the relative energy of the pion–nucleon system may be displaced from the value calculated from the pion kinetic energy and owing to the rather sharp energy dependence of the pion–nucleon cross-section, illustrated in figure 10.6, the ratio (10.30) for scattering inside the nucleus may be substantially modified from the value for free scattering. This effect is certainly apparent in the (p, 2p) reaction (see section 9.3). The relevance of some of these explanations could be assessed if the energy spectrum of the outgoing pions were measured.

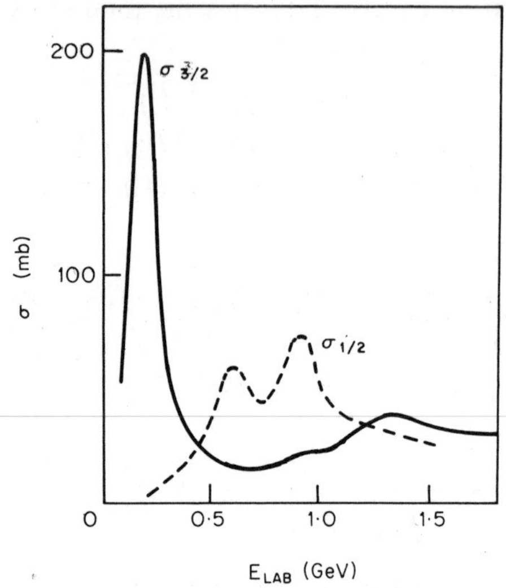

Figure 10.6 The energy dependence of the pion-nucleon total cross-sections for the $T=\frac{3}{2}$ and $T=\frac{1}{2}$ states.

Measurements of the total cross-sections and absorption cross-sections for pion–nucleus scattering have also been used to study the pion–nucleon interaction in the nucleus. Crozon *et al.* [151] have measured σ_{tot} and σ_{abs} for π^- scattering from Be, C, and Al, at energies from 500 MeV to 1300 MeV. Using a π^- optical potential of the form

$$V^-(r) \propto [Zf^-(0) + Nf^+(0)]\,\rho(r)\,,$$

where f^- is the forward scattering amplitude for π^- on protons and f^+ is the scattering amplitude for π^+ on protons (or π^- on neutrons), and varying f^+ to obtain a best fit to the measured cross-sections they determine the quantities

$$\bar{\sigma} = \tfrac{1}{2}\frac{4\pi}{k}(\text{Im}\,f^+ + \text{Im}\,f^-)\,,$$

$$\bar{D} = \tfrac{1}{2}(\text{Re}\,f^+ + \text{Re}\,f^-)\,,$$

They find that $\bar{D}$ is not significantly different from the value obtained using the free pion-nucleon scattering amplitudes whereas $\bar{\sigma}$ is substantially different in magnitude from the free value and does not show the same marked energy dependence due to resonances in the pion–nucleon system. A simple calculation suggests that the smoothing out of the energy dependence is due to the effect of the momentum distribution of the nucleons in the nucleus. Auerbach *et al.* [142] have extended this approach to calculate the absorption cross-sections for $\pi^{\pm}$ scattering on Pb at 700 MeV for comparison with the data of Abashian *et al.*, using optical potentials of the form

$$V^{\pm}(r) \propto [Zf^{\pm}(0)\,\rho_{\mathrm{p}}(r) + Nf^{\mp}(0)\,\rho_{\mathrm{n}}(r)]\,. \tag{10.32}$$

In this case the scattering amplitudes were taken from the pion–nucleon scattering and the quantity studied was the neutron distribution $\rho_{\mathrm{n}}(r)$. It can be seen from the expression (10.32) for the optical potential that if the pion–nucleon amplitudes f^{+} and f^{-} are identical, the pion–nucleus potential must be the same for positive and negative pions. In fact, these amplitudes are in general not the same, as can be seen from figure 10.6, so that a comparison of π^{+} and π^{-} scattering from the same nucleus provides information about the difference between the proton and neutron distributions. The quantity studied is the ratio

$$q = (\sigma^{-}_{\mathrm{abs}} - \sigma^{+}_{\mathrm{abs}})/\sigma^{+}_{\mathrm{abs}}\,. \tag{10.33}$$

Calculations suggest that this ratio is insensitive to uncertainties in the optical parameters although the effect of the nucleon momentum distribution is to increase the calculated value of q slightly.

Calculations of the total and absorption cross-sections have also been made using the semi-classical approximation. From equations (3.65), (3.67), (3.68), (3.140), and (10.9) the absorption and total cross-section for pion–nucleus scattering are given by

$$\sigma_{\mathrm{abs}} = 2\int [1 - \mathrm{Re}\, e^{i\chi(\boldsymbol{b})}]\, d^2 b = 2\int [1 - e^{-\frac{1}{2}\sigma T(\boldsymbol{b})}]\, d^2 b$$

$$\sigma_{\mathrm{tot}} = \int [1 - |e^{i\chi(\boldsymbol{b})}|^2]\, d^2 b = \int [1 - e^{-\sigma T(\boldsymbol{b})}]\, d^2 b\,,$$

where we must now write

$$\sigma T(\boldsymbol{b}) = \sigma^{+} Z \int_{-\infty}^{\infty} \rho_{\mathrm{n}}(\boldsymbol{b}, z)\, dz + \sigma^{-} N \int_{-\infty}^{\infty} \rho_{\mathrm{n}}(\boldsymbol{b}, z)\, dz\,, \tag{10.34}$$

and $\sigma^{\pm}$ are the total cross-sections for pion–nucleon scattering which are related to the amplitudes $f^{\pm}$ through the optical theorem. Semi-classical calculations for pion scattering from Pb are consistent with optical model calculations when the Coulomb potential is omitted [152] but the effect of the Coulomb potential is important and should not be neglected. In both cases, interpretation of the data in terms of the proton and neutron distributions depends on the validity of equation (10.32), i.e. on the validity of the impulse approximation in the presence of the pion–nucleon resonances and of the large A approximation.

Further information about the neutron distribution in nuclei may be obtained from a study of pion production by fast protons on nuclei. If it is assumed that this process proceeds through formation of the (3, 3) isobar, the possible reactions which lead to the production of charged pions can be written as follows

$$\begin{aligned} \mathrm{p}+\mathrm{p} &\rightarrow \Delta^{++}+\mathrm{n} \rightarrow \pi^{+}+\mathrm{p}+\mathrm{n} \qquad && T_Z=\tfrac{3}{2} \\ &\rightarrow \Delta^{+}\;\;+\mathrm{p} \rightarrow \pi^{+}+\mathrm{n}+\mathrm{p} && T_Z=\tfrac{1}{2} \\ \mathrm{p}+\mathrm{n} &\rightarrow \Delta^{+}\;\;+\mathrm{n} \rightarrow \pi^{+}+\mathrm{n}+\mathrm{n} && T_Z=\tfrac{1}{2} \\ &\rightarrow \Delta^{0}\;\;+\mathrm{p} \rightarrow \pi^{+}+\mathrm{p}+\mathrm{p} && T_Z=-\tfrac{1}{2} \end{aligned}$$

where the symbol Δ represents the isobar, and the values of the Clebsch-Gordan coefficients for addition of the pion and nucleon I-spins to form a total I-spin of $\frac{3}{2}$ then lead to ratios of these primary interactions of $1:\frac{1}{9}:\frac{1}{9}:\frac{1}{9}$. Thus the ratio of π^+ production to π^- production on a nucleus A should be $(10Z+N)/N$, and π^+ production is strongly favoured. Margolis [140] has used the first term of the semi-classical multiple scattering expansion and has shown that the effective number of nucleons participating in this process is consistent with the number $N(A)$ (equation (10.18)) determined from high energy proton scattering. It may be noted that within the framework of the isobar model, π^- production by protons occurs only through interactions with neutrons in the nucleus, so that a study of this process could yield the most direct information on the distribution of these neutrons. This information may be obscured by the occurence of secondary reactions, and by kinematical corrections of the type discussed in connection with the $(\pi^-, \pi^- n)$ reaction [153].

10.3 Interactions of other particles with nuclei

The interaction of other strongly interacting particles with nucleons in nuclei can also provide information about nuclear structure. At the present time the particles of interest, besides the pion which we have already discussed, are the kaon and the Λ and Σ hyperons, some of whose properties are given in table 10.1. The hyperons belong, with the nucleons and other heavier particles of spin $\frac{1}{2}$, to the family of baryons, and decay in such a way that the baryon number N is conserved. The kaon belongs, with the pion and other heavier particles of spin zero, to the family of mesons. There is apparently no conservation law for meson number since processes such as those represented by equations (10.21) and (10.22) are observed. For an understanding of the interactions and decays of these particles it is necessary to include among their properties the *strangeness* quantum number S which is given by [3]

$$S = 2Q - 2T_Z - N \tag{10.35}$$

where Qe is the charge of the particle and we again use convention A for I-spin. The experimental evidence indicates that strangeness is conserved in strong and electromagnetic interactions but not in weak interactions, and

Table 10.1 Properties of some strong interacting particles.

	Mass (MeV)	Spin J	Strangeness S	I-spin T	I-spin T_Z
Baryons					
p n	938·2 939·5	$\frac{1}{2}$	0	$\frac{1}{2}$	$\frac{1}{2}$ $-\frac{1}{2}$
Λ^0	1115·5	$\frac{1}{2}$	-1	0	0
Σ^+ Σ^0 Σ^-	1189·4 1193·2 1197·6	$\frac{1}{2}$	-1	1	1 0 -1
Mesons					
π^+ π^0 π^-	139·7 135·0 139·7	0	0	1	1 0 -1
K^+ K^0	439·9 497·8	0	$+1$	$\frac{1}{2}$	$\frac{1}{2}$ $-\frac{1}{2}$
$\bar{K}^0$ K^-	497·8 439·9	0	-1	$\frac{1}{2}$	$\frac{1}{2}$ $-\frac{1}{2}$

there are many consequences of this conservation law which are important in elementary particle physics [40]. It is sufficient here to note that with the exception of the Σ^0 the K, Λ, Σ particles can decay only through the weak interaction and therefore have half-lives of the order 10^{-8} to 10^{-10} secs which are long by nuclear standards. (Compare for example the nuclear times discussed at the beginning of Chapter 5.)

Because of the conservation of strangeness, the positive kaon can interact with a nucleon to give only elastic scattering or charge exchange, i.e.

$$K^+ + p \to K^+ + p$$
$$K^+ + n \to K^+ + n$$
$$\to K^0 + p$$

Similarly the negative kaon can interact with a nucleon to give elastic scattering and production of $\overline{K}^0$ by charge-exchange, but it can also interact to produce a Λ or Σ hyperon and one or more pions. The processes which produce one pion are the following

$$\begin{aligned} K^- + p &\to \Lambda^0 + \pi^0 + 181{\cdot}8 \text{ MeV} \\ &\to \Sigma^- + \pi^+ + 96{\cdot}6 \text{ MeV} \\ &\to \Sigma^0 + \pi^0 + 105{\cdot}6 \text{ MeV} \\ &\to \Sigma^+ + \pi^- + 103{\cdot}1 \text{ MeV} \end{aligned} \tag{10.36}$$

$$\begin{aligned} K^- + n &\to \Lambda^0 + \pi^- + 178{\cdot}5 \text{ MeV} \\ &\to \Sigma^- + \pi^0 + 102{\cdot}4 \text{ MeV} \\ &\to \Sigma^0 + \pi^- + 102{\cdot}3 \text{ MeV} \end{aligned} \tag{10.37}$$

The negative kaon can also form a mesic atom. The K^- is first captured into an extra-nuclear orbit of high principal quantum number from which it makes successive transitions to lower orbits accompanied by the emission of x-rays and Auger electrons, until it is sufficiently close to the nucleus for the probability of nuclear capture to dominate over the other processes.

Some data on K^- interactions with the deuteron and the α-particle have been used to study the KN interaction. There are as yet very few data available on K-mesic x-rays, but if such data do become available they can be studied to obtain information on the relation between the K-nucleon and K-nucleus interaction, in a similar way to the study of π-mesic x-rays. In addition, a determination of the quantum numbers corresponding to the x-ray transitions would indicate the 'cut-off' orbit for which the probability for nuclear capture exceeds the probability for x-ray transition [154]. At present, interpretations of K^- interactions with nuclei rely on calculations of these transition probabilities which indicate that for heavy nuclei the nuclear capture processes occur well outside the half-way radius of the nuclear matter distribution.

In order to examine the information on nuclear structure which might be obtained from K^- capture we construct a simple expression for the transition matrix element by treating the process as a direct nuclear reaction. If we consider the primary interaction

$$K + N \to \Sigma + \pi \tag{10.38}$$

and take the KN interaction to be of zero range, the nuclear matrix element has the form

$$T_{if} = \langle \Phi_f \chi_\Sigma \chi_\pi | V_{KN} | \Phi_i \phi_K \rangle \tag{10.39}$$

where Φ_i, Φ_f are the initial and final nuclear wavefunctions, $\chi_\Sigma \chi_\pi$ are the distorted wavefunctions of the outgoing $\Sigma\pi$ pair and ϕ_K is the wavefunction of the K^- in an extra-nuclear bound state. Thus if the process leaves the residual nucleus of $A-1$ nucleons in one of a limited group of final states which can be identified, the relevant matrix element can be written as

$$T_{if} \propto \int \chi_\Sigma^{-*}(\boldsymbol{k}_\Sigma, \boldsymbol{r}) \chi_\pi^{-*}(\boldsymbol{k}_\pi, \boldsymbol{r}) \psi(\boldsymbol{r}) \phi_K(\boldsymbol{r}) \, d\boldsymbol{r} \tag{10.40}$$

where $\psi(\boldsymbol{r})$ is the overlap integral defined in equation (9.21). We note also that since the K^- is captured essentially at rest, the conservation of momentum gives

$$\boldsymbol{k}_N = \boldsymbol{k}_\Sigma + \boldsymbol{k}_\pi$$

so that measurement of the summed momenta of the $\Sigma\pi$ pair provides a direct measurement of the distorted momentum distribution of the nucleon in the nucleus. It is found experimentally that stars arising from K^- capture with emission of a charged $\Sigma\pi$ pair occur predominantly for light nuclei, and some data exist for capture in ^{12}C leading to low-lying states in ^{11}B.

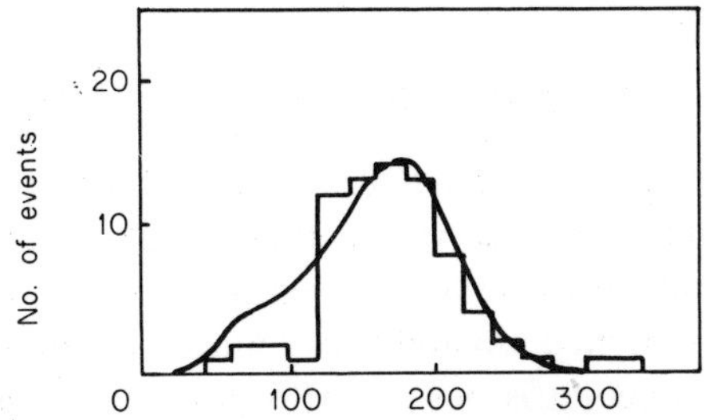

Figure 10.7 The distribution of recoil momenta measured for kaon capture in ^{12}C compared with curve predicted by a simple direct reaction calculation. [After Adair, this version from *Advances in Physics* **17**, 481 (1968).]

The distribution as a function of recoil momentum $-k_N$ is shown in figure 10.7 and indicates a most probable momentum of ~170 MeV/c compared with a value of ~100 MeV/c obtained from the (p, 2p) reaction. The theoretical curve shown in the figure is that obtained by Adair [155] using the form (10.40) for the matrix element, a form for the overlap integral which has the correct separation energy, and including the distortion of the outgoing $\Sigma\pi$ pair by the residual nucleus. These results can be interpreted to suggest that the effect of the distortion and the localization of the reactions are such that K^- capture and the (p, 2p) reaction are sensitive to distinctly different regions of the distorted momentum distribution, and that the (p, 2p) reaction is more sensitive to the low momentum components, although such an interpretation can only be tentative in view of the limited amount of data on K^- capture. It is clear, however, that a study of K^- capture on light nuclei could provide information complementary to that obtained from transfer and knock-out reactions.

In the more general situation the final states of the residual nucleus will not be identified and, if it is assumed that the residual nucleus may be left in any state up to a high level of excitation, it is permissible to sum the matrix elements T_{if} over all final states and to use the closure relation for the final state wavefunctions. This gives

$$\sum_f |T_{if}|^2 \propto \int \rho(\boldsymbol{r})|\phi_K(\boldsymbol{r})|^2 d\boldsymbol{r}$$

where $\rho(\boldsymbol{r})$ is the nuclear matter distribution defined in equation (2.42). The localization of the capture process in configuration space can be studied by calculating the function

$$P(r) = r^2 \rho(r)|\phi_K(r)|^2 ,$$

and some typical results are shown in figure 10.8. Wilkinson [154] has drawn attention to the localization observed in these results and has suggested that K^- capture should provide a very sensitive means of studying the extreme surface region of medium and heavy nuclei. In addition, by identifying the reaction products it is possible to distinguish between reactions with protons and with neutrons and thereby to compare the density distributions of protons and neutrons in nuclei. A comparison of the ratio of events leading to production of $\Sigma^- + \pi^0$ to events leading to production of $\Sigma^\pm + \pi^\mp$ has shown that the ratio of interactions with neutrons to interactions with protons is approximately five times greater in heavy nuclei than in light nuclei, and this result has been interpreted by Burhop [156] as evidence for the existence of a neutron-rich skin in heavy nuclei. These results can be reproduced by increasing the surface thickness of the neutron distribution in heavy nuclei by 45–50 per cent or increasing the halfway radius by ~ 15 per cent, or by some combined increase in these two quantities. Such an

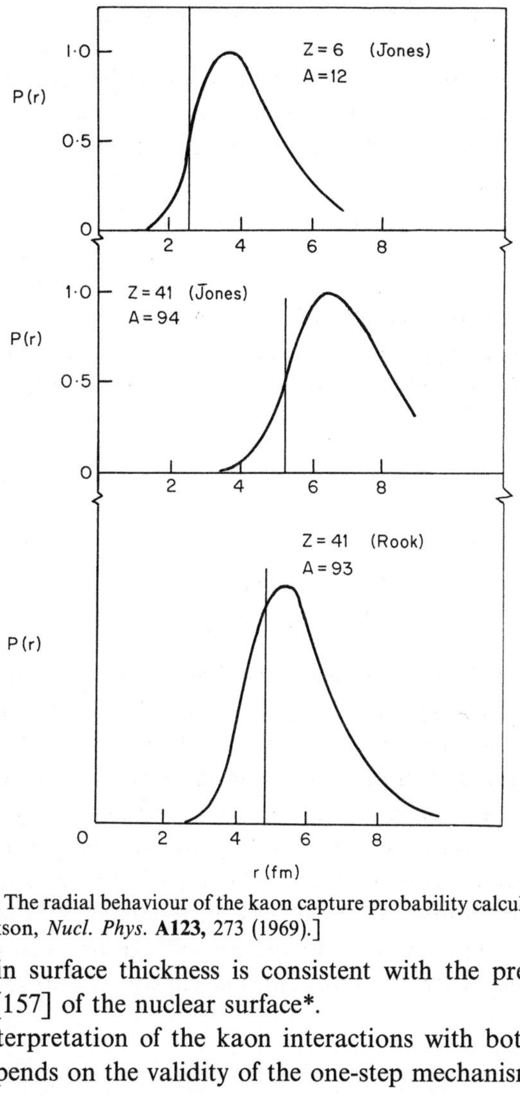

Figure 10.8 The radial behaviour of the kaon capture probability calculated for various nuclei. [D. F. Jackson, *Nucl. Phys.* **A123**, 273 (1969).]

increase in surface thickness is consistent with the predictions given by theories [157] of the nuclear surface*.

The interpretation of the kaon interactions with both light and heavy nuclei depends on the validity of the one-step mechanism, and is therefore

* Present information on the difference between the halfway radii of the neutron and proton distributions for heavy nuclei is conflicting. Analyses of low energy proton scattering [79] and pion production [140] and also calculations using single-particle potentials [46,123] suggest that the neutron radii are greater than the proton radii by as much as 0.6 fm, whereas studies of pion scattering and absorption [142] and of Coulomb energy differences between isobaric analogue states and their parent states [158] indicate that the proton radii are equal to or very slightly greater than the neutron radii.

subject to uncertainties similar to those discussed for pion–nucleus interactions.

In some of the events due to K^- capture in nuclei a Λ^0 particle is formed and replaces a nucleon in the nucleus giving rise to a bound system called a *hyperfragment* or *hypernucleus.* Thus, K^- capture in helium can give rise to the following reactions

$$\begin{aligned} K^- + {}^4\mathrm{He} &\rightarrow {}^4\mathrm{He}_\Lambda + \pi^- \\ &\rightarrow {}^4\mathrm{H}_\Lambda + \pi^0 , \end{aligned}$$

where the notation for the hypernuclei indicates that they are composed of three nucleons and one hyperon. Since the hyperon is distinguishable from the nucleons it can go into a 1*s* state. It subsequently decays, either through a mesonic mode

$$\begin{aligned} \Lambda^0 &\rightarrow \mathrm{p} + \pi^- \\ &\rightarrow \mathrm{n} + \pi^0 , \end{aligned}$$

or through a non-mesonic mode

$$\begin{aligned} \Lambda^0 + \mathrm{p} &\rightarrow \mathrm{p} + \mathrm{n} \\ \Lambda^0 + \mathrm{n} &\rightarrow \mathrm{n} + \mathrm{n} . \end{aligned}$$

The experimental data so far consist of binding energies of *s*-shell and *p*-shell hypernuclei, transition rates of excited hypernuclei, branching ratios for various decay modes, and energy spectra of the decay products [159]. The existence of certain double hypernuclei has also been demonstrated. The task in hypernuclear physics is essentially the same as in ordinary nuclear structure physics, namely to interpret the hypernuclear properties in terms of the two-body interactions using a suitable nuclear model as a basis. In this case of single and double hypernuclei the basic interactions to be studied are the ΛN and ΛΛ interactions, but reliable calculations also demand a sound understanding of the structure of the nuclear core of the hypernucleus.

Bibliography

Chapter 1

For this chapter, references are given to standard introductory texts on quantum mechanics and nuclear physics. The list is far from exhaustive. See also, references [9], [12], [16], [31], and [32].

[1] W. E. Burcham, *Nuclear Physics* (Longmans, London, 1963).
[2] R. M. Eisberg, *Fundamentals of Modern Physics* (Wiley, New York, 1961).
[3] E. Segre, *Nuclei and Particles*, (Benjamin, New York, 1964).
[4] D. Bohm, *Quantum Theory* (Constable, London, 1954).
[5] L. I. Schiff, *Quantum Mechanics* (McGraw-Hill, New York-London, second edition 1955, third edition 1969).
[6] L. Eisenbud and E. P. Wigner, *Nuclear Structure* (Princeton University Press; Oxford University Press, 1958).
[7] J. M. Blatt and V. F. Weisskopf, *Theoretical Nuclear Physics* (Wiley, New York, 1952).

Chapter 2

Most of the references in this chapter are to standard advanced texts in nuclear physics with special reference to nuclear models. Reference [9] gives a valuable list of references up to 1963. References [17] together with reference [5] provide a sufficient background in special functions. The remaining references relate to specific points in sections 2.5 and 2.6.

[8] D. M. Brink, *Nuclear Forces* (Pergamon, Oxford, 1965).
[9] *Selected Reprints on Nuclear Structure* (American Institute of Physics, New York, 1965).
[10] J. P. Davidson, *Collective Models of the Nucleus* (Academic Press, New York-London, 1968).
[11] A. E. S. Green, T. Sawada, and D. Saxon, *The Nuclear Independent Particle Model* (Academic Press, New York-London, 1968).
[12] M. A. Preston, *Physics of the Nucleus* (Addison-Wesley, Reading, Mass., 1962).

Chapter 2 (contd)

[13] G. E. Brown, *Unified Theory of Nuclear Models and Forces* (North-Holland, Amsterdam, 1967).

[14] A. Bohr and B. R. Mottelson, *Mat. Fys. Medd. Dan. Vid. Selsk.*, **27**, No. 16 (1953).
A. Bohr and B. R. Mottelson, article in *Beta and Gamma-Ray Spectroscopy*, (ed. Siegbahn, North-Holland, Amsterdam, 1955).
B. R. Mottelson, *Proceedings of the International Conference on Nuclear Structure, Kingston*, 1960 (University of Toronto Press, 1960).

[15] Articles in *Encyclopedia of Physics*, Vol. 39, (Springer, Berlin, 1957), by J. P. Elliott and A. M. Lane, and by S. A. Moskowski.

[16] M. G. Mayer and H. G. Jensen, *Elementary Theory of Nuclear Shell Structure* (Wiley, New York, 1955).

[17] I. N. Sneddon, *Special Functions of Mathematical Physics and Chemistry*, (Oliver and Boyd, London, 1956).
E. Jahnke and F. Emde, *Tables of Functions* (Constable, London: Dover, New York, 1945).

[18] A. de Shalit and I. Talmi, *Nuclear Shell Theory* (Academic Press, New York-London, 1963).

[19] O. Nathan and S. G. Nilsson, article in *Alpha-, Beta-, and Gamma-Ray Spectroscopy* (ed. Siegbahn, North-Holland, Amsterdam, 1965).

[20] S. G. Nilsson, *Mat. Fys. Medd. Dan. Vid. Selsk.* **29**, No. 16 (1955).

[21] A. M. Lane, *Nuclear Theory* (Benjamin, New York, 1964).

[22] R. E. Peierls, article in *Lectures in Theoretical Physics*, Vol. I, (Interscience, New York-London, 1959).

[23] H. A. Bethe and M. E. Rose, *Phys. Rev.* **51**, 283 (1937).

[24] J. P. Elliott and T. H. R. Skyrme, *Proc. Roy. Soc.* **A232**, 561 (1955).

[25] H. J. Lipkin, article in *The Many Body Problem* (Methuen, London, 1959).

[26] E. U. Baranger and C. W. Lee, *Phys. Rev.* **108**, 482 (1957).

[27] L. J. Tassie and F. C. Barber, *Phys. Rev.* **111**, 940 (1958); *Nuo. Chim.* **19**, 1211 (1961).

[28] K. Gottfried, *Annals of Phys.* **21**, 29 (1963).

[29] K. W. McVoy and L. Van Hove, *Phys. Rev.* **125**, 1034 (1962).

[30] See also, references [36] and [82].

Chapter 3

For this chapter, references are given to standard works on advanced scattering theory and on angular momentum in quantum theory. These references are also used in later chapters.

[31] A. Messiah, *Quantum Mechanics* (North-Holland, Amsterdam, 1965).

[32] N. F. Mott and H. S. W. Massey, *The Theory of Atomic Collisions*, (Clarendon Press, Oxford, second edition 1949, third edition 1968).

[33] M. E. Rose, *Elementary Theory of Angular Momentum* (Wiley, New York: Chapman and Hall, London, 1957).

[34] A. R. Edmonds, *Angular Momentum in Quantum Mechanics* (Princeton University Press, 1957).

[35] L. Wolfenstein, *Annual Reviews of Nuclear Science*, **6**, 43 (1956).
[36] A. K. Kerman, H. McManus and R. M. Thaler, *Annals of Phys.* **8**, 551 (1959).
[37] W. E. Frahn, article in *Fundamentals of Nuclear Theory* (International Atomic Energy Agency, Vienna, 1967).
[38] M. L. Goldberger and K. M. Watson, *Collision Theory* (Wiley, New York-London, 1964).
[39] R. J. Glauber, article in *Lectures in Theoretical Physics* Vol. I (Interscience, New York-London, 1959).
See also, references [131], [132], [133].

Chapter 4

References [41] and [44] are reprinted in *Nuclear and Nucleon Structure* (ed. Hofstadter, Benjamin, New York, 1963), which contains most of the important papers in this field up to 1962.

[40] J. Hamilton, *The Theory of Elementary Particles* (Clarendon Press, Oxford, 1959).
[41] D. R. Yennie, D. G. Ravenhall, and R. N. Wilson, *Phys. Rev.* **95**, 500 (1954).
[42] G. Rawitscher, *Phys. Rev.* **151**, 846 (1966).
[43] K. Alder, A. Bohr, T. Haas, B. Mottelson and A. Winther, *Rev. Mod. Phys.* **28**, 432 (1956).
[44] L. I. Schiff, *Phys. Rev.* **96**, 765 (1954).
[45] Landolt-Börnstein, New Series, Group I, Vol. 2 (ed. Schopper, Springer, Berlin.)
[46] L. R. B. Elton, *Proceedings of the International Conference* on *Electromagnetic Sizes of Nucleons*, Ottawa 1967 (Carleton University, Ottawa 1967), and earlier references quoted therein.
[47] D. S. Onley, J. T. Reynolds and L. E. Wright, *Phys. Rev.* **134**, B945 (1964) and earlier references quoted therein.
T. Schucan, *Nucl. Phys.* **61**, 417 (1965).
[48] L. I. Schiff, *Nuo. Cim.* **5**, 1223 (1957).
T. de Forest and J. D. Walecka, *Advances in Physics* **15**, 1 (1966).
[49] U. Meyer-Berkhout, *Phys. Rev.* **115**, 1300 (1959).
[50] T. de Forest, *Nucl. Phys.* **A132**, 305 (1969).
[51] U. Amaldi, article in *Proceedings of the International School of Physics-E. Fermi*, (ed. Ericson, Academic Press, New York-London, 1967) and references given therein.
[52] A. Watt, *Phys. Lett.* **27B**, 190 (1968).
C. D. Epp and T. Griffy, *Bull. Am. Phys. Soc.* **14**, 572 (1969).
[53] See, for example, section 5.3.4 of reference [86].
[54] P. H. Stelson and L. Grodzins, *Nuclear Data* **A1**, No. 1 (1966).
[55] C. S. Wu, *Proceedings of the International Nuclear Physics Conference, Gatlinburg* 1966 (Academic Press, New York-London, 1967).
[56] R. C. Barrett, *Proceedings of the International Conference on Electromagnetic Sizes of Nucleons, Ottawa* 1967 (Carleton University, Ottawa, 1967).
[57] R. C. Barrett, S. J. Brodsky, G. W. Erikson, and M. H. Goldhaber, *Phys. Rev.* **166**, 1589 (1968).

Chapter 5

References [6], [7] deal extensively with the subject of this chapter. References [6], [7], [61], [62], [63], [64] contain extensive lists of references to the early papers.

[58] N. Bohr, *Nature* **137**, 344 (1936).
[59] G. Breit and E. P. Wigner, *Phys. Rev.* **49**, 519 (1936).
[60] R. P. Feynman, *Lectures on Physics*, Vol. I (Addison-Wesley, Reading, Mass., 1963).
[61] R. G. Sachs, *Nuclear Theory* (Addison-Wesley, Reading, Mass., 1953).
[62] V. F. Weisskopf and E. Friedman, article in *Niels Bohr and the Development of Physics*, (McGraw-Hill, New York, 1955).
[63] E. Vogt, article in *Advances in Nuclear Physics*, Vol. I (Plenum Press, New York, 1968).
[64] Articles in *Nuclear Reactions* Vol. I (ed. Endt and Demeur, North-Holland, Amsterdam, 1959), by E. Vogt, H. E. Gove and K. Le Couteur.
[65] D. Bodansky, *Annual Reviews of Nuclear Science* **12**, 79 (1962).
[66] A. M. Lane and D. Robson, *Phys. Rev.* **151**, 774 (1966).
[67] E. P. Wigner and L. Eisenbud, *Phys. Rev.* **72**, 29 (1947).
[68] P. L. Kapur and R. E. Peierls, *Proc. Roy. Soc.* **A166**, 277 (1938).

Chapter 6

References [13] and [39] contain important material relevant to this chapter.

[69] W. Hauser and H. Feshbach, *Phys. Rev.* **87**, 366 (1952).
[70] A. M. Lane, R. G. Thomas and E. P. Wigner, *Phys. Rev.* **98**, 693 (1955).
[71] H. Feshbach, *Annals of Physics* **5**, 357 (1958); **19**, 287 (1962).
R. H. Lemmer, *Reports on Progress in Physics*, **29**, 131 (1966).
[72] J. J. Griffin, *Phys. Rev. Lett.* **17**, 478 (1966).
[73] A. K. Kerman, L. S. Rodberg and J. E. Young, *Phys. Lett.* **11**, 422 (1963).
[74] P. E. Hodgson, *The Optical Model of Elastic Scattering* (Clarendon Press, Oxford, 1963).
[75] P. B. Jones, *The Optical Model in Nuclear and Particle Physics* (Interscience, New York-London, 1963).
[76] Articles in *Isospin in Nuclear Physics* (ed. Wilkinson, North-Holland, Amsterdam, 1969) by G. R. Satchler, J. Jänecke and G. M. Temmer.
[77] F. G. Perey and B. Buck, *Nucl. Phys.* **32**, 353 (1962).
[78] G. Rawitscher, *Phys. Rev. Lett.* **20**, 673 (1968).
[79] G. W. Greenlees, G. J. Pyle and Y. C. Tang, *Phys. Rev.* **171**, 1115 (1968).
[80] C. G. Morgan and D. F. Jackson, *Phys. Rev.* **188** (1970). To be published.
[81] I. E. McCarthy, *Introduction to Nuclear Theory* (Wiley, New York-London, 1968).
[82] H. McManus, article in *Nuclear Forces and the Few Nucleon Problem* (Pergamon, Oxford, 1960).

Chapter 7

References [37], [81], [87] and [88] contain detailed discussions of the material in this chapter and extensive lists of references to original papers. Reference [36] gives the basis of the formalism described in section 7.2.

[83] A helpful discussion of these approximations is given in an appendix by J. S. Blair and E. M. Henley, *Phys. Rev.* **112**, 202 (1958).

[84] R. M. Haybron, *Phys. Rev.* **160**, 756 (1967).

[85] R. M. Haybron and H. McManus, *Phys. Rev.* **136B**, 1730 (1964).

[86] The procedures used in DWBA calculations is described by R. H. Bassel, R. M. Drisko, and G. R. Satchler, *The Distorted Wave Theory of Direct Reactions*, Oak Ridge National Laboratory Report, ORNL 3240 (1962).

[87] N. Austern, article in *Selected Topics in Nuclear Theory* (International Atomic Energy Agency, Vienna, 1963).
G. R. Satchler, *Proceedings of the International Nuclear Physics Conference, Gatlinburg*, 1966 (Academic Press, New York-London, 1967).

[88] G. R. Satchler, *Nucl. Phys.* **A77**, 481 (1966); **A95**, 1 (1967).

[87] N. Austern, article in *Selected Topics in Nuclear Theory* (International Atomic Energy Agency, Vienna, 1963).
G. R. Satchler, *Proceedings of the International Nuclear Physics Conference, Gatlinburg*, 1966 (Academic Press, New York-London, 1967).

[88] G. R. Satchler, *Nucl. Phys.* **A77**, 481 (1966); **A95**, 1 (1967).

[89] F. G. Perey, *Proceedings of the Conference on Direct Interactions and Nuclear Reaction Mechanisms, Padua* 1962 (Gordon and Breach, New York, 1963).

[90] B. Buck, A. P. Stamp, and P. E. Hodgson, *Phil.* Mag. **8**, 1805 (1963).

[91] J. D. Anderson, *Proceedings of the Symposium on Nuclear Spectroscopy with Direct Reactions, Argonne* 1964, Argonne National Laboratory Report ANL 6878, (1964).
See also, reference [157].

[92] P. E. Hodgson and J. R. Rook, *Nucl. Phys.* **37**, 632 (1962).

[93] C. J. Batty, B. E. Bonner, E. Friedman, C. Tschalär, L. E. Williams, A. S. Clough and J. B. Hunt, *Nucl. Phys.* **116**, 643 (1968).

Chapter 8

This topic is discussed at length in references [37] and [94], which contain extensive lists of references.

[94] J. S. Blair, article in *Lectures in Theoretical Physics* Vol. VIII-C (University of Colorado Press, Boulder, 1966).

[95] N. Austern, *Annals of Phys.* **15**, 299 (1961).

[96] These models are reviewed in reference [37]. See also Chapter 10.

[97] R. M. Drisko, G. R. Satchler and R. H. Bassel, *Phys. Lett.* **5**, 347 (1963).

[98] J. R. Rook, *Nucl. Phys.* **61**, 219 (1965).

[99] J. S. Blair and B. Fernandez, *Phys. Rev.* (1970), to be published.

[100] S. D. Baker and J. A. McIntyre, *Phys. Rev.* **161**, 1200 (1967).

Chapter 8 (contd)

[101] G. Joos, *Theoretical Physics* (Blackie, London, 2nd edition 1951).
R. S. Longhurst, *Geometrical and Physical Optics* (Longmans, London, 1957).

[102] N. K. Glendenning, *Phys. Rev.* **114**, 1291 (1959).
A. J. Kromminga and I. E. McCarthy, *Phys. Rev. Lett.* **6**, 69 (1961).

[103] J. S. Blair, *Proceedings of the Conference on Direct Interactions and Nuclear Reaction Mechanisms, Padua* 1962 (Gordon and Breach, New York, 1963).

[104] E. Rost, *Phys. Rev.* **128**, 2708 (1962).

[105] N. Austern and J. S. Blair, *Annals of Phys.* **33**, 15 (1965).

Chapter 9

References [81] and [87] are relevant to this chapter, and reference [86] gives details of the procedures used in DWBA calculations.

[106] G. R. Satchler, *Proceedings of the Symposium on Nuclear Spectroscopy with Direct Reactions, Argonne* 1964, Argonne National Laboratory Report ANL 6878 (1964).

[107] N. Austern, R. M. Drisko, E. C. Halbert and G. R. Satchler, *Phys. Rev.* **133**, B4 (1964).
J. K. Dickens, R. M. Drisko, F. G. Perey and G. R. Satchler, *Phys. Lett.* **15**, 337 (1965).

[108] P. J. A. Buttle and L. J. B. Goldfarb, *Proc. Phys. Soc.* **83**, 701 (1964).

[109] M. H. Macfarlane and J. B. French, *Rev. Mod. Phys.* **32**, 567 (1960).
M. H. Macfarlane, *Proceedings of the Symposium on Nuclear Spectroscopy with Direct Reactions, Argonne* 1964, Argonne National Laboratory Report ANL 6878 (1964).

[109] M. H. Macfarlane and J. B. French, *Rev. Mod. Phys.* **32**, 567 (1960).
M. H. Macfarlane, *Proceedings of the Symposium on Nuclear Spectroscopy with Direct Reactions, Argonne* 1964, Argonne National Laboratory Report ANL 6878 (1964).

[110] A very useful account of the calculation of spectroscopic factors and sum rules for single nucleon transfer reactions is given by I. S. Towner, *Nuclear Physics Theoretical Group Report* No. 31, University of Oxford, 1967.

[111] L. J. B. Goldfarb and R. C. Johnson, *Nucl. Phys.* **18**, 353 (1960).

[112] R. C. Johnson and F. D. Santos, *Phys. Rev. Lett.* **19**, 364 (1967).

[113] Reviews of this subject have been given by N. K. Glendenning, *Annual Reviews of Nuclear Science* **13**, 191 (1963), and by
R. L. Jaffe and W. J. Gerace, *Nucl. Phys.* **A125**, 1 (1969).

[114] I. S. Towner and J. C. Hardy, *Advances in Physics* **18**, 401 (1969).

[115] I. E. McCarthy, *Proceedings of the Conference on Correlations of Particles Emitted in Nuclear Reactions, Rev. Mod. Phys.* **37**, No. 3, (1965).
See also, reference [81].

[116] D. F. Jackson and T. Berggren, *Nucl. Phys.* **62**, 353 (1965).

[117] T. Berggren and G. Jacob, *Nucl. Phys.* **47**, 481 (1963).

[118] K. L. Lim and I. E. McCarthy, *Phys. Rev.* **133**, 1006 (1964); *Nucl. Phys.* **88**, 433 (1966).

B. K. Jain, *Nucl. Phys.* **A129**, 145 (1969).
[119] D. F. Jackson, *Nuo. Cim.* **51B**, 49 (1967).
[120] T. Berggren, *Nucl. Phys.* **72**, 337 (1965).
W. T. Pinkston and G. R. Satchler, *Nucl. Phys.* **72**, 641 (1965).
A review of more recent work is given by R. J. Philpott, W. T. Pinkston and G. R. Satchler, *Nucl. Phys.* **A119**, 241 (1968).
[121] R. Huby and J. R. Mines, *Proceedings of the Conference on Correlations of Particles Emitted in Nuclear Reactions, Rev. Mod. Phys.* **37**, No. 3 (1965).
T. Berggren, *Nucl. Phys.* **A109**, 265 (1968); *Nucl. Phys.* (1970) to be published.
[122] J. P. Schiffer, *Proceedings of the International Conference on Nuclear Structure, Tokyo* 1967, Supplement to the *Journal of the Physical Society of Japan*, **24** (1968).
[123] Calculations of neutron states in single-particle potentials have been carried out by L. R. B. Elton, *Phys. Rev.* **158**, 970 (1967); E. Rost, *Phys. Lett.* **26B**, 184 (1968); C. J. Batty and G. W. Greenlees, *Nucl. Phys.* **A133**, 673 (1969); and comparisons of these wavefunctions and potentials with those required for the analysis of (d,p) and (d,t) reactions have been made by M. Dost, W. R. Hering and W. R. Smith, *Nucl. Phys.* **A93**, 357 (1967); W. C. Parkinson, D. L. Hendrie, H. H. Duhm, J. Mahoney, J. Saudinos and G. R. Satchler, *Phys. Rev.* **178**, 1976 (1969).
[124] See, for example, papers in the *Proceedings of the Conference on Correlations of Particles Emitted in Nuclear Reactions, Rev. Mod. Phys.* **37**, No. 3 (1965).
[125] H. G. Pugh, D. L. Hendrie, M. Chabre and E. Boschitz, *Phys. Rev. Lett.* **14**, 434 (1965).
Y. Dupont and M. Chabre, *Phys. Lett.* **26B**, 362 (1968).
[126] R. Bock, H. H. Duhm, R. Rüdel and R. Stock, *Phys. Lett.* **13**, 151 (1965).
[127] S. K. Penny and G. R. Satchler, *Nucl. Phys.* **53**, 145 (1964).
B. Kozlowsky and A. de Shalit, *Nucl. Phys.* **77**, 215 (1966).
P. J. Iano and N. Austern, *Phys. Rev.* **151**, 853 (1966).
[128] W. Tobocman, *Theory of Direct Nuclear Reactions* (Oxford University Press, 1961).
S. Edwards, *Nucl. Phys.* **47**, 652 (1963).

Chapter 10

For the topic of this chapter, there are as yet few standard works, but references [131], [146], [154], and [159] provide a summary of work up to 1968. References [3] and [40] provide an introduction to the necessary background in elementary particle physics.

[129] C. J. Batty, *Nucl. Phys.* **23**, 562 (1961).
H. Palevksy *et al.*, *Phys. Rev. Lett.* **18**, 1200 (1967).
[130] W. E. Frahn and G. Wiechers, *Phys. Rev. Lett.* **16**, 810 (1966).
[131] R. J. Glauber, *Proceedings of the Conference on High Energy Physics and Nuclear Structure, Rehovoth*, 1967 (ed. Alexander, North-Holland, Amsterdam, 1967).
[132] G. P. McCauley and G. E. Brown, *Proc. Phys. Soc.* **71**, 893 (1958).
[133] The range of validity of the small-angle formalism and its extension to larger angles is discussed by D. K. Ross, *Phys. Rev.* **173**, 1965 (1968); L. I. Schiff, *Phys. Rev.* **176**, 1390 (1968).
[134] H. K. Lee and H. McManus, *Phys. Rev. Lett.* **20**, 337 (1968).

Chapter 10 (contd)

[135] R. H. Bassel and C. Wilkin, *Phys. Rev.* **174**, 1179 (1968).
[136] D. F. Jackson and B. K. Jain, *Phys. Lett.* **27B**, 147 (1968).
[137] R. Sutler *et al*, *Phys. Rev. Lett.* **19**, 1189.
H. G. Pugh, *Phys. Rev.* **20**, 601 (1968).
[138] G. Belletini *et al.*, *Nucl. Phys.* **79**, 609 (1966).
[139] G. Matthiae, *Nucl. Phys.* **87**, 809 (1967).
[140] B. Margolis, *Nucl. Phys.* **B4**, 433 (1968).
W. Hirt, *Nucl. Phys.* **B9**, 447 (1969).
D. F. Jackson, *Nuo. Cim.* **63A**, 343 (1969).
[141] M. Ericson and T. Ericson, *Annals of Phys.* **36**, 323 (1966).
[142] E. H. Auerbach, D. M. Fleming and M. M. Sternheim, *Phys. Rev.* **162**, 1683 (1967).
E. H. Auerbach, H. M. Qureshi, and M. M. Sternheim, *Phys. Rev. Lett.* **21**, 1 (1968).
[143] A. Richter and R. A. Chatwin, *Phys. Lett.* **27B**, 181 (1968).
[144] D. T. Chivers *et al.*, *Phys. Lett.* **26B**, 573 (1968).
[145] R. A. Lundy, I.A. Pless., D. R. Rust, D. D. Yovanovitch and V. Kistiakowsky, *Phys. Rev. Lett.* **20**, 283 (1968).
[146] T. Ericson, *Proceedings of the International Nuclear Physics Conference, Gatlinburg*, 1966 (Academic Press, New York-London, 1967).
[147] K. A. Brueckner, R. Serber, and K. M. Watson, *Phys. Rev.* **84**, 258 (1951).
[148] G. Charpak *et al.*, *Phys. Lett.* **16**, 54 (1965).
H. Davies, H. Muirhead and J. N. Woulds, *Nucl. Phys.* **78**, 633 (1966).
F. I. Kopaleishvili, I. Z. Machbeli, G. Sh. Goksadze and N. B. Krupennikova, *Phys. Lett.* **22**, 181 (1966).
F. Pellegrini, *Nuo. Cim.* **54B**, 335 (1968).
[149] V. M. Kolybasev, *Soviet J. of Nucl. Phys.* **2**, 101 (1966).
See also A. O. Aganyants *et al.*, *Phys. Lett.* **27B**, 590 (1968).
[150] O. D. Dalkarov, *Phys. Lett.* **26B**, 610 (1968).
[151] M. Crozon *et al.*, *Nucl. Phys.* **64**, 567 (1965).
[152] Private communications from C. J. Batty and V. K. Kembhavi.
[153] F. Selleri, *Phys. Rev.* **164**, 1475 (1967).
[154] D. H. Wilkinson, *Proceedings of the International Conference on Nuclear Structure, Tokyo* 1967, Supplement to the *Journal of the Physical Society of Japan* **24**, (1968); *Proceedings of the Symposium on the Use of Nimrod for Nuclear Structure Physics*, Rutherford Laboratory 1968, Rutherford Laboratory Report RHEL/R166 (1968); and earlier references quoted therein.
[155] R. K. Adair, *Phys. Lett.* **6**, 86 (1963).
[156] E. H. S. Burhop, *Nucl. Phys.* **B1**, 438 (1967).
[157] L. Wilets, *Phys. Rev.* **101**, 1805 (1956).
[158] J. A. Nolen, J. P. Schiffer, and N. Williams, *Phys. Lett.* **27B**, 1 (1968).
H. A. Bethe and P. J. Siemens, *Phys. Lett.* **27B**, 549 (1968).
[159] See for examples articles in the *Proceedings of the Conference on High Energy Physics and Nuclear Structure, Rehovoth*, 1967 (ed. Alexander, North-Holland, Amsterdam, 1967).

Index